# FLAVONOIDS IN THE LIVING SYSTEM

# ADVANCES IN EXPERIMENTAL MEDICINE AND BIOLOGY

Editorial Board:
NATHAN BACK, *State University of New York at Buffalo*
IRUN R. COHEN, *The Weizmann Institute of Science*
DAVID KRITCHEVSKY, *Wistar Institute*
ABEL LAJTHA, *N. S. Kline Institute for Psychiatric Research*
RODOLFO PAOLETTI, *University of Milan*

Recent Volumes in this Series

Volume 433
RECENT ADVANCES IN PROSTAGLANDIN, THROMBOXANE, AND LEUKOTRIENE RESEARCH
Edited by Helmut Sinzinger, Bengt Samuelsson, John R. Vane, Rodolfo Paoletti, Peter Ramwell, and Patrick Y-K Wong

Volume 434
PROCESS-INDUCED CHEMICAL CHANGES IN FOOD
Edited by Fereidoon Shahidi, Chi-Tang Ho, and Nguyen van Chuyen

Volume 435
GLYCOIMMUNOLOGY 2
Edited by John S. Axford

Volume 436
ASPARTIC PROTEINASES: Retroviral and Cellular Enzymes
Edited by Michael N. G. James

Volume 437
DRUGS OF ABUSE, IMMUNOMODULATION, AND AIDS
Edited by Herman Friedman, John J. Madden, and Thomas W. Klein

Volume 438
LACRIMAL GLAND, TEAR FILM, AND DRY EYE SYNDROMES 2: Basic Science and Clinical Relevance
Edited by David A. Sullivan, Darlene A. Dartt, and Michele A. Meneray

Volume 439
FLAVONOIDS IN THE LIVING SYSTEM
Edited by John A. Manthey and Béla S. Buslig

Volume 440
CORONAVIRUSES AND ARTERIVIRUSES
Edited by Luis Enjuanes, Stuart G. Siddell, and Willy Spaan

Volume 441
SKELETAL MUSCLE METABOLISM IN EXERCISE AND DIABETES
Edited by Erik A. Richter, Bente Kiens, Henrik Galbo, and Bengt Saltin

Volume 442
TAURINE 3: Cellular and Regulatory Mechanisms
Edited by Stephen Schaffer, John B. Lombardini, and Ryan J. Huxtable

A Continuation Order Plan is available for this series. A continuation order will bring delivery of each new volume immediately upon publication. Volumes are billed only upon actual shipment. For further information please contact the publisher.

# FLAVONOIDS IN THE LIVING SYSTEM

Edited by

## John A. Manthey

United States Department of Agriculture
Winter Haven, Florida

and

## Béla S. Buslig

State of Florida Department of Citrus
Lake Alfred, Florida

PLENUM PRESS • NEW YORK AND LONDON

Library of Congress Cataloging in Publication Data

Flavonoids in the living system / edited by John A. Manthey and Béla S. Buslig.
    p.    cm.—(Advances in experimental medicine and biology; v. 439)
"Based on a symposium on flavonoids and related compounds held during the 212th National Meeting of the American Chemical Society held in Orlando, Florida, on August 28–29, 1996, under the sponsorship of the Division of Agricultural and Food Chemistry"—Pref.
Includes bibliographical references and index.
ISBN 0-306-45905-1
    1. Bioflavonoids—Congresses. I. Manthey, John A. II. Buslig, Béla S. III. American Chemical Society. Meeting (212th: 1997: Orlando, Fla.) IV. American Chemical Society. Division of Agricultural and Food Chemistry. V. Series.
QP772.B5F566 1998
572'.5—dc21                                                                                             98-8317
                                                                                                                          CIP

Proceedings of a Symposium on Flavonoids in the Living System,
held August 28–29, 1996, in Orlando, Florida

ISBN 0-306-45905-1

© 1998 Plenum Press, New York
A Division of Plenum Publishing Corporation
233 Spring Street, New York, N.Y. 10013

http://www.plenum.com

10 9 8 7 6 5 4 3 2 1

All rights reserved

No part of this book may be reproduced, stored in a retrieval system, or transmitted in any form or by any means, electronic, mechanical, photocopying, microfilming, recording, or otherwise, without written permission from the Publisher

Printed in the United States of America

# PREFACE

The presence of contaminant flavonoids in vitamin C preparations from citrus fruits initially led Szent-Györgyi and his collaborators to suggest that a flavonoid compound, with biological activity for the prevention of capillary fragility, was vitamin P. Later research, although not disproving biological activity, discontinued the use of the vitamin classification for these compounds. However, the ubiquitous distribution of flavonoids in living organisms, and the continued discovery of various activity in biological systems makes these compounds targets of wide ranging investigation.

This volume is primarily based on a Symposium on *Flavonoids and related compounds* held during the 212th National Meeting of the American Chemical Society held in Orlando, Florida on August 28–29, 1996 under the sponsorship of the Division of Agricultural and Food Chemistry. While the book is not intended to be a comprehensive volume on flavonoid research, the papers provide various approaches to exploring the biological functions of flavonoids in plants and animals, their chemical modifications for enhanced activity, some analytical techniques, as well as their use in food classification. A significant portion is devoted to medicinal implications of these compounds.

The organizers would like to express their appreciation to Tropicana Products, Inc., Bradenton, Florida, Coca-Cola Foods Division, Plymouth, Florida and the American Chemical Society's Division of Agricultural and Food Chemistry for financial support. Of course, the book could not be produced without the authors, whose cooperation and patience is greatly appreciated.

John A. Manthey
Citrus and Subtropical
Products Laboratory
U.S. Department of Agriculture
P. O. Box 1909
Winter Haven, Florida 33883

Béla S. Buslig
Florida Department of Citrus
700 Experiment Station Road
Lake Alfred, Florida 33850

# CONTENTS

1. Flavonoids in the Living System: An Introduction .......................... 1
   John A. Manthey and Béla S. Buslig

2. Flavonoids and Arbuscular-Mycorrhizal Fungi .......................... 9
   Horst Vierheilig, Berta Bago, Catherine Albrecht, Marie-Josée Poulin, and Yves Piché

3. The Role of Glycosylation in Flavonol-Induced Pollen Germination .......... 35
   Loverine P. Taylor, Darren Strenge, and Keith D. Miller

4. Expression of Genes for Enzymes of the Flavonoid Biosynthetic Pathway in the Early Stages of the *Rhizobium*-Legume Symbiosis .................. 45
   H. I. McKhann, N. L. Paiva, R. A. Dixon, and A. M. Hirsch

5. Prospects for the Metabolic Engineering of Bioactive Flavonoids and Related Phenylpropanoid Compounds .................................... 55
   Richard A. Dixon, Paul A. Howles, Chris Lamb, Xian-Zhi He, and J. Thirupathi Reddy

6. Flavonoid Accumulation in Tissue and Cell Culture: Studies in *Citrus* and Other Plant Species ................................................. 67
   Mark A. Berhow

7. Flavonoids of the Orange Subfamily Aurantioideae ....................... 85
   John A. Manthey and Karel Grohmann

8. Citrus Flavonoids: A Review of Past Biological Activity Against Disease: Discovery of New Flavonoids from Dancy Tangerine Cold Pressed Peel Oil Solids and Leaves ......................................... 103
   Antonio Montanari, Jie Chen, and Wilbur Widmer

9. Differentiation of Soy Sauce Types by HPLC Profile Pattern Recognition: Isolation of Novel Isoflavones ................................... 117
   Emiko Kinoshita, Yoshinori Ozawa, and Tetsuo Aishima

10. Induction of Oxidative Stress by Redox Active Flavonoids .................. 131
    William F. Hodnick, Sami Ahmad, and Ronald S. Pardini

11. Flavonoids in Foods as *in Vitro* and *in Vivo* Antioxidants .................... 151
    Joe A. Vinson

12. Antithrombogenic and Antiatherogenic Effects of Citrus Flavonoids:
    Contributions of Ralph C. Robbins ................................. 165
    John A. Attaway and Béla S. Buslig

13. Effect of Plant Flavonoids on Immune and Inflammatory Cell Function ....... 175
    Elliott Middleton, Jr.

14. Flavonoids: Inhibitors of Cytokine Induced Gene Expression ............... 183
    Mary E. Gerritsen

15. Recent Advances in the Discovery and Development of Flavonoids and Their
    Analogues as Antitumor and Anti-HIV Agents ...................... 191
    Hui-Kang Wang, Yi Xia, Zheng-Yu Yang, Susan L. Morris Natschke, and
    Kuo-Hsiung Lee

16. Inhibition of Mammary Cancer by Citrus Flavonoids ..................... 227
    N. Guthrie and K. K. Carroll

17. Inhibition of Neoplastic Transformation and Bioavailability of Dietary
    Flavonoid Agents .............................................. 237
    Adrian A. Franke, Robert V. Cooney, Laurie J. Custer, Lawrence J. Mordan,
    and Yuichiro Tanaka

18. Flavonoids as Hormones: A Perspective from an Analysis of
    Molecular Fossils ............................................. 249
    Michael E. Baker

Index ................................................................ 269

# FLAVONOIDS IN THE LIVING SYSTEM

## An Introduction

John A. Manthey[1] and Béla S. Buslig[2]

[1]USDA, ARS, SAA
U.S. Citrus and Subtropical Products Laboratory
600 Avenue S, NW
Winter Haven, Florida 33881
[2]Florida Department of Citrus
700 Experiment Station Road
Lake Alfred, Florida, 33850

Flavonoids and related phenols are ubiquitous in land plants (Mabry and Ulubelen, 1980), and as a common element of plants they have taken their place in importance in far ranging arrays of biological systems, being integral to the function of most living systems. The study of these compounds has attracted the attention of generations of chemists. In fact, as one of the first flavonoids described, hesperidin, the primary flavanone glycoside in citrus, was first reported by Lebreton nearly 170 years ago (Lebreton, 1828). Although there has been a long history of the use of medicinal plants rich in flavonoids, our recent interest in the biological properties of the flavonoids began with the findings of Szent-Györgi who found that citrus flavonoids were important to proper capillary function (Rusznyák and Szent-Györgyi, 1936). Although the vitamin status that was briefly attributed to flavonoids was discontinued in 1950, subsequent research has clearly shown an ever increasing list of biological systems over which flavonoids exert significant control. The pace of this research has rapidly accelerated, and recently, a much clearer understanding has emerged of the significance of these compounds, not only in plants, but also in animal systems, and ultimately pertaining to human health. This volume is based primarily on an American Chemical Society symposium held in Orlando, FL in August, 1996. Some of the material was added later to highlight new findings regarding the biological properties of flavonoids. It is not intended to be a comprehensive treatment of flavonoid research, nevertheless it is hoped that it shows new perspectives and encourages new research into this broad subject area at the cellular and molecular levels.

*Flavonoids in the Living System*, edited by Manthey and Buslig
Plenum Press, New York, 1998.

## 1. PLANTS

In this book, the functions of these compounds are explored in the simplest organisms, and then continued to higher plants, and finally to biological systems in animals. In the first chapters the roles of flavonoids in plant/microbe communication and signaling are discussed. This topic is first discussed in the chapter by Vierheilig *et al.*, where flavonoids are shown to impact many aspects of the symbiotic associations between arbuscular mycorrhizal fungi (AMF) and plants, including signaling between plant and fungi, spore germination, hyphal growth, and fungal differentiation. Regulation of AMF-plant associations is still poorly understood in terms of the communication between the two partners, but the coexistence and interactions between soil microbes and plant roots in the rhizosphere most likely extend from the rise of land plants and has been intimately linked to chemical communication between plants and soil microorganisms. Plant phenolics, and flavonoids in particular, are seen in recent studies to be central to these associations (Fisher and Long, 1992). Key to understanding the roles of flavonoids in this ancient microbe-plant communication is an understanding of the nature of the binding sites in the AMF. In acting as chemical signals in these early plant-microbe symbioses, the binding of these compounds to key receptor molecules is crucial, and it is in this receptor-ligand binding influencing cell-cell interaction that we gain insights into some of the most fundamental properties of these flavonoids.

The following chapters by Dixon *et al.* and McKhann *et al.* deal with many of the recent findings regarding the enzymes of the flavonoid biosynthetic pathways in the early stages of the *Rhizobium*-legume symbiosis. In this symbiosis, flavonoids attract rhizobia chemotactically to plant roots, enhance rhizobial growth, induce rhizobial *nod* genes, and are produced as part of the increase in *nod* gene-inducing flavonoid response. The McKhann *et al.* chapter explores the time and spacial expressions of chalcone synthase and isomerase enzymes involved in the *nod* gene-inducing flavonoid response that occurs several days after root inoculation. As stated in this chapter, certain parallels exist in this response and that of plant responses (phytoalexin biosynthesis) to pathogenic bacteria (Buffard *et al.*, 1996, McKhann and Hirsch, 1994), and findings reported in previous studies suggest that flavonoids may modify endogenous hormone levels (Hirsch, 1992, Yang *et al.*, 1992), which results in the triggering of cell division beginning nodule formation. Research findings discussed in the chapter by Dixon *et al.* deal with the metabolic engineering of bioactive flavonoids, and the recent discoveries of the regulatory architecture for flavonoid flux control, substrate channelling, and of the biosynthesis, storage, and function of flavonoids as part of plant responses, including elicitor/pathogen recognition, and the flavonoid nodulation signals involved in the *Rhizobium*-legume symbiosis. It is emphasized that the availability of the complete set of genes and knowledge of the mechanisms underlying their transcriptional regulation are in themselves insufficient to guarantee success in the genetic manipulation of complex pathways. Also discussed are several transgenic approaches for determining the control points for flux into the branch pathways of flavonoid biosynthesis, and the potential metabolic channelling in the flavonoid biosynthesis pathway.

The influence of flavonoids on pollen tube growth (Taylor and Jorgensen, 1992; Vogt and Taylor, 1995; Vogt *et al.*, 1995) provides us with further insights into the potential interactions between these compounds and higher plant development and evolution. Origins of land plants may have involved an endocellular mutualism between green algae and a tip-growing, fungus-like organism (Atsatt, 1988; Jorgensen, 1996), and this mutualism may have used, in part, the tip-growing symbiont's capacity for invasive growth. A role for the flavonoids in this development may be reflected in the dependence of pollen tube tip growth on flavonoids as previously reported (Mo et al., 1992; Pollak et al., 1992). The chapter by Taylor et al. deals with the recent findings of the enzymatic conversions of flavonol glycosides during pollen tube growth and development. Glycosidations are of widespread importance in cell-cell recognition, and this study addresses the role of glycosidation in regulating the biological activity of the flavonols in this particular biological system. As shown in this chapter, part of the regulation is controlled by localized pH and by the metabolic fluxes of the different substrates. These discussions are pertinent to understanding the more general physiological roles of flavonoids in plant tissue. As shown in the chapters by Montanari et al., Manthey and Grohmann, and Berhow, certain plant species, including the many varieties of citrus, contain remarkably large numbers of different flavonoids, some at very high concentrations. The questions concerning transport and storage are extremely important to the function and subsequent bioconversions of plant flavonoids, and central to much of this are the glycosidation reactions, which strongly influence cytotoxicity, solubility within the plant vacuoles, and ability to cross biological membrane systems. Considerable literature exists regarding the chemotaxonomic applications of flavonoids. Extending this approach to plant derived food products is discussed in the chapter of Kinoshita et al. where flavonoids are used in the classification of production of various types of soy sauces.

These elements of the bioregulation of flavonoid biosynthesis in plant tissue are shown to play pivotal roles in the ecochemical functionality of the plant. As discussed in the chapter by Berhow, flavonoids are also involved as growth regulators, modulators of gene expression, stress responses, and intracellular signaling. This chapter summarizes the important contributions of cell culture studies to flavonoid research. As discussed in the chapter by Dixon et al. the enzymes of flavonoid biosynthesis pathways are often present as isomeric forms, indicating groups of tightly controlled metabolic chains which are driven by the formation of very specific end products.

## 2. ANIMALS

In addition to the biochemistry of flavonoids in plants, this book also explores the physiological properties of flavonoids in animals. Much of this work has been done with *in vitro* human cell lines, but increasing numbers of *in vivo* animal feeding studies are being reported. In studying these properties, it has become apparent that flavonoids are critical to many aspects of human physiology, and as important components to the human diet, are critical to long term health maintenance. The scope of the actions of these compounds in the control of inflammation, cancer, thrombogenesis, cell damage by oxidation, viral infection and erythrocyte aggregation suggests an importance of these compounds that we are only beginning to understand. An extensive review of many antitumor and anti-HIV flavonoids and their analogues is presented by Wang et al. A number of naturally occurring flavonoids as well as synthetic 2-phenyl-4-quinolones and substituted 2-phenylthiochromen-4-ones are shown to have potent anti-HIV and antitumor activities against a number of cancer cell lines.

In this book the discussion of the roles of flavonoids in animal systems centers on a number of important studies of the anticancer, antioxidant, and antiinflammatory properties of the flavonoids on activated cell lines. Common to many of these reports are the potent anticancer and antiinflammatory activities of the citrus polymethoxylated flavones, which, due to their significantly lower polarities compared to the more common flavonol glycosides, are considered to be more easily transported across biological membranes and thus, are more likely to have access to key enzyme systems essential for cell proliferation (Kandaswami et al., 1991). As discussed in these chapters, flavonoids are inhibitory to a number of enzymes needed for cell activation in cancer and also in the inflammation and allergic responses. These enzymes include: protein kinase C, protein tyrosine kinase, lipoxygenases, phosphatase A2, phospholipases, cyclooxygenase, topoisomerase, NADH oxidase, and others. In these discussions, it is pointed out that the biological activities of flavonoids are, for the most part, specific towards activated cells and that flavonoids have very little effect on normal, control cells.

As discussed by Middleton, and later in the chapter by Gerritsen, one of the most prevalent activities of flavonoids is the effect that these compounds have on cell adhesion. Cell adhesion is critical to many aspects to human physiology, including, in part: inflammation, the allergic response, cancer cell metastasis, platelet adhesion during thrombosis, and erythrocyte adhesion. Flavonoids also have strong inhibitory effects on the function of a number of secretory cell types (for example, neutrophils, mast cells, eosinophils, macrophages, B and T lymphocytes, monocytes, and platelets) involved in inflammation (for a review, see Middleton and Ferriola, 1988). In the study by Gerritsen, flavonoids are shown to affect the induction of the biosynthesis of key adhesion molecules during the inflammation response. The flavone apigenin inhibited the cytokine induced upregulation of several other inflammatory genes including IL-6, IL-8, and cyclooxygenase-2. These cytokine regulated genes are central to the initiation and propagation of immune and inflammatory responses. This provides an example of flavonoids regulating gene expression in activated cells, and these results provide an explanation of how flavonoids affect the adhesion of cells essential to the inflammation response. Pertaining to cancer cells, the polymethoxylated flavone tangeretin has been shown to have strong anti-invasive activity towards invasive cancer cell lines, and that this activity was blocked by an adhesion molecule inhibitor (Bracke et al., 1989). Addition of tangeretin to nonaggregating cancer cells led to rapid aggregation. Additionally, several of the flavanones and polymethoxylated flavones discussed in these chapters have been shown in other studies to inhibit *in vitro* and *in vivo* leukocyte binding to sites of injury at the endothelium (Middleton and Kandaswami, 1992; Bouskela and Donyo, 1997). Inhibition of leukocyte binding directly affects the levels of tissue damage during the inflammation response.

The effects of flavonoids on cell adhesion are also critical to two other important physiological responses affecting human health. As summarized by Attaway and Buslig, the work by Robbins extended the much earlier findings by Szent-Györgyi, and showed that flavonoids, and in particular the polymethoxylated flavones, significantly affect red blood cell adhesion and aggregation (Robbins, 1976,1977). This is particularly important in injured tissue, and by affecting blood viscosity, erythrocyte aggregation influences microcirculation. Additionally, these flavonoids were shown to have strong antithrombogenic effects. Thrombogenesis, caused initially in many cases by oxidative injury to the endothelium, involves platelet binding to the injured site. Studies have shown that flavonoids inhibit platelet binding and thus decrease thrombi formation (for reviews, see Cazenave et al., 1986; Beretz and Cazenave, 1988).

Several of the studies reported here showed the strong synergy that occurs when different cancer cell lines were exposed to combinations of flavonoids. As reported in the chapter by Franke, several of the citrus flavanone glycosides strongly inhibited cell transformations, yet, when two of the polymethoxylated flavones (tangeretin and nobiletin) were run in combination, the activities were far greater than when the compounds were run singly. Similarly, results in the chapter by Guthrie et al. show that when these compounds were run in combinations with different dietary tocotrienols, significant decreases in the $IC_{50}$ occurred for these flavonoid compounds. The chapter by Franke also discusses findings relevant to the uptake and bioavailability of dietary flavonoids. The understanding of the metabolism and bioavailability of flavonoids in the human body is far from complete, yet these factors, along with the synergistic effects of combinations of these compounds may profoundly affect the levels of the pharmacological properties of the flavonoid compounds.

In addition to the effects that flavonoids have on cell adhesion and gene expression of activated cells, the ability of most dietary flavonoids to scavenge reactive oxygen species is another key element of the physiological roles of these compounds in the human body (Pignol *et al.*, 1988). The production of the various deleterious reactive oxygen species is involved in a number of key physiological processes, including in part, the inflammation and immune responses, atherogenesis, and thrombogenesis. Damage caused by these reactive species is a major source of tissue damage during chronic inflammation. In addition to the modulation by the flavonoids of these processes by cell adhesion and gene expression, the flavonoids as antioxidants are able to remove these reactive oxygen species, and thus inhibit the ensuing cascades of biochemical events. As a class of compounds the flavonoids are some of the most potent dietary antioxidants, and thus are of critical importance to human health. The structure/activity relationships of the antioxidant properties of flavonoids in *in vitro* and *in vivo* studies are discussed in the chapter by Vinson. A review is also given of the epidemiological evidence for their health benefits in heart disease. Results suggest that the hydrophobic flavonoids bind to low density lipoproteins and the amphophilic flavonoids in the aqueous fluid may be powerful *in vivo* antioxidants, critical to the prevention of lipid oxidation leading to atherogenesis (Steinberg *et al.*, 1989). Included in the discussion by Vinson are results of studies of *in vivo* absorption of dietary flavonoids, and the ability of these compounds to increase the levels of antioxidant activity *in vivo*.

In contrast to the antioxidant properties of most flavonoids, the chapter by Hodnick *et al.* deals with the potential, under certain conditions, to induce oxidative stress by certain redox active flavonoids. Mitochondrial uncoupling activities have been reported for flavonoids (Ravanel, 1986), and the studies summarized in this chapter show the abilities of redox active flavonoids, mainly those with catechol and pyrogallol structures in the B ring, to inhibit two mitochondrial oxidative enzymes. In addition to this, numerous flavonoids are shown to be capable of autooxidation, a result of which is the production of reactive oxygen species. The effects of differences in the levels of dietary redox active flavonoids on changes in the *in vivo* levels of detoxifying enzymes were studied with insect model systems.

Beyond the redox nature of flavonoids, the interactions between flavonoids and the living cell are functions of binding, and hence, of structure. Flavonoid binding is critical to the physiological activities of flavonoids in living systems, and in order to understand these activities at a molecular level, we need to consider the sites of interactions between flavonoids and a host of potential receptor binding sites. The interactions between flavonoids and living systems are ancient, and in the final chapter, Baker evaluates the possible links between flavonoids and the evolution of steroid hormone receptors in animal systems, and discusses the possible hormonal actions of plant flavonoids during evolutionary divergence. The studies reviewed in this chapter show the occurrence of significant sequence homologies between mammalian proteins involved in hormone action and with flavonoid-binding proteins in plants, insects, and bacteria. The similarities are sufficient in some cases to suggest that molecules that recognize a plant, insect, or bacterial protein may interact with a mammalian relative. As shown by this work, certain chemical themes have been taken and reworked for construction of intercellular signals. Amino acid sequence homologies between *R. meliloti* NodG (flavonoid binding protein receptor) and human 17-β-hydroxysteroid dehydrogenase—the enzyme that metabolizes estrogens and androgens in humans—suggest the occurrence of common ancestors for enzymes that regulate steroid, retinoid, and prostaglandin action in humans and signals between plants, bacteria, and insects. In searching for flavonoid binding sites within the cell, we begin to uncover the modes of action of these compounds in all living systems. The effects of flavonoids on the cell adhesion and gene expression of activated cells involved in a number of physiological processes critical to maintaining good human health show us the extreme importance of these compounds obtained through the human diet.

## REFERENCES

Beretz, A.; Cazenave, J.P. The effect of flavonoids on blood-vessel wall interactions. In *Plant Flavonoids in Biology and Medicine II: Biochemical, Cellular, and Medicinal Properties*. Cody, V., Middleton, E. Jr., Harborne, J.B., Beretz, A., Eds.; Alan R. Liss: New York, 1988; pp 187–200.

Bouskela, E.; Donyo, K.A. Effects of oral-administration of purified micronized flavonoid fraction on increased microvascular permeability induced by various agents and on ischemia/reperfusion in the hamster-cheek pouches. *Angiology*, **1997**, *48*, 391–399.

Bracke, M.E.; Vyncke, B.M.; Van Larebeke, N.A.; Bruyneel, E.A.; De Bruyne, G.K.; De Pestel, G.H.; De Coster, W.J.; Espeel, M.; Mareel, M.M. The flavonoid tangeretin inhibits invasion of MO4 mouse cells into embryonic chick heart in vitro. *Clin. Exptl. Metastasis*, **1989**, *7*, 283–300.

Buffard, D.; Esnault, R.; Kondorosi, A. Role of plant defence in alfalfa during symbiosis. *World J. Microbiol. Biotechnol.* **1996**, *12*, 175–188.

Cazenave, J.-P.; Beretz, A.; Anton, R. The effects of flavonoids on human platelet function. In *Flavonoids and Bioflavonoids*. Farkas, L., Gabor, M., Kallay, F., Eds.; Elsevier Science Publishers: Amsterdam, 1985; 373–380.

Fisher, R.F.; Long, S.R. *Rhizobium*-plant signal exchange. *Nature*. **1992**, *387*, 655–660.

Hirsch, A.M. Developmental biology of legume nodulation. *New Phytol.* **1992**, *122*, 211–237.

Jorgensen, R. The origin of land plants: a union of alga and fungus advanced by flavonoid? *Biosystems*, **1993**, *31*, 193–207.

Kandaswami, C.; Perkins, E.; Solonuik, D.S.; Drzewiecki, G.; Middleton, E. Jr. Antiproliferative effects of citrus flavonoids on human squamous cell carcinoma *in vitro*. *Cancer Letters* **1991**, *56*, 147–152.

Lebreton, P. *J. Pharm. Chim. Paris*, **1828**, *14*, 377.

Mabry, T.J.; Ulubelen, A. Chemistry and utilization of phenylpropanoids including flavonoids, coumarins, and lignans. *J. Agric. Food Chem.* **1980**, *28*, 188–196.

McKhann, H.I.; Hirsch, A.M. Does *Rhizobium* avoid the host response? In *Current Topics in Microbiology and Immunology, Vol. 192, Bacterial Pathogenesis of Plants and Animals*. Dangl, J., Ed.; Springer Verlag: Berlin, 1994; pp 139–162.

Middleton, E. Jr.; Ferriola, P. Effect of flavonoids on protein kinase C: relationship to inhibition of human basophil histamine release. In *Plant Flavonoids in Biology and Medicine II: Biochemical, Cellular, and Medicinal Properties*. Cody, V., Middleton, E. Jr., Harborne, J.B., Beretz, A., Eds.; Alan R. Liss: New York, 1988; pp 251–266.

Middleton, E. Jr.; Kandaswami, C. Effects of flavonoids on immune and inflammatory cell functions. *Biochemical Pharmacology*, **1992**, *43*, 1167–1179.

Mo, Y.; Nagel, C.; Taylor, L.P. Biochemical complementation of chalcone synthase mutants defines a role for flavonols in functional pollen. *Proc. Natl. Acad. Sci. USA*, **1992**, *89*, 7213–7217.

Pignol, B.; Etienne, A.; Crastes de Paulet, A.; Deby, C.; Mencia-Huerta, J.M.; Braquet, P. Role of flavonoids in the oxygen-free radical modulation of the immune response. In *Plant Flavonoids in Biology and Medicine II: Biochemical, Cellular, and Medicinal Properties.*, Cody, V., Middleton, E. Jr., Harborne, J.B., Beretz, A., Eds.; Alan R. Liss: New York, 1988; 173–182.

Pollak, P.E.; Hansen, K.; Astwood, J.D.; Taylor, L.P. Chalcone synthase and flavonol accumulation in stigmas and anthers of *Petunia hybrida*. *Plant Physiol.* **1993**, *102*, 925–932.

Ravanel, P. Uncoupling activity of a series of flavones and flavonols on isolated plant mitochondria. *Phytochemistry* **1986**, *25*, 1015–1020.
Robbins, R.C. Regulatory action of phenylbenzo-γ-pyrone (PBP) derivatives on blood constituents affecting rheology in patients with coronary heart disease (CHD). *Internat. J. Vit. Nutr. Res.* **1976**, *46*, 338–347.
Robbins, R.C. Stabilization of flow properties of blood with phenylbenzo-γ-pyrone derivatives (flavonoids) *Internat. J. Vit. Nutr. Res.* **1977**, *47*, 373–382.
Rusznyák, S.; Szent-Györgyi, A. Vitamin P: flavonols a vitamins. *Nature* **1936**, *27*, 138.
Steinberg, D.; Parathasaraghy, S.; Carew, T.E.; Khoo, J.C.; Witzum, J.L. Beyond chloresterol; modification of low-density lipoprotein that increases its atherogenicity. *New Engl. J. Med.* **1989**, *320*, 915.
Taylor, L.P.; Jorgensen, R. Conditional male fertility in chalcone synthase-deficient petunia. *J. Heredity*, **1992**, *83*, 11–17.
Vogt, T.; Taylor, L.P. Flavonol 3-*O*-glycosyltransferases associated with petunia pollen produce gametophyte-specific flavonol diglycosides. *Plant Physiol.*, **1995**, *108*, 903–911.
Vogt, T.; Wollenweber, E.; Taylor, L.P. The structural requirements of flavonols that induce pollen germination of conditionally male fertile petunia. *Phytochemistry*, **1995**, *38*, 589–592.
Yang, W.-C.; Cremers, H.C.J.C.; Hogendijk, P.; Katinakis, P.; Wijffelman, C.A.; Franssen, H.; van Kammen, A.; Bisseling, T. *In-situ* localization of chalcone synthase mRNA in pea root nodule development. *Plant J.* **1992**, *2*, 143–151.

# FLAVONOIDS AND ARBUSCULAR-MYCORRHIZAL FUNGI

Horst Vierheilig, Berta Bago, Catherine Albrecht, Marie-Josée Poulin, and Yves Piché

Université Laval
Faculté de Foresterie et de Géomatique
CRBF, Pavillon C.-E.- Marchand
Québec GlK 7P4, Canada

## 1. ABSTRACT

Arbuscular mycorrhizal fungi (AMF) are ancient Zygomycetes forming the most widespread plant-fungus symbiosis. The regulation of this association is still poorly understood in terms of the communication between the two partners. Compounds inside the root and released by the root, such as flavonoids, are hypothesized to play a role in this plant-fungus communication, as already demonstrated in other symbiotic associations (e.g. *Rhizobium*-leguminoseae). Here we give a general overview of the research concerning this question.

## 2. WHAT IS AN ARBUSCULAR MYCORRHIZA?

This term refers to the symbiotic association between roots of over 80% of all land plants and a small group of fungi (Glomales) commonly found in the soil. In general, this association is beneficial for both partners. The host plant receives mineral nutrients via the fungal mycelium, while the heterotrophic fungus obtains photosynthetically produced carbon compounds from the host (Smith and Read, 1997).

The AM colonization of the root can be divided into three different stages. These are i) the precolonization stage, in which the plant and the fungus are not in contact yet; ii) the cell-to-cell stage, in which the hyphae attach to the root, forming appressoria; and iii) the colonization stage, in which the fungus, after penetrating the root, forms inter- and intracellular hyphae, arbuscules, thought to be the place for nutrient exchange between both partners, and vesicles, thought to be storage organs of the fungus (Bonfante-Fasolo, 1984).

One of the fundamental difficulties in the study of the arbuscular mycorrhizal (AM) symbiosis is the inability of the AMF to complete their life cycle without a host plant. It can be assumed that mycorrhiza formation is a continuous exchange of signals between the plant and the fungus. These signals may: i) be exuded by the roots and play a role during the precolonization stage and/or, ii) be intraradical, and act on the fungus at the colonization stage. Attempts have been made to determine the signals necessary to overcome the obligate status of AMF.

Interestingly, some plant families such as the Chenopodiaceae and the Brassicaceae do not form the AM association (AM nonhost plants). Two principal hypotheses have been advanced to explain the inability of these plant families to form the AM symbiosis. Either i) AM nonhost plants lack signals necessary for the formation of the symbiosis, or ii) in AM nonhost plant roots certain compounds inhibitory to AMF are present. In order to elucidate the signals involved in AM symbiosis, the effect of root extracts and exudates of nonhost plants on AMF has been compared to the effect of root extracts and exudates of AM host plants on AMF. Moreover, the effect of flavonoids, known to play a signaling role in the development of the *Rhizobium*-plant symbiosis, has been studied on AMF.

## 3. EFFECT OF ROOT EXTRACTS AND ROOT EXUDATES ON AM FUNGAL SPORE GERMINATION AND HYPHAL GROWTH

### 3.1. Effect of Root Extracts

Root extracts, a mixture of all root compounds, were obtained either with organic solvents or with aqueous solutions. Extracts of AM host roots exhibited an inhibitory or no effect on AM fungal spore germination (Table 1). The different effects obtained seemed to depend on the differential solubility of the extracted compounds. In the case of extracts from AM nonhost roots this differential effect disappeared. Both, organic solvent and aqueous solution extracted compounds, generally inhibited spore germination.

Virtually no data are available on the effect of root extracts on hyphal growth. Vierheilig et al. (1996a) showed that application of aqueous root extract of the AM host plant tobacco *(Nicotiana tabacum)* to the hyphal tip of the AM fungus *Glomus mosseae* showed no effect, whereas the aqueous root extract of the AM nonhost plant stinging nettle *(Urtica dioica)* exhibited an inhibitory effect. A similar effect of aqueous root extracts of AM host and nonhost plants was observed on spore germination. Aqueous root extracts of AM host plants showed no effect, whereas the extracts from AM nonhost plants were inhibitory (Table 1).

It is difficult to draw a conclusion about the biological relevance of the effect of root extracts on AMF. Roots contain a large variety of compounds and it may be assumed that in intact roots not necessarily all of them come into contact with the AM fungus. Some of the compounds present in the extracts, showing activity but not normally coming in contact with the AM fungus, thus playing no role in such cellular communication, may mask the effect of other compounds which do play a role in the formation of AM symbiosis. In general, it seems that inhibitory compounds are prevalent in root extracts of AM nonhost plants.

### 3.2. Effect of Root Exudates

More biologically relevant as signals in the plant-fungus communication are root exudates and volatiles, as these compounds can act on the fungus during the precolonization stage and the cell-to-cell stage. Root exudates have been shown to be involved in AM

Table 1. Effect of root extracts on spore germination of different AMF

| Plant | Fungus | Extraction | Spore germination | Reference |
|---|---|---|---|---|
| **AM host plants** | | | | |
| Abuliton theophrasti | Glomus etunicatum | organic solvent | - | Schreiner and Koide 1993a |
| Daucus carota | G. etunicatum | organic solvent | - | Schreiner and Koide 1993a |
| Lactuca sativa | G. etunicatum | organic solvent | - | Schreiner and Koide 1993a |
| Lycopersicon esculentum | G. mosseae | buffer | 0 | Vierheilig and Ocampo 1990a; 1990b; |
| Medicago sativa | G. mosseae | buffer | 0 | Vierheilig and Ocampo 1990a; |
| | | buffer | 0 | Ocampo et al. 1986 |
| Sorghum bicolor | G. etunicatum | organic solvent | 0 | Schreiner and Koide 1993a |
| **AM non-host plants** | | | | |
| Amaranthus retroflexus | G. etunicatum | organic solvent | - | Schreiner and Koide 1993a |
| Beta vulgaris | G. etunicatum | organic solvent | - | Schreiner and Koide 1993a |
| Brassica campestris | G. etunicatum | organic solvent | - | Schreiner and Koide 1993a |
| B. kaber | G. etunicatum | organic solvent | - | Schreiner and Koide 1993a |
| B. napus | G. etunicatum | organic solvent | - | Schreiner and Koide 1993a |
| | G. mosseae | buffer | - | Vierheilig and Ocampo 1990a |
| B. oleracea | G. mosseae | buffer | - | Vierheilig and Ocampo 1990a; 1990b; |
| | | | | Ocampo et al. 1986 |
| Oxalis pes-caprae | G. mosseae | buffer | - | Parra-Garcia et al. 1992 |
| Raphanus raphanistrum | G. etunicatum | organic solvent | - | Schreiner and Koide 1993a |
| Raphanus sativus | G. mosseae | buffer | - | Vierheilig and Ocampo 1990a |
| Spinacea oleracea | G. etunicatum | organic solvent | - | Schreiner and Koide 1993a |
| | G. mosseae | buffer | - | Vierheilig and Ocampo 1990a |
| Thlaspi arvense | G. etunicatum | organic solvent | 0 | Schreiner and Koide 1993a |

Stimulation (+), inhibition (-), no effect (0).

fungal differentiation. Some authors observed that the morphology (e.g. the branching pattern) of the AM fungal hyphae, changed in the presence of AM host roots (Mosse and Hepper, 1975; Powell, 1976; Graham, 1982; Mosse, 1988; Giovannetti et al., 1993a,b; 1996). This induction of branching has been proposed to be a prerequisite for a successful root colonization, as the presence of AM nonhost roots was not effective in inducing similar differentiations of AM fungal hyphae.

Many studies have been conducted on the effects of exuded and volatile compounds released by roots on spore germination and/or on hyphal growth (see Tables 2, 3). Different experimental approaches were used to determine their effect on AMF: Root exudates (E) were applied to AMF; roots were grown together with AMF (exudates and volatiles (E+V)), or physically separated from AMF (volatiles (V)). The experiments were performed either in axenic culture (AMF in "*in vitro*" conditions), in monoxenic culture (root organ culture + AMF in "*in vitro*" conditions) or in greenhouse experiments with whole plants.

Curiously, results from root organ cultures differed from results from other experimental systems (Tables 2 and 3). Exudates from root organ cultures of a mycorrhizal resistant (Myc-) *Pisum sativum* mutant were inhibitory on hyphal growth (Balaji et al., 1995), whereas the roots of whole *Pisum sativum* (Myc-) plants exhibited a stimulatory effect (Giovannetti et al., 1993b). Moreover, roots of *Pisum sativum* (Myc+) plants stimulated hyphal growth when experiments were carried out with whole plants under greenhouse conditions (Giovannetti et al., 1993b), but root exudates from root organ cultures of this plant were inhibitory (Balaji et al., 1995)(Table 2). Exudates from root organ cultures of the AM host plant *Daucus carota* in one experiment stimulated hyphal growth (Bécard and Piché, 1990) whereas in another experiment no effect was observed (Bécard and Piché, 1989a)

It is interesting to note that contradictory results were also obtained with root organ cultures of AM nonhost plants (Table 3). Exudates from root organ cultures of lupin stimulated hyphal growth (Balaji 1996), whereas exudates from roots of whole lupin plants grown in the greenhouse showed no effect (Gianinazzi-Pearson et al, 1989). These results remind us that although root organ cultures provide an easily controllable system to study the effect of roots on AMF, they are highly artificial systems and not necessarily comparable to the complex conditions found in the rhizosphere of plants in the soil.

In general it was found that spore germination was either stimulated or not affected by root exudates of AM host plants, whereas hyphal growth was stimulated (Table 2). Volatiles alone always stimulated hyphal growth, but nearly never showed an effect on spore germination (Table 2). Bécard and Piché, (1989a) showed that the combination of exudates and volatiles stimulated hyphal growth and this stimulation was synergistical. The stimulatory effect of volatiles has been attributed, at least partially, to the release of $CO_2$ by the root (Bécard and Piché, 1989a). Carbon dioxide has been shown to promote highest hyphal growth of *Gigaspora margarita* at concentrations around 2% (Bécard and Piché, 1989a; Poulin et al., 1993), the usual concentration found in the soil. However, even if it has been suggested that $CO_2$ is the major factor responsible for this stimulation other yet unidentified root volatiles could also play a role (Bécard and Piché, 1989a; Balaji et al., 1995).

The effect of exudates and volatiles of AM nonhost root on spore germination and fungal growth does not seem clear, however, looking at the experimental systems used in the different experiments, we find a hyphal growth stimulation with root organ cultures of AM nonhost plants and no effect or an inhibition in all the other systems used (Table 3).

The results presented in Tables 2 and 3 indicate that root exudates and volatiles from AM nonhost plants exhibit a different effect on AMF than root exudates and volatiles of

AM host plants. With AM host root exudates and volatiles a stimulatory effect was predominant, suggesting their role in AM formation, while AM nonhost plant exudates and volatiles (except from root organ cultures) had no effect or were inhibitory.

From the data presented, it is not clear whether roots of AM nonhost plants lack factors necessary for AM symbiosis, or in AM nonhost plants compounds are present which impede the symbiosis. However, the inability of AM nonhost plant to form the AM symbiosis may not result from a general mechanism. Some AM nonhost roots may lack the necessary signals, while others may contain and/or release inhibitory compounds.

## 4. PHENOLIC COMPOUNDS AS SIGNALS IN AM SYMBIOSIS

AMF establish a symbiotic association with many different plant families. Until now, however, no compounds acting as common signal molecules involved in AM root colonization in the different plant families have been found. Roots produce a wide range of exudates including phenolic compounds. Phenolic compounds may act as antimicrobial agents (Bailey, 1982) or as signal molecules modulating plant-microorganism interactions (Lynn and Chang, 1990; Phillips and Tsai, 1992). Phenolic compounds, specifically flavonoids, appear to be nearly universal signaling molecules and are known to play an important role in the communication between roots of legumes and *Rhizobium* spp.(Firmin et al., 1986; Peters et al., 1986; Redmond et al., 1986; Kosslak et al., 1987). However, the role of these metabolites in the AM symbiosis remains unclear. Recently an overview about phenolics in AM interactions, and their potential role in biological control was given (Morandi 1996). In the following part of our work we try to discuss aspects related with the possible role of these compounds in the signaling between plants and AM fungi.

### 4.1. Effect of AM Symbiosis on Root Phenolic Compounds

A wide range of phenolic compounds is present in roots, most of them are constitutively produced. After AM fungal root colonization, the level of some of these compounds changes (Table 4). Sometimes similar changes could be observed within a plant family. The level of coumestrol and daidzein increased in colonized roots of *Glycine max*, *Medicago sativa* and *M. truncatula* (Harrison and Dixon, 1993; Morandi et al, 1984), all belonging to the leguminoseae family and the level of blumenin was enhanced in colonized roots of several members of the Gramineae (Maier et al., 1995). Sometimes changes of a compound seemed to depend on the plant family. Coumaric acid and ferulic acid showed increases in colonized roots of *Allium cepa* (Liliaceae)(Grandmaison et al., 1993), but not in colonized roots of several Gramineae (Maier et al., 1995). Changes of a compound level seem to depend as well on the AM fungal species. Grandmaison et al. (1993) inoculated *A. cepa* with *G. versiforme* and *G. intraradices*. Several weeks (21 w) after inoculation the levels of the cell wall-bound *N*-feruloyltyramine and *p*-coumaric acid were significantly higher in roots colonized by *G. intraradices* compared to roots colonized by *G. versiforme*. This indicates that either the two AMF have different requirements for their development and thus induce different levels of these compounds or that the plant recognizes the two fungi differently. However, it has to be kept in mind that initial changes occur at a very early stage of root colonization (Harrison and Dixon, 1993), therefore studies of phenolic root compounds several months after root inoculation do not necessarily reflect significant events involved in the formation of the AM symbiosis.

Table 2. Effect of roots (E+V), root exudates (E) and root volatiles (V) of AM host plants on spore germination and hyphal growth of different AMF

| Plant | Fungus | Spore germination | Hyphal growth | Experimental system | Reference |
|---|---|---|---|---|---|
| Abutilon theophrasti | Glomus intraradices | + | ND | AMF + roots in soil | Schreiner and Koide 1993b |
| Alnus glutinosa | G. mosseae | ND | + | AMF + roots in quartz grit | Giovannetti et al. 1994 |
| Allium cepa | Gigaspora margarita | 0 | + | AMF in axenic culture + E | Tawaraya et al. 1996 |
| Ambrosia artemisiifolia | G. intraradices | + | ND | AMF + roots in soil | Schreiner and Koide 1993b |
| Asparagus officinalis | G. mosseae | 0 | ND | AMF + roots in soil | Daniels and Trappe 1980 |
| | Gi. gigantea | 0 | ND | AMF + roots in soil | Daniels and Trappe 1980 |
| Daucus carota | G. etunicatum | + | + | AMF + root organ culture (monoxenic culture) | Schreiner and Koide 1993c |
| | Gi. margarita | ND | + | AMF in axenic culture + E | Poulin et al. 1993 |
| | | ND | + | AMF in axenic culture + E from root organ cultures | Bécard and Piché 1990 |
| | | ND | + | AMF (in axenic culture) + V from root organ cultures | Bécard and Piché 1990 |
| | | ND | 0 | AMF in axenic culture + E from root organ cultures | Bécard and Piché 1989a |
| | | ND | + | AMF (in axenic culture) + V from root organ cultures | Bécard and Piché 1989a |
| | | ND | + | AMF + root organ culture (monoxenic culture) | Bécard and Piché 1989a |
| | | ND | + | AMF + root organ culture (monoxenic culture) | Bécard and Piché 1989b |
| Lactuca sativa | G. intraradices | 0 | ND | AMF + roots in soil | Schreiner and Koide 1993b |
| Lavandula spica | G. geosporus | 0 | ND | AMF + roots in soil | Azcon and Ocampo 1984 |
| Lycopersicon esculentum | G. mosseae | ND | + | AMF + roots in soil | Vierheilig et al. 1995 |

| Plant | Fungus | | | Description | Reference |
|---|---|---|---|---|---|
| Medicago sativa | G. mosseae | 0 | ND | AMF in axenic culture + E from axenically cultured plants | El-Atrach et al. 1989 |
| | | 0 | ND | AMF (in axenic culture) + V from root of axenically cultured plants | El-Atrach et al. 1989 |
| | G. geosporus | ND | + | AMF + roots in quartz grit | Giovannetti et al. 1993b |
| | | 0 | ND | AMF + roots in soil | Azcon and Ocampo 1984 |
| Ocimum basilicum | G. mosseae | ND | + | AMF + roots in quartz grit | Giovannetti et al. 1994 |
| Phaseolus vulgaris | Gi. gigantea | 0 | ND | AMF (in axenic culture) + V from root of axenically cultured plants | Koske 1982 |
| Pisum sativum (Myc+) | G. mosseae | ND | + | AMF + roots in quartz grit | Giovannetti et al. 1993b |
| | Gi. margarita | ND | − | AMF in axenic culture + E from root organ cultures | Balaji et al. 1995 |
| | | ND | + | AMF (in axenic culture) + V from root organ cultures | Balaji et al. 1995 |
| Pisum sativum (Myc−) | G. mosseae | ND | + | AMF + roots in quartz grit | Giovannetti et al. 1993b |
| | Gi. margarita | ND | − | AMF in axenic culture + E from root organ cultures | Balaji et al. 1995 |
| | | ND | + | AMF (in axenic culture) + V from root organ cultures | Balaji et al. 1995 |
| Poncirus trifoliata X | | | | | |
| Citrus sinensis | G. epigaeum | + | + | AMF in axenic culture + E | Graham 1982 |
| Trifolium pratense | Gi. margarita | + | + | AMF in axenic culture + E | Gianinazzi et al. 1989 |
| | G. mosseae | ND | + | AMF + root organ culture (monoxenic culture) | Hepper and Mosse 1975 |
| T. repens | G. fasciculatus | 0 | + | AMF in axenic culture + E | Elias and Safir 1987 |
| | G. mosseae | 0 | + | AMF in axenic culture + E | Elias and Safir 1987 |
| Zea mays | G. geosporus | 0 | ND | AMF + roots in soil | Azcon and Ocampo 1984 |
| | Gi. gigantea | + | ND | AMF (in axenic culture) + V from root organ cultures | Suriyapperuma and Koske 1995 |

Stimulation (+), inhibition (−), no effect (0). E= collected root exudates; V= volatiles.

**Table 3.** Effect of roots (E+V), root exudates (E) and root volatiles (V) of AM nonhost plants on spore germination and hyphal growth of different AMF

| Plant | Fungus | Spore germination | Hyphal growth | Experimental system | Reference |
|---|---|---|---|---|---|
| Amaranthus retroflexus | Glomus intraradices | 0 | ND | AMF + roots in soil | Schreiner and Koide 1993b |
| Beta vulgaris | G. etunicatum | + | + | AMF + root organ culture (monoxenic culture) | Schreiner and Koide 1993c |
| | G. mosseae | ND | 0 | AMF + roots in quartz grit | Giovannetti et al. 1994 |
| | Gigaspora margarita | ND | 0 | AMF (in axenic culture) + E from root organ cultures | Bécard and Piché 1990 |
| | Gi. margarita | ND | + | AMF (in axenic culture) + V from root organ cultures | Bécard and Piché 1990 |
| Brassica kaber | G. intraradices | - | ND | AMF + roots in soil | Schreiner and Koide 1993b |
| | G. etunicatum | + | + | AMF + root organ culture (monoxenic culture) | Schreiner and Koide 1993c |
| | | - | ND | AMF + roots in soil | Schreiner and Koide 1993b |
| B. napus | G. mosseae | ND | - | AMF + roots in soil | Vierheilig et al. 1995 |
| | | ND | 0 | AMF + roots in quartz grit | Giovannetti et al. 1994 |
| B. nigra | G. intraradices | - | ND | AMF + roots in soil | Schreiner and Koide 1993b |
| | G. etunicatum | - | ND | AMF + roots in soil | Schreiner and Koide 1993b |
| | | + | + | AMF + root organ culture (monoxenic culture) | Schreiner and Koide 1993c |
| B. oleracea | G. mosseae | 0 | ND | AMF (in axenic culture) + E from axenically cultured plants | El-Atrach et al. 1989 |
| | | - | ND | AMF (in axenic culture) + V from roots of axenically cultured plants | El-Atrach et al. 1989 |
| | G. geosporus | 0 | ND | AMF + roots in soil | Azcon and Ocampo 1984 |
| | G. mosseae | ND | 0 | AMF + roots in quartz grit | Giovannetti et al. 1994 |
| Dianthus caryophyllus | G. mosseae | ND | 0 | AMF + roots in quartz grit | Giovannetti et al. 1994 |
| Eruca sativa | G. mosseae | ND | 0 | AMF + roots in quartz grit | Giovannetti et al. 1994 |
| Lupinus albus | Gi. margarita | 0 | 0 | AMF (in axenic culture) + E | Gianinazzi et al. 1989 |
| | | ND | 0 | AMF (in axenic culture) + E from root organ cultures | Balaji unpublished results |
| | | ND | + | AMF (in axenic culture) + V from root organ cultures | Balaji unpublished results |
| | G. mosseae | ND | 0 | AMF + roots in quartz grit | Giovannetti et al. 1993b |
| | | ND | 0 | AMF + roots in soil | Vierheilig et al. 1995 |
| Nasturtium officinale | G. mosseae | ND | 0 | AMF + roots in quartz grit | Giovannetti et al. 1994 |
| Raphanum raphanistrum | G. geosporus | 0 | ND | AMF + roots in soil | Azcon and Ocampo 1984 |
| Spinacea oleracea | G. mosseae | ND | - | AMF + roots in soil | Vierheilig et al. 1995 |
| | | ND | 0 | AMF + roots in quartz grit | Giovannetti et al. 1994 |
| | G. intraradices | 0 | ND | AMF + roots in soil | Schreiner and Koide 1993b |
| Urtica dioica | G. mosseae | ND | - | AMF + roots in soil | Vierheilig et al. 1996 |

Stimulation (+), inhibition (-), no effect (0). E= collected root exudates; V=volatiles.

Roots of the mycorrhizal resistant *Medicago sativa* mutant (Myc-) do not become colonized, but hyphae grow attached to the root surface forming appressoria. In presence of *G. versiforme* the levels of several flavonoids (daidzein, formononetin, formononetin malonyl glucoside, medicarpin, medicarpin malonyl glucoside) increased in the *Medicago sativa* (Myc-) roots in the same way as in colonized root of the wild type (Myc+) *M. sativa* (Harrison and Dixon, 1993). Volpin et al. (1994) also found an increase of formononetin in inoculated, but still uncolonized roots of *M. sativa* (Myc+). These results suggest that formononetin and the other flavonoids mentioned above do play a role in the plant-fungus signaling during precolonization and the cell-to-cell stage.

In contrast, some compounds are newly induced after AM fungal root colonization. Coumestrol and 4',7-dihydroxyflavone were only present in colonized roots of (Myc+) *M. sativa* and *M. truncatula*, but were not detected in uncolonized roots of the two plants and in roots of inoculated (Myc-) *M. sativa* (Harrison and Dixon, 1993)(Table 4). These results suggest that coumestrol and 4',7-dihydroxyflavone play a signaling role during the intraradical phase of the hyphae in the AM symbiosis.

Root exudates are known to contain flavonoids, which play an important role in the formation of the *Rhizobium* symbiosis (Firmin et al., 1986; Peters et al., 1986; Redmond et al., 1986; Kosslak et al., 1987). Recently in *Vicia sativa* inoculated with *Rhizobium leguminosarum* enhanced levels of flavonoids have been found in root exudates (Recourt et al., 1992). Xie et al, (1995) found that the inoculation of *Glycine max* with *Bradyrhizobium japonicum* increased the colonization by *G. mosseae* and an increased flavonoid concentration in root exudates was suggested to be responsible for this effect. Surprisingly, although there is some information about flavonoids in root exudates of nonmycorrhizal plants (Kape at al., 1992; Phillips and Tsai, 1992), no data are available about flavonoids in exudates of AM colonized roots. Studies about possible changes of the flavonoid pattern would be extremely interesting.

## 4.2. Effect of Phenolic Compounds on AMF *in Vitro*

Different phenolic compounds found in AM fungal roots or those which are known as signal compounds in other mutualistic symbiosis, have been tested for their effect on AM fungal development under axenic conditions. A large variety of sometimes contradictory responses was obtained (see Table 5), which could be attributed mostly to different causes: i) different experimental conditions (such as $CO_2$ levels or compound concentrations); ii) possible fungal species and genera specificities for a given compound; and iii) different developmental events of the symbiosis triggered by different compounds. In any case, as Harrison and Dixon (1993) emphasize, it is difficult to extrapolate from the observed effect of a single compound *in vitro* to the symbiotic situation in whole living root, where complex combinations of metabolites may have synergistic or antagonistic effects.

*4.2.1. Flavonoid Activity and Chemical Structure.* Flavonoids are divided into structurally distinct groups, such as flavonols, flavones, flavanones and isoflavones (Figure 1). In general flavonols have been shown to be stimulatory compounds for hyphal growth of *Gi. margarita,* and at least one hydroxyl group on the B aromatic ring has been found to be necessary for this effect (Bécard et al., 1992; Chabot et al., 1992). Besides this, a hydroxyl on position 3 has also been hypothesized to be essential to confer stimulatory activity to the flavonol molecule, as the flavones luteolin and apigenin (Table 5; Figure 1), lacking this hydroxyl showed no effect (Bécard et al., 1992). Glycosylation at position 3

**Table 4.** Changes of phenolic compound levels in roots induced by different AMF

| Compound | Plant | AMF | Changes in colonized roots | Comments | Reference |
|---|---|---|---|---|---|
| Blumenol C 9-O-(2'-O-ß-glururonosyl)-ß-glucoside (so called blumenin) | Hordeum vulgare | Glomus intraradices | + | | Maier et al. 1995 |
| | Triticum aestivum | G. intraradices | + | | Maier et al. 1995 |
| | Secale cereale | G. intraradices | + | | Maier et al. 1995 |
| | Avena sativa | G. intraradices | + | | Maier et al. 1995 |
| Coumaric acid | Allium cepa | G. intraradices | + | cell wall bound | Grandmaison et al. 1993 |
| | | G. versiforme | + | cell wall bound | Grandmaison et al. 1993 |
| | A. porrum | G. versiforme | 0 | cell wall bound | Codignola et al. 1989 |
| | H. vulgare | G. intraradices | 0 | cell wall bound | Maier et al. 1995 |
| | T. aestivum | G. intraradices | 0 | cell wall bound | Maier et al. 1995 |
| | S. cereale | G. intraradices | 0 | cell wall bound | Maier et al. 1995 |
| | A. sativa | G. intraradices | 0 | cell wall bound | Maier et al. 1995 |
| Coumestrol | Medicago truncatula | Glomus versiforme | + | only detected in AM roots | Harrison and Dixon 1993 |
| | M. sativa | G. versiforme | + | only detected in AM roots | Harrison and Dixon 1993 |
| | M. sativa (Myc-) | G. versiforme | 0 | roots not colonized, but appressoria formation | Harrison and Dixon 1993 |
| | Glycine max | G. mosseae | + | | Morandi et al. 1984 |
| | | G. fasciculatus | + | | Morandi et al. 1984 |
| | | G. intraradices | + | | Morandi 1989 |
| Daidzein | G. max | G. mosseae | + | | Morandi et al. 1984 |
| | | G. fasciculatus | + | | Morandi et al. 1984 |
| | | G. intraradices | + | | Morandi 1989 |
| | M. truncatula | G. versiforme | + | | Harrison and Dixon 1993 |
| | M. sativa | G. versiforme | + | | Harrison and Dixon 1993 |
| | M. sativa (Myc-) | G. versiforme | + | roots not colonized, but appressoria formation | Harrison and Dixon 1993 |
| 4',7-dihydroxyflavone | M. truncatula | G. versiforme | + | only detected in AM roots | Harrison and Dixon 1993 |
| | M. sativa | G. versiforme | + | only detected in AM roots | Harrison and Dixon 1993 |
| | M. sativa (Myc-) | G. versiforme | 0 | roots not colonized, but appressoria formation | Harrison and Dixon 1993 |
| Ferulic acid | A. cepa | G. intraradices | + | cell wall bound | Grandmaison et al. 1993 |
| | | G. versiforme | + | cell wall bound | Grandmaison et al. 1993 |
| | A. porrum | G. versiforme | 0 | cell wall bound | Codignola et al. 1989 |
| | H. vulgare | G. intraradices | 0 | cell wall bound | Maier et al. 1995 |
| | T. aestivum | G. intraradices | 0 | cell wall bound | Maier et al. 1995 |
| | S. cereale | G. intraradices | 0 | cell wall bound | Maier et al. 1995 |
| | A. sativa | G. intraradices | 0 | cell wall bound | Maier et al. 1995 |

# Flavonoids and Arbuscular-Mycorrhizal Fungi

| Compound | Plant | Fungus | Response | Notes | Reference |
|---|---|---|---|---|---|
| Formononetin | M. sativa | G. intraradices | + | | Volpin et al. 1994 |
| | M. truncatula | G. versiforme | +/0 | beginning increase/ later like control | Harrison and Dixon 1993 |
| | M. sativa | G. versiforme | + | weak increase | Harrison and Dixon 1993 |
| | M. sativa (Myc-) | G. versiforme | + | roots not colonized, but appressoria formation | Harrison and Dixon 1993 |
| Formononetin -7-O-glucoside | M. sativa | G. intraradices | 0 | | Volpin et al. 1994 |
| Formononetin malonyl glucoside | M. truncatula | G. versiforme | + | | Harrison and Dixon 1993 |
| | M. sativa | G. versiforme | 0 | | Harrison and Dixon 1993 |
| | M. sativa (Myc-) | G. versiforme | + | roots not colonized, but appressoria formation | Harrison and Dixon 1993 |
| Glyceollin | G. max | G. mosseae | + | late stage of association | Morandi et al. 1984 |
| | | G. fasciculatus | + | late stage of association | Morandi et al. 1984 |
| | | G. intraradices | + | late stage of association | Morandi 1989 |
| | | G. mosseae | 0 | early stage of association | Wyss et al. 1990;1991 |
| Medicarpin | M. truncatula | G. versiforme | +/0 | beginning increase/ later decrease | Harrison and Dixon 1993 |
| | M. sativa | G. versiforme | + | weak increase | Harrison and Dixon 1993 |
| | M. sativa (Myc-) | G. versiforme | + | roots not colonized, but appressoria formation | Harrison and Dixon 1993 |
| Medicarpin malonyl glucoside | M. truncatula | G. versiforme | + | | Harrison and Dixon 1993 |
| | M. sativa | G. versiforme | + | | Harrison and Dixon 1993 |
| | M. sativa (Myc-) | G. versiforme | + | roots not colonized, but appressoria formation | Harrison and Dixon 1993 |
| N-feruloyltyramine | A. cepa | G. intraradices | - | soluble | Grandmaison et al. 1993 |
| | A. cepa | G. versiforme | - | soluble | Grandmaison et al. 1993 |
| N-feruloyltyramine | A. cepa | G. intraradices | + | cell wall bound | Grandmaison et al. 1993 |
| | A. cepa | G. versiforme | + | cell wall bound | Grandmaison et al. 1993 |
| Protocathecuic acid | A. porrum | G. versiforme | 0 | cell wall bound | Codignola et al. 1989 |
| Syringic acid | A. porrum | G. versiforme | 0 | cell wall bound | Codignola et al. 1989 |
| Tyrosol | A. porrum | G. versiforme | 0 | cell wall bound | Codignola et al. 1989 |
| Vanillic acid | A. porrum | G. versiforme | 0 | cell wall bound | Codignola et al. 1989 |

(+) higher in mycorrhizal roots; (-) lower in mycorrhizal roots; (0) no changes.

Table 5. Effect of phenolic compounds on spore germination and hyphal growth of AM fungi

| Compound | Fungus | Carbondioxide concentration(%) | Tested concentrations (effective conc.) μM | Spore germination | Hyphal growth | Reference |
|---|---|---|---|---|---|---|
| Apigenin | Gigaspora margarita | 0.03 | 0.015-1.5 (0.15) | + | + | Gianinazzi-Pearson et al.1989 |
|  | Gi. margarita | 2 | 10 | ND | - | Bécard et al. 1992 |
|  | Gi. gigantea | 0.03 | 25-200 | 0 | ND | Baptista and Siqueira 1994 |
|  |  | liquid medium | 1-10 | ND | 0 | Baptista and Siqueira 1994 |
| Biochanin A | Gi. margarita | 2 | 10 | - | - | Chabot et al. 1992 |
|  |  | 2 | 10 | ND | - | Bécard et al. 1992 |
|  |  | 2 | 10 | ND | - | Poulin unpublished results |
|  | Gi. gigantea | 0.03 | 50-400 (50<400) | - | ND | Baptista and Siqueira 1994 |
|  |  | liquid medium | 1-10 (1<10) | ND | - | Baptista and Siqueira 1994 |
|  | Glomus intraradices | 2 | 0.01-10 (0.1<10) | ND | + | Poulin unpublished results |
|  | Glomus sp. | 0.03 | 17.6 | ND | + | Nair et al. 1991 |
| Chalcone | Gi. margarita | 2 | 10 | ND | - | Bécard et al. 1992 |
| Coumestrol | Gi. margarita | 0.03 | 0.05-50 (50) | 0 | + | Morandi et al. 1992 |
| Chrysin | Gi. margarita | 2 | 10 | - | - | Chabot et al. 1992 |
|  |  | 2 | 10 | ND | - | Bécard et al. 1992 |
| Fisetin | Gi. gigantea | 0.03 | 6.25-50 (6.25<30>50) | - | ND | Baptista and Siqueira 1994 |
|  |  | liquid medium | 1-10 (1<4>5) | ND | + | Baptista and Siqueira 1994 |
| Flavone | Gi. margarita | 2 | 10 | ND | - | Bécard et al. 1992 |
|  | Gi. gigantea | 0.03 | 12.5-100(12.5<100) | - | ND | Baptista and Siqueira 1994 |
|  |  | liquid medium | 1-10 | ND | 0 | Baptista and Siqueira 1994 |
| Formononetin | G. etunicatum | 0.03 | 2.5 | - | ND | Tsai and Phillips 1991 |
|  | G. macrocarpum | 0.03 | 2.5 | - | ND | Tsai and Phillips 1991 |
|  | Glomus sp. | 0.03 | 18.6 | ND | + | Nair et al. 1991 |
|  | Gi. gigantea | 0.03 | 50-400 | 0 | ND | Baptista and Siqueira 1994 |
|  |  | liquid medium | 0.25-10 (0.25=2.5) | ND | + | Baptista and Siqueira 1994 |

| Compound | Fungus | Conc. | Range | Effect | Reference |
|---|---|---|---|---|---|
| Daidzein | G. mosseae | 0.03 | 2.0; 5.0 (2.0>5.0) | + | Kape et al. 1992 |
| | Gi. gigantea | liquid medium | 1-10 | ND | Baptista and Siqueira 1994 |
| 4,4'-dihydroxy-2'-methoxychalcone | G. etunicatum | 0.03 | 2.5 | 0 | Tsai and Phillips 1991 |
| | G. macrocarpum | 0.03 | 2.5 | - | Tsai and Phillips 1991 |
| 4',7-dihydroxyflavone | G. etunicatum | 0.03 | 2.5 | + | Tsai and Phillips 1991 |
| | G. macrocarpum | 0.03 | 2.5 | 0 | Tsai and Phillips 1991 |
| N-feruloyltyramine | G. intraradices | 0.03 | 0.14 | ND | Grandmaison et al. 1993 |
| Galangin | Gi. margarita | 2 | 10 | ND | Chabot et al. 1992 |
| Genistein | Gi. margarita | 2 | 10 | - | Chabot et al. 1992 |
| | Gi. gigantea | 0.03 | 25-200 | 0 | Baptista and Siqueira 1994 |
| | | liquid medium | 1-10 | ND | Baptista and Siqueira 1994 |
| Glyceollin | Gi. margarita | 0.03 | 0.05-50 (0.5=5.0) | 0 | Morandi et al. 1992 |
| Hesperitin | Gi. margarita | 0.03 | 0.015-15 (15) | + | Gianinazzi-Pearson et al. 1989 |
| | | 0.03 | 0.015-15 (0.015<0.15=1.5) | ND | Gianinazzi-Pearson 1989 |
| | | 2 | 10 | - | Chabot et al. 1992 |
| | | 2 | 10 | - | Bécard et al. 1992 |
| | Gi. gigantea | 0.03 | 50-400 (50<252>400) | ND | Baptista and Siqueira 1994 |
| | | liquid medium | 1-10 (1<2) | + | Baptista and Siqueira 1994 |
| Kaempferol | Gi. margarita | 2 | 10 | + | Chabot et al. 1992 |
| | | 2 | 10 | + | Bécard et al. 1992 |
| | Gi. gigantea | liquid medium | 1-10 (10) | - | Baptista and Siqueira 1994 |

Table 5. (continued)

| Compound | Fungus | Carbondioxide concentration(%) | Tested concentrations (effective conc.) µM | Spore germination | Hyphal growth | Reference |
|---|---|---|---|---|---|---|
| Liquiritigenin (4';7-dihydroxyflavanone) | G. etunicatum | 0.03 | 2.5 | + | ND | Tsai and Phillips 1991 |
| | G. macrocarpum | 0.03 | 2.5 | 0 | ND | Tsai and Phillips 1991 |
| Luteolin | G. etunicatum | 0.03 | 2.5 | 0 | ND | Tsai and Phillips 1991 |
| | G. macrocarpum | 0.03 | 2.5 | 0 | ND | Tsai and Phillips 1991 |
| | Gi. margarita | 2 | 10 | ND | 0 | Chabot et al. 1992 |
| | Gi. margarita | 2 | 10 | ND | 0 | Bécard et al. 1992 |
| Luteolin-7-O-glucoside | G. etunicatum | 0.03 | 2.5 | + | ND | Tsai and Phillips 1991 |
| | G. macrocarpum | 0.03 | 2.5 | 0 | ND | Tsai and Phillips 1991 |
| Morin | Gi. margarita | 2 | 10 | ND | + | Chabot et al. 1992 |
| | Gi. gigantea | 0.03 | 50-200 (50<200) | - | ND | Baptista and Siqueira 1994 |
| | | liquid medium | 1-10 (1<4.1>5) | ND | + | Baptista and Siqueira 1994 |
| Myricetin | G. mosseae | 0.03 | 0.8-5.0 (2.0) | ND | + | Kape et al. 1992 |
| | Gi. margarita | 2 | 10; 20 (10=20) | ND | + | Bécard et al. 1992 |
| | | 0.03<2.0 | 10 | ND | + | Poulin et al. 1993 |
| | Gi. gigantea | liquid medium | 1-10 | ND | 0 | Baptista and Siqueira 1994 |
| Naringenin | Gi. margarita | 0.03 | 0.015-1.5 (1.5) | 0 | + | Gianinazzi-Pearson et al.1989 |
| | | 2 | 10 | ND | 0 | Bécard et al. 1992 |
| | | 0.03>2.0 | 10 | ND | - | Poulin et al. 1993 |
| | Gi. gigantea | 0.03 | 50-400 (50<400) | - | ND | Baptista and Siqueira 1994 |
| | | liquid medium | 1-10 (1<2) | ND | + | Baptista and Siqueira 1994 |

# Flavonoids and Arbuscular-Mycorrhizal Fungi

| Compound | Fungus | Conc. | Range | Effect 1 | Effect 2 | Reference |
|---|---|---|---|---|---|---|
| Quercetin | G. mosseae | 0.03 | 0.8-5.0 (0.8>2) | ND | + | Kape et al. 1992 |
| | G. etunicatum | 0.03 | 0.05-10.0 (1.0<2.5) | + | + | Tsai and Phillips 1991 |
| | G. intraradices | 2 | 10 | ND | + | Bécard et al. 1992 |
| | | 2 | 10 | ND | + | Bécard et al. 1992 |
| | | 2 | 5 | ND | 0 | Poulin unpublished results |
| | G. macrocarpum | 0,03 | 2,5 | + | ND | Tsai and Phillips 1991 |
| | Gi. margarita | 0.03<2.0 | 10 | ND | + | Chabot et al. 1992 |
| | | 0.0<2.0 | 10; 20 (10=20) | ND | + | Bécard et al. 1992 |
| | | 2 | 10 | ND | + | Bécard et al. 1995 |
| | | 0.03 | 0.1-10 | 0 | 0 | Morandi et al. 1992 |
| | | 2 | 5 | ND | + | Poulin unpublished results |
| | | 0.03<2.0 | 10 | ND | + | Poulin et al. 1993 |
| | Gi. gigantea | 0.03 | 25-200 (25<60>200) | - | ND | Baptista and Siqueira 1994 |
| | | liquid medium | 5-50 (5<19) | ND | + | Baptista and Siqueira 1994 |
| Quercetin-3-O-galactoside | G. etunicatum | 0.03 | 2.5 | + | ND | Tsai and Phillips 1991 |
| | G. macrocarpum | 0.03 | 2.5 | + | ND | Tsai and Phillips 1991 |
| Quercitrin (Quercetin-3-rhamnoside) | Gi. margarita | 2 | 10 | ND | 0 | Bécard et al. 1992 |
| Rutin | Gi. margarita | 2 | 10 | ND | 0 | Chabot et al. 1992 |
| | | 2 | 10 | ND | 0 | Bécard et al. 1992 |
| Taxifolin | Gi. gigantea | 0.03 | 25-200 (25<200) | - | ND | Baptista and Siqueira 1994 |
| | | liquid medium | 1-10 | ND | 0 | Baptista and Siqueira 1994 |

Stimulation (+), inhibition (-), no effect (0), not determined (ND); (x>y) means x has a higher effect compared to y.

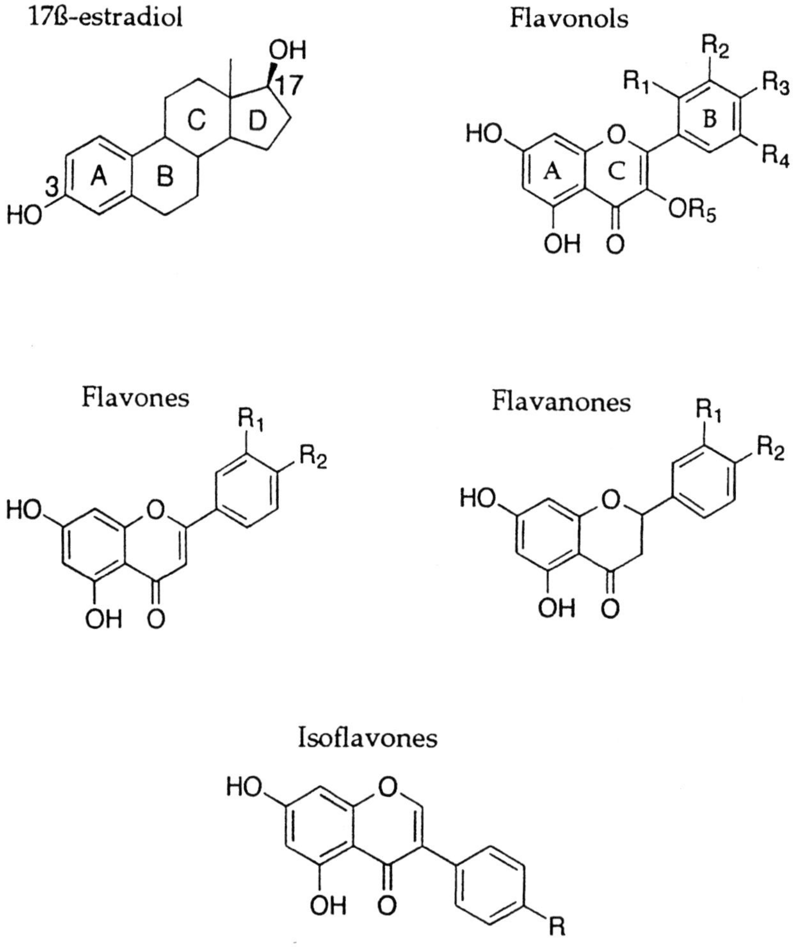

**Figure 1.** Chemical structure of estrogen and flavonoids. *Estrogen*: 17 β-estradiol. *Flavonols*: Quercetin $R_1=R_4=R_5=H$, $R_2=R_3=OH$; Kaempferol $R_1=R_2=R_4=R_5=H$, $R_3=OH$; Morin $R_2=R_4=R_5=H$, $R_1=R_3=OH$; Galangine $R_1=R_2=R_3=R_4=R_5=H$; Rutin $R_1=R_4=H$, $R_2=R_3=OH$, $R_5=$rutinose. *Flavones*: Luteolin $R_1=R_2=OH$; Chrysin $R_1=R_2=H$. *Flavanone*: Hesperetin $R_1=OH$, $R_2=OCH_3$. *Isoflavones*: Biochanin A $R=OCH_3$; Genistein $R=OH$.

also promoted the loss of the stimulatory activity, as in quercitrin and rutin (Bécard et al., 1992; Chabot et al., 1992) (Figure 1).

Moreover, it has been suggested that saturation of the 2,3 double bond in flavonols, as seen in flavanones, promoted the loss of activity on hyphal growth of *Gi. margarita* (Chabot et al., 1992). This loss of activity was attributed to the loss of the planar configuration of the flavonol molecule when the double bond disappears. Experiments with dihydroquercetin and dihydrokaempferol, which have the same structure as the corresponding flavonols quercetin and kaempferol, except for the saturated 2,3 double bond, showed no effect on hyphal growth on *Gi. margarita* and it was suggested that the presence of the 2,3 double bond is essential for the stimulatory effect on hyphal growth. However, these results are in contrast with those of Gianinazzi-Pearson et al. (1989) and Baptista and Siqueira (1994). Naringenin, lacking the 2,3 double bond, was stimulatory with *Gi. mar-*

*garita* and *Gi. gigantea*, whereas apigenin, differing only by the presence of the 2,3 double bond, showed no effect with *Gi. gigantea*.

The data listed in Table 5 and 6 suggest that, in general, the flavonols are the effective flavonoid group on hyphal growth of the *Gigaspora* genera, whereas *Glomus* spp. is stimulated by flavonols as well as by isoflavones (Tables 5 and 6). The flavonols might be more general signal molecules as they showed effect in *Gigaspora* spp. and *Glomus* spp.. Contradictory results have been reported on the effect of isoflavones, flavones and flavanones on *Gigaspora* spp.. Compounds of these groups depending on the experiment either stimulated, inhibited, or showed no effect on AM fungal growth (Tables 5 and 6). Data about the effect of flavonoids on other AM fungal genera as *Acaulospora* spp. or *Scutellospora* spp. are not available.

*4.2.2. Effect of Flavonoids on Spore Germination.* Few data are available on the effect of phenolic compounds on *in vitro* AM fungal spore germination and these data are sometimes contradictory. For example, apigenin stimulated spore germination of *Gi. margarita* (Gianinazzi-Pearson et al., 1989) and showed no effect with *Gi. gigantea* (Baptista and Siqueira, 1994)(Table 5). Formononetin was inhibitory at low concentrations (2,5 µM) on two *Glomus* species (Tsai and Phillips, 1991), but exhibited no effect at high concentrations (50–400 µM) on *Gi. gigantea* (Baptista and Siqueira, 1994)(Table 5). Even two *Glomus* species seemed to differ in their response to the same concentration of 4,4'-dihydroxy-2'-methoxychalcone, 4',7-dihydroxyflavone and 4',7-dihydroxyflavanone (Tsai and Phillips, 1991). These compounds seemed to be fungal species specific (Table 5).

*4.2.3. Effect of Flavonoids on Hyphal Growth.* In most experiments with flavonoids their effect on hyphal growth of AMF was studied. From the overview of the data (Table 5) it can be concluded, that the effect of some flavonoids seemed to be dependent on the $CO_2$ concentration, the compound concentration, and the AM fungal species.

Table 6. Groups of flavonoids and their effect on hyphal growth of different AMF

| | Effect on hyphal growth | | | Effect on hyphal growth | |
|---|---|---|---|---|---|
| | Gigaspora spp. | Glomus spp. | | Gigaspora spp. | Glomus spp. |
| **ISOFLAVONES** | | | **FLAVONOLS** | | |
| Biochanin A | - | + | Fisetin | + | ND |
| Coumestrol | + | ND | Galangin | - | ND |
| Daidzein | 0 | ND | Kaempferol | +/- | ND |
| Formononetin | + | + | Morin | + | ND |
| Genistein | -/0 | ND | Myricetin | +/0 | + |
| Glyceollin | + | ND | Quercetin | +/0 | +/0 |
| | | | Quercitrin | 0 | ND |
| **FLAVONES** | | | Rutin | 0 | ND |
| Apigenin | +/-/0 | ND | **FLAVANONES** | | |
| Chrysin | - | ND | | | |
| 4',7-dihydroxyflavone | ND | ND | Hesperetin | +/- | ND |
| Flavone | -/0 | ND | 4',7-dihydroxyflavanone | ND | ND |
| Luteolin | 0 | ND | Naringenin | +/-/0 | ND |
| | | | Taxifolin | 0 | ND |

Stimulation (+); inhibition (-); no effect (0); no data (ND), contradictory data (+/-/0). For more details about each compound see Table 5.

High $CO_2$ levels have been shown to enhance the effect of flavonoids on AM fungal growth. For example, the effect of myricetin and quercetin on hyphal growth was more pronounced at $CO_2$ concentrations higher than 0.03% (ambiental $CO_2$ level) and was highest at a $CO_2$ concentration around 2% (usual level in the soil)(Bécard et al., 1992; Chabot et al., 1992; Poulin et al., 1993) (Table 5).

Different concentrations of fisetin (1–5μM), morin (1–5μM)(Baptista and Siqueira, 1994) and quercetin (Tsai and Phillips, 1991 (0.25–2.5μM); Kape et al., 1992 (0.8–2μM); Baptista and Siqueira, 1994 (5–10μM)) showed different stimulation levels on hyphal growth. At higher concentrations fisetin (10μM), morin (10μM)(Baptista and Siqueira, 1994) and quercetin (Kape et al., 1992 (5μM); Baptista and Siqueira, 1994 (50μM)) exhibited an inhibitory effect.

Different AM fungal genera did not always show the same response to certain flavonoids (see as well "4.2.5. Flavonoid binding sites of AM fungi"). Biochanin A inhibited hyphal growth of the *Gigaspora* genus and stimulated hyphal growth of the *Glomus* genus (Tables 5 and 6). Other compounds, such as flavone, genistein, hesperetin, naringenin, kaempferol, liquiritigenin and luteolin-7-*O*-glucoside showed interspecific differences, e.g. flavone, which inhibits hyphal growth of *Gi. margarita* showed no effect on *Gi. gigantea*, and naringenin and hesperetin, which stimulated hyphal growth of *Gi. gigantea*, showed inhibition or no effect on *Gi. margarita*. However, in these experiments the $CO_2$ levels and/or the concentrations of the tested compound were different, therefore, it is difficult to attribute these differences to a single cause. Regardless of the fungal genera, *Glomus* or *Gigaspora*, myricetin and quercetin in general stimulated hyphal growth (Tables 5 and 6), suggesting their role as general AM signal molecules in plant-AM fungus communication. It would be interesting if their hyphal growth stimulating effect is also present on other AM genera; however, no data are available.

*4.2.4. Effect of Flavonoids on Fungal Differentiation.* Spore germination and hyphal growth are the responses most frequently used for studying the effect of phenolic compounds on AMF. However, these compounds may also play a role in other processes, such as hyphal differentiation during precolonization and/or colonization. Changes in the morphology of AMF induced by compounds released by AM host roots have been reported above (see "3.2. Effect of root exudates"). This hyphal differentiation effect may be caused by flavonoids in the root exudates. Enhanced hyphal branching of *G. etunicatum* was found *in vitro* after application of quercetin (Tsai and Phillips, 1991) and 4',7-dihydroxyflavone (Phillips and Tsai, 1992). 4',7-dihydroxyflavone has been suggested to play a signaling role during the intraradical phase of the fungus (Harrison and Dixon, 1993), as it was only detected in colonized roots. The enhanced hyphal branching observed *in vitro* in presence of 4',7-dihydroxyflavone might be an indication for its role in the formation of the highly ramified arbuscules. Naringenin found in root exudates of bean (Hungria et al., 1991) enhanced in axenic culture of *Gi. margarita* the number of vesicle clusters per germinated spore. A similar effect was observed with apigenin and hesperetin (Gianinazzi-Pearson et al., 1989). Luteolin, found in the seed rinse of alfalfa (Hartwig et al., 1990), although showing no stimulation of hyphal growth, stimulated the production of auxiliary cells in *Gi. margarita* (Bécard et al., 1992). Hyphal differentiation possibly requires a different stimulatory mechanism or induction than does hyphal growth. Some flavonoids may be involved in only one of these mechanisms and not in the other. However, quercetin and myricetin, which exhibit a hyphal growth stimulating effect on different AM fungal genera, also enhanced the formation of auxiliary cells (Bécard et al., 1992), suggesting again a more general role of these two compounds compared to other flavonoids.

*4.2.5. Flavonoid Binding Sites on AMF.* Flavonoids have been shown to have the capability to bind to estrogen receptors and to exhibit certain steroid-related functions in mammalian systems, but to be less active than estrogens (Markaverich et al., 1988; Whitten and Naftolin, 1991; Miksicek, 1993). When 17β-estradiol was applied to the AMF *G. intraradices* it stimulated hyphal growth, although to a lower extent than the flavonoid biochanin A (Poulin et al. 1997; Table 7). There is a strong structural resemblance between these two molecules (Figure 1). The enhanced fungal growth with 17β-estradiol suggested that the fungus possesses estrogen like-binding sites.

Antiestrogens are estrogen derivatives. They bind specifically to estrogen receptors without the transformation of the bound receptor to the active form and the following gene activation (Brann et al., 1995). Interestingly the antiestrogen, EM-652 decreased the stimulatory effect of biochanin A on hyphal growth of *G. intraradices* (Poulin et al., 1997; Table 7). Since EM-652 alone, at the same concentrations, had no significant effect on hyphal growth, it was suggested that the blocking of the stimulatory effect results from the capability of EM-652 to block the active binding sites for biochanin A in *G. intraradices.*

Similarly, hyphal growth stimulation of *Gi. margarita* by quercetin, decreased significantly when the fungus was grown in the presence of quercetin and the antiestrogen EM-170, indicating a competition between the flavonoid and the antiestrogen for the same binding site (Poulin et al., 1997; Table 7).

Biochanin A showed a stimulatory effect with *G. intraradices,* but was inhibitory for *Gi. margarita,* and quercetin showed hyphal growth stimulation with *Gi. margarita,* but

Table 7. Effect of flavonoids, estrogen, antiestrogens and their combinations on hyphal growth of two AMF

| Compounds and compound combinations | Effect on hyphal growth of | |
|---|---|---|
| | *G. intraradices* | *Gi. margarita* |
| **Flavonoids** | | |
| Biochanin A | +++ | - |
| Quercetin | 0 | +++ |
| **Estrogen** | | |
| 17β-estradiol | ++ | 0 |
| **Antiestrogens** | | |
| EM 139 | 0 | 0 |
| EM 170 | ND | 0 |
| EM 652 | 0 | ND |
| **Combinations** | | |
| Biochanin A + EM 652 | + | ND |
| Biochanin A + EM 139 | +++ | ND |
| Quercetin + EM 170 | ND | + |
| Quercetin + EM 139 | ND | + |

Stimulation of hyphal growth compared to control treatment(+), (+++)=high stimulation; inhibition of hyphal growth compared to control treatment (-); hyphal growth as control treatment (0); not determined (ND).

not with *G. intraradices* (Table 7). 17β-estradiol stimulated hyphal growth of *G. intraradices*, but not of *Gi. margarita* (Poulin et al., unpublished results; Table 7). The antiestrogen EM-139 alone showed no effect on the two fungi tested. Applied in combination with quercetin to *Gi. margarita*, hyphal growth was reduced, thus indicating the blocking of the quercetin binding site. However, when EM-139 was applied to *G. intraradices,* in combination with biochanin A the fungal growth was stimulated to the same degree as with biochanin A alone (Poulin et al., unpublished results; Table 7). Apparently, EM-139 was able to block the binding site for quercetin in *Gi. margarita,* but not for biochanin A in *G. intraradices.* These results suggest different binding sites for the two flavonoids on the two fungi.

## 4.3. Effect of Flavonoids on AM *in Vivo* (Root Colonization)

Flavonoids have been shown to exhibit certain effects on AM fungal spore germination and hyphal growth, but their actual role in the formation of the AM symbiosis is not known. In a recent study it has been shown that inoculation of *Glycine max* with *Bradyrhizobium japonicum* enhanced root AM colonization and this effect has been attributed to the increase of certain flavonoids in the root exudates (Xie at al., 1995). Direct application of different flavonoids to the soil in which plants were growing in the presence of AM fungi resulted in an increase in root colonization (Table 8).

For example, apigenin, coumestrol and daidzein increased root colonization of *G. max* by *G. mosseae* (Xie at al., 1995; Table 8). The regulatory mechanism for this enhanced colonization is still unclear. Biochanin A and formononetin are known to stimulate hyphal growth of different *Glomus* species *in vitro* (Nair et al., 1991; Poulin et al., 1997; Table 5). The two compounds applied to the soil enhanced root colonization by *Glomus* spp. in many plant species (Siqueira et al., 1991a,b; Nair et al., 1991; Vierheilig unpublished results; Table 8), suggesting that these two compounds, isolated from clover (Nair et al. 1991), are general stimulants for root colonization by *Glomus* spp.. Application of biochanin A and formononetin to soil in which the AM nonhost plants, *Lupinus polyphyllus* and *Spinacea oleracea* grew in the presence of *G. mosseae*, resulted in hyphal attachment and more hyphae around these roots (Vierheilig and Piché, 1995). These results could indicate a simple hyphal growth stimulation to be responsible for the enhanced root colonization in AM host plants, however, as soil application also resulted in a colonization of *L. polyphyllus* roots by *G. mosseae* (Vierheilig and Piché, 1995; Vierheilig et al., 1996b), a role of these compounds in the plant-fungus communication during the colonization stage is suggested.

## 5. CONCLUSION

The presented data confirm the hypothesis that root exudates play an important role in the plant-fungus communication and signaling during the precolonization stages of AM symbiosis. Phenolic compounds, or more precisely, flavonoids, are active components in root exudates, therefore they may play a fundamental role in the symbiont's communication, before and during AM colonization, as already demonstrated with other mutualistic symbioses. Specific flavonoids do not cause similar effects on different AMF. This could be attributed either to differences in experimental conditions, or to a specificity of each compound for a fungal genera or even species, as well as to the implication of each compound in a different event of the establishment of AM symbiosis and functioning. It also

# Flavonoids and Arbuscular-Mycorrhizal Fungi

Table 8. Effect of flavonoid application to the soil on AM formation

| Plant | Fungus | Flavonoid | Tested concentration (effective conc.) μM | Effect on colonization | Frequency of application | Reference |
|---|---|---|---|---|---|---|
| **AM host plants** | | | | | | |
| Cucumis sativus | Glomus mosseae | Biochanin A | 18 | + | twice during 4 weeks | Vierheilig unpublished results |
| Glycine max | G. mosseae | Apigenin | 0.1 | + | daily during 7 days | Xie et al. 1995 |
| | G. mosseae | Coumestrol | 0.1 | + | daily during 7 days | Xie et al. 1995 |
| | G. mosseae | Daidzein | 0.1 | + | daily during 7 days | Xie et al. 1995 |
| | G. mosseae | Genistein | 0.1 | 0 | daily during 7 days | Xie et al. 1995 |
| Lycopersicon esculentum | G. mosseae | Biochanin A | 18 | + | twice during 4 weeks | Vierheilig unpublished results |
| Phaseolus vulgaris | G. mosseae | Biochanin A | 18 | + | twice during 4 weeks | Vierheilig unpublished results |
| Trifolium repens | G. intraradices | Biochanin A | 9-70 (19>37) | + | once during 4 weeks | Siqueira et al. 1991a |
| | G. mosseae | Biochanin A | 18 | + | twice during 4 weeks | Vierheilig unpublished results |
| | Glomus sp. | Biochanin A | 18 | + | once during 4 weeks | Nair et al. 1991 |
| | G. intraradices | Formononetin | 9-76 (19>37) | + | once during 4 weeks | Siqueira et al. 1991a |
| | Glomus sp. | Formononetin | 19 | + | once during 4 weeks | Siqueira et al. 1991b; Nair et al 1991 |
| | Glomus sp. | Chrysin | 20-240 (120<240) | + | once during 4 weeks | Siqueira et al. 1991a |
| | G. intraradices | Luteolin | 18 | 0 | once during 4 weeks | Siqueira et al. 1991a |
| | G. intraradices | Genistein | 19 | 0 | once during 4 weeks | Siqueira et al. 1991a |
| | G. intraradices | Naringenin | 18 | 0 | once during 4 weeks | Siqueira et al. 1991a |
| | G. intraradices | 7-8, Dihydroxyflavone | 20 | 0 | once during 4 weeks | Siqueira et al. 1991a |
| | G. intraradices | Hesperetin | 17 | 0 | once during 4 weeks | Siqueira et al. 1991a |
| Triticum vulgare | G. mosseae | Biochanin A | 18 | + | twice during 4 weeks | Vierheilig unpublished results |
| Zea mays | G. intraradices | Biochanin A | 18 | + | once during 4 weeks | Siqueira et al. 1991b |
| | G. intraradices | Formononetin | 19 | + | once during 4 weeks | Siqueira et al. 1991b |
| **AM nonhost plants** | | | | | | |
| Brassica campestris | Glomus sp. | Formononetin | 19 | 0 | once during 4 weeks | Siqueira et al. 1991b |
| B. napus | G. mosseae | Biochanin A | 18 | 0 | twice during 4 weeks | Vierheilig unpublished results |
| Spinacea oleracea | G. mosseae | Biochanin A | 18 | 0 | twice during 4 weeks | Vierheilig unpublished results |
| Lupinus albus | G. mosseae | Biochanin A | 18 | 0 | twice during 4 weeks | Vierheilig unpublished results |
| L. polyphyllus | G. mosseae | Biochanin A | 18 | + | twice during 4 weeks | Vierheilig and Piché 1995 |
| | G. mosseae | Formononetin | 19 | + | twice during 4 weeks | Vierheilig and Piché 1995 |

Stimulation (+), no effect (0); (x<y) means the concentration x showed a lower effect than the concentration y.

should not be forgotten that in all *in vitro* experiments AMF were exposed to a single flavonoid and not to a complex combination of metabolites as found in the root and exuded by the root. Standardization of the experimental conditions and additional data about the effects of flavonoids on other AM fungal genera and in soil are further challenges for future research in order to clarify the importance and the precise role of flavonoids in arbuscular mycorrhizae.

## ACKNOWLEDGMENTS

This work was supported by a grant of the "Deutsche Forschungsgemeinschaft" (DFG) Germany, to H.V. Thanks to Eric Langlois for Figure 1 and Christine Juge for helpful discussion.

## REFERENCES

Azcon, R., Ocampo, J. A. Effect of root exudation on VA mycorrhizal infection at early stages of plant growth. Plant and Soil **1984**, 82, 133–138.
Bailey, J.A. Mechanisms of phytoalexin accumulation. In *Phytoalexins*; Bailey, J. A., Mansfield, J. W. Eds. Blackie and Son; Glasglow, UK. **1982**, pp. 289–318.
Balaji, B. Use of Ri T-DNA transformed roots of pea mutants and a non-host (lupin) in studying precolonization and colonization stages of the arbuscular mycorrhizal (AM) symbiosis. PhD Thesis, Université Laval, Québec City, Canada, **1996**, pp. 132.
Balaji, B., Poulin, M. J., Vierheilig, H., Piché, Y. Responses of an arbuscular mycorrhizal fungus, *Gigaspora margarita*, to exudates and volatiles from the Ri T-DNA-transformed roots of nonmycorrhizal and mycorrhizal mutants of *Pisum sativum* L. Sparkle. Exp. Mycol. **1995**, 19, 275–283.
Baptista, M. J., Siqueira, J. O. Efeito de flavonóides na germinação de esporos e no crescimento assimbiótico do fungo micorrízico arbuscular *Gigaspora gigantea*. R. Bras. Fisiol. Veg. **1994**, 6, 127–134.
Bécard, G., Piché, Y. Fungal growth stimulation by $CO_2$ and root exudates in vesicular-arbuscular mycorrhizal symbiosis. Appl. Environ. Microbiol. **1989**a, 55, 2320–2325.
Bécard, G., Piché, Y. New aspects on the acquisition of biotrophic status by a vesicular-arbuscular mycorrhizal fungus, *Gigaspora margarita*. New Phytol. **1989**b, 112, 77–83.
Bécard, G., Piché, Y. Physiological factors determining vesicular-arbuscular mycorrhizal formation in host and nonhost Ri T-DNA transformed roots. Can. J. Bot. **1990**, 68, 1260–1264.
Bécard, G., Douds, D. D., Pfeffer, P. E. Extensive *in vitro* hyphal growth of vesicular-arbuscular mycorrhizal fungi in the presence of $CO_2$ and flavonols. Appl. Environ. Microbiol. **1992**, 58, 821–825.
Bécard, G., Taylor, L. P., Douds, D. D., Pfeffer, P. E., Doner, L. W. Flavonoids are not necessary plant signal compounds in arbuscular mycorrhizal symbiosis. MPMI **1995**, 8, 252–258.
Bonfante-Fasolo, P. Anatomy and morphology of VA mycorrhizae. In *VA Mycorrhizas*; Powell, C. L., Bajyarai, D. J., Eds. **1984**, CRC Press, Boca Raton, Fl., pp. 5–33.
Brann, D. W., Hendry, L. B., Mahesh, V. B. Emerging diversities in the mechanism of action of steroid hormones. J. Steroid Biochem. Mol. Biol. **1995**, 52, 113–133.
Chabot, S., Bel-Rhlid, R., Chênevert, R., Piché, Y. Hyphal growth promotion *in vitro* of the VA mycorrhizal fungus, *Gigaspora margarita* Becker & Hall, by the activity of structurally specific flavonoid compounds under $CO_2$-enriched conditions. New Phytol. **1992**, 122, 461–467.
Codignola, A., Verotta, L., Spanu, P., Maffei, M., Scannerini, S., Bonfante-Fasolo, P. Cell wall bound-phenols in roots of vesicular-arbuscular mycorrhizal plants. New Phytol. **1989**, 112, 221–228.
Daniels, B. A., Trappe, J. M. Factors affecting spore germination of the vesicular-arbuscular mycorrhizal fungus *Glomus epigaeus*. Mycologia **1980**, 72, 457–471.
El-Atrach, F., Vierheilig, H., Ocampo, J. A. Influence of non-host plants on vesicular-arbuscular mycorrhizal infection of host plants and on spore germination. Soil Biol. Biochem. **1989**, 21, 161–163.
Elias, K. S., Safir, G. R. Hyphal elongation of *Glomus fasciculatus* in response to root exudates. Appl. Environ. Microbiol. **1987**, 53, 1928–1933.
Firmin, J. L., Wilson, K. E., Rossen, L., Johnston, A. W. B. Flavonoid activation of nodulation genes in *Rhizobium* reversed by other compounds in plants. Nature **1986**, 324, 90–92.

Gianinazzi-Pearson, V., Branzanti, B., Gianinazzi, S. *In vitro* enhancement of spore germination and early hyphal growth of a vesicular-arbuscular mycorrhizal fungus by host root exudates and plant flavonoids. Symbiosis **1989**, 7, 243–255.

Giovannetti, M., Avio, L., Sbrana, C., Citernesi, A. S. Factors affecting appressorium development in the vesicular-arbuscular mycorrhizal fungus *Glomus mosseae* (Nicol. & Gerd.) Gerd. & Trappe. New Phytol. **1993**a, 123, 115–122.

Giovannetti, M., Sbrana, C., Avio, L., Citernesi, A. S., Logi, C. Differential hyphal morphogenesis in arbuscular mycorrhizal fungi during preinfection stages. New Phytol. **1993**b, 125, 587–593.

Giovannetti, M., Sbrana, C., Citernesi, A. S., Avio, L. Analysis of factors involved in fungal recognition responses to host-derived signals by arbuscular mycorrhizal fungi. New Phytol. **1996**, 133, 65–77.

Giovannetti, M., Sbrana, C., Logi, C. Early processes involved in host recognition by arbuscular mycorrhizal fungi. New Phytol. **1994**, 127, 703–709.

Graham, J. H. Effect of citrus root exudates on germination of chlamydospores of the vesicular-arbuscular mycorrhizal fungus, *Glomus epigaeum*. Mycologia **1982**, 74, 831–835.

Grandmaison, J., Olah, G. M., Van Calsteren, M. R., Furlan, V. Characterization and localization of plant phenolics likely involved in the pathogen resistance expressed by endomycorrhizal roots. Mycorrhiza **1993**, 3, 155–164.

Harrison, M., Dixon, R. A. Isoflavonoid accumulation and expression of defense gene transcripts during the establishment of vesicular-arbuscular mycorrhizal associations in roots of *Medicago truncatula*. MPMI **1993**, 6, 643–654.

Hartwig, U.A., Maxwell, C.A., Joseph, C.M., Phillips, D.A. Chrysoeriol and luteolin released from alfalfa seeds induce *nod* genes in *Rhizobium meliloti*. Plant Physiol. **1990**, 92, 116–122.

Hepper, C. M., Mosse, B. Techniques used to study the interaction between *endogone* and plant roots. In *Endomycorrhizas*, Sanders, F. E., Mosse, B., Tinker, P. B., Eds., **1975**, Academic Press, London, pp. 65–75.

Hungria, M., Joseph, C.M., Phillips, D.A. *Rhizobium nod*-gene inducers exuded naturally from roots of common bean (*Phaseolus vulgaris* L.). Plant Physiol. **1991**, 97, 759–764.

Kape, R., Wex, K., Parniske, M., Görge, E., Wetzel, A., Werner, D. Legume root metabolites and VA-mycorrhiza development. J. Plant Physiol. **1992**, 141, 54–60.

Koske, R. E. Evidence for a volatile attractant from plant roots affecting germ tubes of a VA mycorrhizal fungus. Trans. Br. Mycol. Soc. **1982**, 79, 305–310.

Kosslak, R. M., Bookland, R., Barkei, J., Paaren, H. E., Appelbaum, E. R. Induction of *Bradyrhizobium japonicum* common *Nod* genes by isoflavones isolated from *Glycine max*. Proc. Natl. Acad. Sci. USA, **1987**, 84, 7428–7432.

Lynn, D. G., Chang, M. Phenolic signals in cohabitation: implications for plants development. Annu. Rev. Plant Physiol. Plant Mol. Biol. **1990**, 41, 497–526.

Maier, W., Peipp, H., Schmidt, J., Wray, V., Strack, D. Levels of a terpenoid glycoside (blumenin) and cell wall-bound phenolics in some cereal mycorrhizas. Plant Physiol. **1995**, 109, 465–470.

Markaverich, B., Roberts, R. R., Alejandro, M.A., Gregory, A., Middleditch, B. S., Clark, J. H. Bioflavonoid interaction with rat uterine type II binding sites and cell growth inhibition. J. Steroid Biochem. Mol. Biol. **1988**, 36, 71–78.

Miksicek, R. J. Commonly occurring plant flavonoids have estrogenic activity. Mol. Pharmacol. **1993**, 44, 37–43.

Morandi, D. Effect of endomycorrhizal infection and biocides on phytoalexin accumulation in soybean roots. Agriculture, Ecosystems and Environment. **1989**, 29, 303–305.

Morandi, D. Occurrence of phytoalexins and phenolic compounds in endomycorrhizal interactions, and their potential role in biological control. Plant and Soil. **1996**, 185, 241–251.

Morandi, D., Bailey, J. A., Gianinazzi-Pearson, V. Isoflavonoid accumulation in soybean roots infected with vesicular-arbuscular mycorrhizal fungi. Physiol. Plant Pathol. **1984**, 24, 357–364.

Morandi, D., Branzanti, B., Gianinazzi-Pearson, V. Effect of some plant flavonoids on *in vitro* behaviour of an arbuscular mycorrhizal fungus. Agronomie **1992**, 12, 811–816.

Mosse, B. Some studies relating to "independent" growth of vesicular-arbuscular endophytes. Can. J. Bot. **1988**, 66, 2533–2540.

Mosse, B., Hepper, C. M. Vesicular-arbuscular mycorrhizal infections in root organ cultures. Physiol. Plant Pathol. **1975**, 5, 215–223.

Nair, M. G., Safir, G. R., Siqueira, J. O. Isolation and identification of vesicular-arbuscular mycorrhiza stimulatory compounds from clover *(Trifolium repens)* roots. Appl. Environ. Microbiol. **1991**, 57, 434–439.

Ocampo, J. A., Cardona, F. L., El-Atrach, F. Effect of root extracts of non host plants on VA mycorrhizal infection and spore germination. In *Mycorrhizae: physiology and genetics*. Gianinazzi-Pearson, V., Gianinazzi, S., Eds., **1986**, Dijon INRA, Paris p.721–724.

Parra-Garcia, M. D., Lo Giudice V., Ocampo, J. A. Absence of VA colonization in *Oxalis pes-caprae* inoculated with *Glomus mosseae*. Plant and Soil **1992**, 145, 298–300.

Peters, N. K., Frost, J. W., Long, S. R. A plant flavone, luteoline, induces expression of *Rhizobium meliloti* nodulation genes. Science **1986**, 233, 977–980.

Phillips, D. A., Tsai, S. M. Flavonoids as plant signals to rhizosphere microbes. Mycorrhiza **1992**, 1, 55–58.

Poulin, M. J., Bel-Rhlid, R., Piché, Y., Chênevert, R. Flavonoids released by carrot *(Daucus carota)* seedlings stimulate hyphal development of vesicular-arbuscular mycorrhizal fungi in the presence of optimal $CO_2$ enrichment. J. Chem. Ecol. **1993**, 19, 2317–2327.

Poulin, M.J., Simard, J., Catford, J.G., Labrie, F., Piché, Y. Response of symbiotic endomycorrhizal fungi to estrogens and antiestrogens. MPMI **1997**, 10, 481–487.

Powell, C. L. Development of mycorrhizal infection from *Endogene* spores and infected root fragments. Trans. Brit. Mycol. Soc. **1976**, 66, 439–445.

Recourt, K., van Tunen, A. J., Mur, L. A., van Brussel, A. A. N., Lugtenberg, B. J. J., Kijine, J. W. Activation of flavonoid biosynthesis in roots of *Vicia sativa* subsp. *nigra* plants by inoculation with *Rhizobium leguminosarum* biovar *viciae*. Plant Mol. Biol. **1992**, 19, 411–420.

Redmond, J. W., Batley, M., Djordjevic, M. A., Innes, R. W., Kuempel, P. L., Rolfe, B. G. Flavones induce expression of nodulation genes in *Rhizobium*. Nature **1986**, 323, 632–635.

Schreiner, P., Koide, R. T. Antifungal compounds from the roots of mycotrophic and non-mycotrophic plant species. New Phytol. **1993**a, 123, 99–105.

Schreiner, P., Koide, R. T. Mustards, mustard oils and mycorrhizas. New Phytol. **1993**b, 123, 107–113.

Schreiner, P., Koide, R. T. Stimulation of vesicular-arbuscular mycorrhizal fungi by mycotrophic and nonmycotrohpic plant root systems. Appl. Environ. Microbiol. **1993**c, 59, 2750–2752.

Siqueira, J. O., Safir, G. R., Nair, M. G. Stimulation of vesicular-arbuscular mycorrhiza formation and growth of white clover by flavonoid compounds. New Phytol. **1991**a, 118, 87–93.

Siqueira, J. O., Safir, G. R., Nair, M. G. VA-mycorrhizae and mycorrhiza stimulating isoflavonoid compounds reduce herbicide injury. Plant and Soil **1991**b, 134, 233–242.

Smith, S. E., Read, D. J. Mycorrhizal Symbiosis. **1997**, Academic Press, London.

Suriyapperuma, S. P., Koske, R. E. Attraction of germ tubes and germination of spores of the arbuscular mycorrhizal fungus *Gigaspora gigantea* in the presence of roots of maize exposed to different concentrations of phosphorus. Mycologia **1995**, 87, 772–778.

Tawaraya, K., Watanabe, S., Yoshida, E., Wagatsuma, T. Effect of onion *(Allium cepa)* root exudates on the hyphal growth of *Gigaspora margarita*. Mycorrhiza **1996**, 6, 57–59.

Tsai, S. M., Phillips, D. A. Flavonoids released naturally from alfalfa promote development of the symbiotic *Glomus* spores *in vitro*. Appl. Environ. Microbiol. **1991**, 57, 1485–1488.

Vierheilig, H., Ocampo, J. A. Role of root extract and volatile substances of non-host plants on vesicular-arbuscular mycorrhizal spore germination. Symbiosis **1990**a, 9, 199–202.

Vierheilig, H., Ocampo, J. A. Effect of isothiocyanates on germination of spores of *Glomus mosseae*. Soil Biol. Biochem. **1990**b, 22, 1161–1162.

Vierheilig, H., Alt, M., Mäder, P., Boller, T., Wiemken, A. Spreading of *Glomus mosseae*, a vesicular-arbuscular mycorrhizal fungus, across the rhizosphere of host and non-host plants. Soil. Biol. Biochem. **1995**, 27, 1113–1115.

Vierheilig, H., Iseli, B., Alt, M., Raikhel, N., Wiemken, A., Boller, T. Resistance of *Urtica dioica* to mycorrhizal colonization: a possible involvement of *Urtica dioica* agglutinin. Plant and Soil **1996**a, 183, 131–136.

Vierheilig, H., Albrecht, C., Bago, B., Piché, Y. Do flavonoids play a role in root colonization by AM fungi ? (Abstract) First International Conference on Mycorrhizae (ICOM) Berkeley, Calif., USA. **1996**b.

Vierheilig, H., Piché, Y. Facteurs biochimiques potentiellement impliqués dans les interactions entre les champignons endomycorhiziens et leurs plantes non-hôtes. In *La Symbiose Mycorhizienne*. Fortin, J. A., Charest, C., Piché, Y. Eds. Éditions Orbis Frelighsburg, Québec, Canada **1995**, pp.109–124.

Volpin, H., Elkind, Y., Okon, Y., Kapulnik, Y. A vesicular-arbuscular mycorrhizal fungus *(Glomus intraradix)* induces a defense response in alfalfa roots. Plant Physiol. **1994**, 104, 683–689.

Whitten, P. L., Naftolin, F. Dietary estrogens: a biologically active background for estrogen action. In *The new biology of steroid hormones*. Hochberg, R., Naftolin, F. Eds. **1991**, Raven Press, New York, pp. 155–167.

Wyss, P., Boller, T., Wiemken, A. Effect of high phosphorus supply on the interaction of soybean roots with *Glomus mosseae* and *Rhizoctonia solani*: degree of infection and accumulation of the phytoalexin glyceollin. Symbiosis **1990**, 9, 383–387.

Wyss, P., Boller, T., Wiemken, A. Phytoalexin response is elicited by a pathogen *(Rhizoctonia solani)* but not by a mycorrhizal fungus *(Glomus mosseae)* in soybean roots. Experientia **1991**, 47, 395–399.

Xie, Z. P., Staehelin, C., Vierheilig, H., Wiemken, A., Jabbouri, S., Broughton, W. J., Voegeli-Lange, R., Boller, T. Rhizobial nodulation factors stimulate mycorrhizal colonization of nodulating and nonnodulating soybeans. Plant Physiol. **1995**, 108, 1519–1525.

# THE ROLE OF GLYCOSYLATION IN FLAVONOL-INDUCED POLLEN GERMINATION

Loverine P. Taylor, Darren Strenge, and Keith D. Miller

Department of Genetics and Cell Biology
Washington State University
Pullman, Washington 99164-4234

## 1. ABSTRACT

Flavonols are small (C15) plant-specific molecules that are required for petunia and maize pollen to germinate. They exist in two chemical forms: the aglycone or glycosyl conjugates. Flavonol-deficient pollen is biochemically complemented by flavonol aglycones but not by the glycosylated forms that accumulate in wild type (WT) pollen. Coincident with the biochemical induction of germination, the added flavonol aglycone is rapidly converted to a galactoside and then to a glucosyl galactoside (diglycoside) that is identical to the compound present in WT pollen. A flavonol 3-*O*-galactosyltransferase (F3GalTase) activity has been identified that controls the formation of glycosylated flavonols in pollen. Importantly, this enzyme also catalyzes the reverse reaction, *i.e.* the production of the flavonol aglycone from the galactoside and UDP (Fig. 1). F3GalTase/RevGalTase therefore has the potential to control the level of the bioactive flavonol species and as a result, pollen germination.

## 2. BACKGROUND

### 2.1. Pollen Development and Germination

Pollen develops within the anther and at maturity contains the products of sporophytic gene expression, arising from the tapetal layer of the anther wall, and gametophytic gene expression from the vegetative and generative nuclei (Mascarenhas, 1993). When a pollen grain falls on a receptive stigma the stored RNA, protein, and bioactive small molecules (*e.g.* flavonols) allow rapid germination and outgrowth of a tube which penetrates

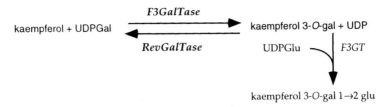

**Figure 1.** Interconversion of flavonol aglycones and glycosides by flavonol glycosyl transferases.

and grows within the style (Mascarenhas, 1993; Taylor and Hepler, 1997). Eventually the pollen tube deposits the two sperm cells in the embryo sac where they fuse with the egg and central cell to form the zygote and endosperm, respectively. Current research is providing a growing appreciation of the critical role that glycomoieties on proteins, lipids, polymers and small molecules play in plant reproduction. S-glycoproteins have long been known to be responsible for self-incompatible (SI) pollinations that prevent inbreeding (Nasrallah and Nasrallah, 1989). Now the glyco-conjugates involved in normal or compatible pollinations are being identified and even at this early stage, a theme is evident. Bioactivity is associated with two properties: the extent of glycosylation and the specificity inherent in sugar usage and inter-glycosidic linkage (Taylor and Hepler, 1997 and references therein).

## 2.2. Biochemical Complementation Defined a Role for Flavonols in Pollen Germination

Our laboratory has established that pollen germination and tube growth in maize and petunia requires flavonols (Mo *et al.*, 1992). Chalcone synthase (CHS) catalyzes the first step in flavonol biosynthesis and CHS mutants not only lacked all flavonols, but they also produced non-functional, but viable, pollen (Coe *et al.*, 1981; Mo *et al.*, 1992; Taylor and Jorgensen, 1992; van der Meer *et al.*, 1992; Ylstra *et al.*, 1994). A naturally occurring maize mutant and a genetically engineered petunia mutant both showed the same reproductive defect: the failure to produce a functional pollen tube (Mo *et al.*, 1992; Taylor and Jorgensen, 1992; Pollak *et al.*, 1995). Although flavonol-deficient pollen never functioned in self-crosses, reciprocal crosses showed that pollen from the mutant plants could germinate on wild-type stigmas, a phenotype we defined as Conditional Male Fertility (CMF) (Taylor and Jorgensen, 1992). *In vitro* germination and HPLC analysis of wild-type stigma extracts identified kaempferol, a flavonol aglycone, as the active element; adding as little as 0.4 µM kaempferol to a suspension of the non-functional CMF pollen completely restored germination and tube growth (pollen rescue assay) (Mo *et al.*, 1992). Biochemical complementation also worked *in vivo*; adding flavonol aglycones to stigmas or pollen at pollination led to full seed set in self-crosses of both the maize and petunia mutants (Mo *et al.*, 1992).

*In vitro* pollen rescue is critically dependent on how efficiently the added compounds are taken up and processed by the intact pollen. The chemical properties of a rescuing compound can dramatically affect uptake, and consequently germination. For a particular flavonol there is a direct correlation between the polarity of the molecule and the efficiency at inducing germination (Vogt *et al.*, 1995).

## 2.3. The Paradox: Flavonol Aglycones Induce Pollen Germination But Only Flavonol Glycosides Are Present in WT Pollen

Even though they function specifically at germination, flavonols accumulate much earlier in development (Mo *et al.*, 1992; Pollak *et al.*, 1993; Ylstra *et al.*, 1994). They are detected following the first pollen mitosis when the tapetal layer of the anther wall is beginning to disintegrate. Flavonol levels rise throughout development eventually reaching a peak of 4000 pmol per anther in mature pollen. This study also showed that all the flavonols are conjugated to sugars; absolutely no flavonol aglycones were detected at any stage during pollen development (Pollak *et al.*, 1993).

The flavonol profile of WT petunia pollen is simple: equal molar quantities of two flavonol glycosides were identified by FAB MS, $^{13}$C NMR and $^{1}$H NMR as kaempferol and quercetin 3-*O*-glucosyl 1→2 galactoside (3-*O*-(2"-*O*-β-D-glucopyranosyl)-β-D-galactopyranoside) (Zerback *et al.*, 1989; Vogt and Taylor, 1995). These compounds are unique to the pollen and are unlike the flavonol 3-*O*-rhamnosyl 1→6 glucosides that accumulate in the adjacent sporophytic tissues of the corolla (Brugliera *et al.*, 1994; Kroon *et al.*, 1994).

## 2.4. Flavonol 3-*O*-Glycosyltransferases Are Site and Substrate Specific

Glycosylation is one of the final steps in flavonol biosynthesis (Heller and Forkmann, 1994). The enzymes responsible for adding sugars to flavonol aglycones are UDP monosaccharide: flavonoid glycosyltransferases and the most common glycosyl donors are UDP sugar nucleotides (Fig. 1). Flavonol glycosyltransferases have precise substrate and site specificity (Kleinehollenhorst *et al.*, 1982; Bar-Peled *et al.*, 1991; Ishikura and Mato, 1993; Ishikura *et al.*, 1993; Heller and Forkmann, 1994) Disaccharide formation occurs by sequential addition of the sugars and the enzymes that form the interglycosidic bond share little or no homology to the glycosyltransferases that attach the sugar to the flavonol moiety (Brugliera *et al.*, 1994; Kroon *et al.*, 1994; L. P. Taylor unpublished results). The most common site for glycosyl addition is the 3-hydroxyl group of the flavonol C-ring although other sites, especially the hydroxyl at carbon 7, are often substituted (Ishikura *et al.*, 1993).

## 2.5. Two Mechanisms for Generating an Aglycone from a Glycoside

If glycosyl conjugates function as latent pools of bioactive molecules then an enzyme activity is required to re-generate the active species. Generation of an aglycone from a glycoside can occur via two mechanisms: hydrolysis by enzymes of the glycosidase class (E.C. 3.2) or transglycosylation mediated by transferases (E.C 2.4) using UDP as the sugar acceptor (Fig. 1). The search for specific glycosidases that release biologically active compounds has focused on enzymes that act on phytohormone conjugates (Sembdner *et al.*, 1994); few if any relevant activities have been identified.

Transglycosylation is reversible (Sutter and Grisbach, 1975). A partially purified flavonol 3-*O*-glucosyltransferase from a parsley suspension culture, catalyzed formation of quercetin and kaempferol from quercetin or kaempferol 3-*O*-glucoside and UDP (Sutter and Grisbach, 1975). The glucosides were stable in solution and formation of the aglycones proceeded only upon addition of UDP. We have found a similar phenomenon in pollen: the forward reaction of F3GalTase converts the aglycone to a galactoside and the reverse reaction liberates the aglycone from the conjugate.

# 3. RESULTS

## 3.1. Gametophytically-Expressed Glycosyltransferases Produce Pollen-Specific Flavonols

To reconcile the contradiction that aglycones induce CMF pollen germination but only diglycosides are present in WT pollen, a search for a ß-glycosidase activity in pollen that could generate kaempferol from kaempferol 3-O-glucosyl 1→2 galactoside was performed. In this case generating the aglycone would require the sequential action of two ß-glycosidase activities or a diglycosidase. However no specific, or even non-specific, hydrolyase activity was detected during pollen development or pollen germination. In addition, the pool of flavonol diglycosides did not appear to decrease during several hours of germination.

However when the metabolic fate of the rescuing flavonol aglycone in CMF pollen was determined, it was found that intact pollen rapidly modified the added aglycone by a series of glycosylation reactions (Fig. 1). Kaempferol glycosides were detected within 1 min of adding a radioactive kaempferol derivative to germinating pollen (Vogt and Taylor, 1995; Xu et al., 1997). Accumulation continued throughout a 4 h incubation period during which the tube grew 300 µm, a rate similar to *in vitro* germinated WT pollen. Kaempferol was not catabolized during germination; all of the radioactivity was recovered in the aglycone or glycosides (Xu et al., 1997). Spectral analysis (FAB MS, $^{13}$C NMR and $^1$H NMR) confirmed that the structure of the major flavonol 3-O-glycoside from CMF pollen was identical to the kaempferol and quercetin glucosyl 1→2 galactoside that accumulated in WT pollen (Zerback et al., 1989; Vogt and Taylor, 1995; Xu et al., 1997). Under certain conditions, the intermediate galactoside also accumulated in rescued CMF pollen although at much lower levels than the diglycoside. The two activities responsible for the formation of pollen flavonol glycosides were isolated and characterized (Vogt and Taylor, 1995), this report). A flavonol 3-O-galactosyl transferase (F3GalTase) adds galactose to the flavonol aglycone at position 3 of the C-ring followed by the action of a glucosyl transferase (F3GT), which links glucose to the 2' carbon of the galactose forming flavonol 3-O-glucosyl 1→2 galactoside.

The substrate requirements of the two transferase activities were determined in (i) a pollen cell free extract and (ii) intact CMF pollen (Vogt and Taylor, 1995) and showed the following:

- F3GalTase has an absolute requirement for UDPGal; it will not use UDPGlu which is at least 10-fold more abundant in pollen (Schlupmann et al., 1994; Vogt and Taylor, 1995).
- F3GT accepts either UDPGal and UDPGlu but *in vitro* it prefers UDPGlu 2:1 (Vogt and Taylor, 1995).
- Mature WT pollen accumulates flavonol disaccharides (Vogt and Taylor, 1995; Xu et al., 1997) and the structure is invariant: glucosyl 1→2 galactoside, never a galactosyl 1→2 galactoside. This suggests that *in vivo* conditions do not favor the use of UDPGal by F3GT.
- In contrast to WT, 3 flavonols can be detected in CMF pollen rescued with a flavonol aglycone: the galactoside, the glucosyl 1→2 galactoside and the aglycone.

## 3.2. CMF Pollen Germinates after Treatment with Flavonol 3-O-Galactoside

Kaempferol is the most active flavonol producing full germination at 0.4 µM; quercetin is about 10-fold less active (Table 1; Mo et al., 1992; Vogt et al., 1995; Xu et al.,

Table 1. Induction of pollen germination by flavonol glycosides

| Compound[*] | Glycosylation status | % germination[†] 1 μm | 10 μm | 100 μm |
|---|---|---|---|---|
| K | aglycone | 63 | 60 | 59 |
| Q | aglycone | 23 | 64 | 48 |
| K 3-galactoside | 3-mono | 30 | 52 | 55 |
| Q 3-galactoside | 3-mono | 4 | 12 | 35 |
| K 3-glucoside | 3-mono | 7 | 8 | 12 |
| Q 3-glucoside | 3-mono | 6 | 10 | 19 |
| Q 3-rhamnoside | 3-mono | 8 | 7 | 2 |
| K 3-rhamnosyl 1→6 glucoside | 3-di | 4 | 8 | 13 |
| K 3-glucosyl 1→2 galactoside | 3-di | 0 | 0 | 0 |
| Q 3-glucosyl 1→2 galactoside | 3-di | 0 | 0 | 0 |
| K 3-rhamnosyl 1→6 galactoside, 7-rhamnoside | 3-di, 7-mono | 2 | 8 | 7 |
| K 7-rhamnosyl 1→2 glucoside | 7-di | 2 | 5 | 43 |

K, kaempferol; Q, quercetin
[*]All compounds HPLC pure.
[†]No se greater than ±5.

1997). A series of flavonol glycosides, including the WT diglycosides and the galactoside intermediates that accumulate in rescued CMF pollen, were tested for the ability to stimulate germination (Table 1). Although the WT diglycosides were inactive, kaempferol-, and to a lesser extent, quercetin 3-*O*-galactoside induced significant levels of germination (Table 1). Significantly, the active molecules were galactosides, not the glucoside isomers or rhamnose conjugates. Thus germination is highly specific for the compounds that are normally produced by pollen but only the mono-glycosylated species is active, higher levels of sugar conjugation, *i.e.* the diglycoside, are inactive.

## 3.3. Reverse F3GalTase Activity Can Generate the Aglycones Required for Pollen Germination

In characterizing the substrate usage of F3GT, it was found that if the galactoside substrate was in excess, two flavonol products were formed. In addition to the expected flavonol 3-*O*-glucosyl 1→2 galactoside, the aglycone was also generated. A plausible explanation was that F3GalTase activity in the cell-free extract was using the UDP generated in the F3GT-catalyzed reaction, together with the galactoside, as substrates for a reverse transferase reaction (Fig. 1). This assumption was strengthened when the UDP concentration was increased 40-fold and all of the galactoside was converted to the aglycone (data not shown). The activity that generated the flavonol aglycone from flavonol 3-*O*-galactoside and UDP is designated reverse galactosyltransferase (RevGalTase).

Significantly intact pollen has intrinsic RevGalTase activity. The flavonol profile of suspensions of CMF pollen incubated for 2 h with 100 μM of the kaempferol 3-*O*-galactoside substrate and three different concentrations of UDP (0, 0.5 and 20 mM) shows that formation of the kaempferol product (peak 3 in HPLC chromatogram) increases from 86 to 486 and to 609 pmol per mg protein with increasing UDP concentrations (Fig. 2, panels A, B & C). This result brings up the possibility that the germination attributed to kaempferol 3-*O*-galactoside (Table 1) might actually be induced by kaempferol generated by RevGalTase activity. This hypothesis was confirmed upon incubation of CMF pollen

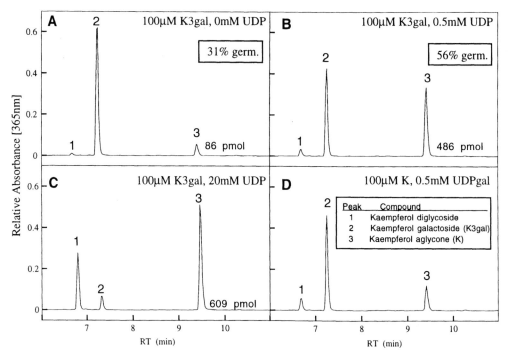

**Figure 2.** Evidence that a reverse galactosyltransferase (RevGalTase) activity can generate kaempferol from kaempferol 3-*O*-galactoside. A suspension of CMF pollen was incubated with the substrates listed in the upper right of each panel. The HPLC profiles in panels A, B, and C show the kaempferol generated (peak 3) with increasing amounts of UDP. The inset in panel A and B indicates the % germination of each suspension after 4 h incubation. Panel D displays the flavonol profile of the forward reaction with kaempferol as the added substrate (peak 3) and the two products, kaempferol 3-*O*-galactoside (peak 2) and the diglycoside (peak 1).

with 1µM kaempferol 3-*O*-galactoside and either 0 or 0.5mM UDP. Germination frequency increased from 31% to 56%, a significant enhancement (Fig. 2, inset in panel A & B). The high baseline germination frequency of 31% is likely due to kaempferol formation stimulated by the UDP released into the germination medium (GM) from the burst pollen grains that are inevitably present (~5%). We were unable to determine the effect of higher (*i. e.* 20 mM) UDP concentrations because at this level pollen germination was inhibited in both CMF and WT pollen. These data show that the RevGalTase activity operates in intact pollen and that the germination frequencies are increased by the same conditions that favor formation of the aglycone from the galactoside. When the same experiment was performed with kaempferol 3-*O*-glucoside as the substrate, no kaempferol was formed and no germination occurred (data not shown). This indicates that the specificity of F3GalTase for galactose in the forward direction is matched by an equivalent specificity in the reverse direction. Thus we have biochemical and physiological evidence that both the forward and reverse galactosyltransferase reactions operate in intact pollen and that RevGalTase can stimulate pollen germination via kaempferol 3-*O*-galactoside.

The results shown in Figure 2 raise questions concerning the location of F3GalTase activity. If we presume that flavonol 3-*O*-galactosides cannot diffuse through the pollen wall, then a surface localized F3GalTase activity must be considered. The aglycones would be generated outside the pollen by the RevGalTase activity and then internalized to

Table 2. Relative amounts of kaempferol glycoside and aglycone in pollen and GM

|  | Germination medium | | Pollen | | |
|---|---|---|---|---|---|
| mM UDP | 3-galactoside | Aglycone | 3-diglycoside | 3-galactoside | Aglycone |
| 0 | 17580* | 560 | 410 | 270 | 280 |
| 0.5 | 9760 | 5830 | 2030 | 160 | 1390 |
| 20 | 2570 | 13640 | 1880 | 80 | 2260 |

*pmol total.

stimulate germination. This is precisely what was observed when the experiment shown in Figure 2 was repeated but the pollen and GM were analyzed separately. Table 2 shows the relative proportions of each flavonol species in the two locations. As expected, kaempferol 3-O-galactoside levels dropped and aglycone levels rose with increased UDP concentrations; this occurred in both the GM and in the pollen fraction. Substantially larger amounts of kaempferol accumulated in the GM as compared to the pollen indicating that the aglycone is generated outside the pollen grain.

## 3.4. Kinetic Properties of F3GalTase and Their Influence on Flavonol-Induced Germination

A single F3GalTase activity was purified by conventional and affinity chromatography. Kinetic analyses were done using kaempferol and UDP-Gal for the forward reaction and kaempferol-3-O-galactoside and UDP for the reverse reaction. All four substrates follow Michaelis-Menten kinetics over the linear portion of the assay (10 min). Initial velocity versus substrate concentration curves of the four substrates were linearized using Lineweaver-Burk or Hanes-Woolf plots to estimate both the $K_m$ and $V_{max}$ values which were calculated as shown in Table 3. The $K_m$ values not only approximate the apparent affinity of the substrate for F3GalTase but may also indicate the approximate concentrations of each of the substrates *in vivo*.

The two reactions have different pH optima: activity of the forward reaction is 1.6-fold higher at pH 7.5 than at 5.5 whereas the activity of the reverse reaction is 8-fold higher at pH 5.5 than at 7.5. The pH effect suggests that the direction of the reaction could be controlled *in vivo* by localized hydronium ion changes. The kinetic analysis showed that F3GalTase catalyzes the formation of kaempferol from kaempferol-3-O-galactoside in a manner that is dependent on increasing concentrations of UDP. One potential mechanism for regulating a reversible reaction is through the metabolic flux of the four substrates. If F3GalTase is located in cellular compartments with different pH and substrate concentrations, the forward reaction might be favored in some locations and the reverse reaction in others. The activity in each compartment could be regulated in response to lo-

Table 3. Kinetic properties of F3Gal Tase and RevGalTase

| Substrate | $K_m$ (µM) | $V_{max}$ (µmol · min$^{-1}$ · mg protein$^{-1}$) |
|---|---|---|
| Kaempferol | 11 | 1.087 |
| UDPGal | 318 | 1.27 |
| Kaempferol-3-gal | 39 | 0.26 |
| UDP | 13 | 0.218 |

calized metabolic perturbations. Our kinetic data suggests that UDP and UDP sugar concentrations in pollen regulate the formation of different flavonol species.

## 4. DISCUSSION

WT pollen develops in the presence of flavonols, CMF does not. Flavonol aglycones added to CMF pollen are metabolized, first to the galactoside and then to the same end product that accumulates in WT pollen. A logical assumption is that the aglycone is the active species. However the WT pollen flavonol profile provides no apparent support for this hypothesis, nor for the alternative idea that the galactoside is a potential source of the aglycone even though we have shown that the aglycone and galactoside are interconvertible and that F3GalTase can mediate this interconversion. The inability to detect the galactoside or the aglycone in WT pollen is puzzling. The abundance of the diglycoside dictates that large amounts of the galactoside are formed, an assumption which is supported by the coordinated rise in F3GalTase activity with flavonol accumulation during pollen development and by the dependence of diglycoside formation on monoglycoside formation (Vogt and Taylor, 1995). Could the diglycoside function as the active molecule? There is no evidence to support this view but we cannot rule out that the diglycoside may act through another germination-inducing agent. Alternatively the aglycone and monoglycoside may escape detection in WT pollen because they are present at very low levels or are rapidly channeled to the glycosylating activity associated with the pollen. Photomicrographs show that the pollen and the tapetal cells are tightly appressed during the period of flavonol formation in the anther (G. Deboo and L. P. Taylor, unpubl. data). Perhaps the flavonol aglycone and/or monoglycoside act before they are converted to the terminal diglycoside. This could occur during uptake from the locule, setting in motion a stimulus that is stored until needed at germination. Formation of this second signal would explain the inability to detect the aglycone or consumption of the diglycoside pool during germination and tube growth.

This study addresses the role of glycosylation in regulating biological activity of flavonols. The result of conjugating bioactive molecules, including flavonol aglycones, to sugars is conversion of a chemically reactive, lipophilic molecule into a more water-soluble, less-reactive form. In animals these conjugates are excreted but in plants they are transported to cellular compartments, chiefly vacuoles, for sequestration. Conjugation also provides a reservoir of potentially active molecules. The alternative roles for conjugation are not mutually exclusive and both may operate within a particular tissue in response to developmental or metabolic signals. In the case of flavonols, there is evidence for both roles: high levels of aglycones are cytotoxic (Vogt *et al.*, 1994) indicating a need for detoxification. Pollen accumulates flavonol glycosides but germinates in response to flavonol aglycones, suggesting a reversible storage mechanism. F3GalTase and its reverse activity, RevGalTase, have the ability to control pollen germination by maintaining pools of active and inactive compounds. In this regard we must consider that F3GalTase/RevGalTase may serve more than one function. Activity in the forward direction suggests a detoxifying mechanism and/or a facilitator of flavonol uptake. If F3GalTase acts as a flavonol receptor the catalytic activity could be a way to inactivate a 'spent' signal molecule. Activity in both directions suggests that in addition to detoxification, the galactosyltransferase acts to provide a pool of potentially active molecules. Knowing the structure of the inducing molecule will facilitate the design and synthesis of "tagged" molecules which mimic the active flavonol and can be used to isolate interacting proteins. This research will also reveal whether glycosylating enzymes represent a potential target for controlling pollen fertility.

A surface localized F3GalTase has important implications for recognition and facilitated uptake of flavonols especially in light of a similar role performed by a mammalian cell surface galactosyltransferase. Unlike plants, animal GalTases have been extensively characterized, especially those involved in the biosynthesis of membrane-bound and secretory glyco-conjugates (Kleene and Berger, 1993). One enzyme, $\beta$ 1→4 GalTase, is unusual among animal transferases in that it is found in two distinct subcellular pools with very different functions associated with each pool. This differential targeting is mediated by a 13 amino acid extension on the surface-localized GalTase (Lopez *et al.*, 1991). In the Golgi apparatus, GalTase catalyzes the transfer of galactose to terminal N-acetylglucosamine residues of proteins and lipids. The surface-localized GalTase is proposed to be involved in sperm-egg recognition, neurite outgrowth and spermatocyte-Sertoli cell interaction (Cooke and Shur, 1994). In these situations, GalTase is supposed to function as a receptor rather than catalytically (*e. g.*, adding UDPGal causes loss of cellular adhesion). Support for a signaling role is provided by the involvement of G-proteins and phosphorylation in the surface interactions (Cooke and Shur, 1994).

# REFERENCES

Bar-Peled, M.; Lewinsohn, E.; Fluhr, R.; Gressel, J. UDP-Rhamnose:flavonone-7-*O*-glucoside-2"-*O*-rhamnosyltransferase. Purification and characterization of an enzyme catalyzing the production of bitter compounds in citrus. *J Biol. Chem.* **1991** 266, 20953–20959.

Brugliera, F.; Holton, T.A.; Stevenson, T.W.; Farcy, E.; Lu, C.; Cornish, E.C. Isolation and characterization of a cDNA clone corresponding to the *Rt* locus of *Petunia hybrida*. *Plant J.* **1994** 5(1), 81–92.

Coe, E.H.; McCormick, S.M.; Modena, S.A. White pollen in maize. *J. Hered.* **1981** 72, 318–320.

Cooke, S.V.; Shur, B.D. Cell surface $\beta$ 1,4-galactosyltransferase: expression and function. *Develop. Growth & Differ.* **1994** 36, 125–132.

Heller, W.; Forkmann, G. Biosynthesis of flavonoids. In *The Favonoids, Advances In Research Since 1986* ; J. B. Harborne, Ed.; Chapman and Hall: London, 1994.

Ishikura, N.; Mato, M. Partial purification and some properties of flavonol 3-*O*-glycosyltransferases from seedlings of *Vigna mungo*, with special reference to the formation of kaempferol 3-*O*-galactoside and 3-*O*-glucoside. *Plant Cell Physiol.* **1993** 34, 329–335.

Ishikura, N.; Yang, Z.-Q.; Teramoto, S. UDP-D-glucose: flavonol 3-*O*- and 7-*O*-glucosyl transferases from young leaves of *Paederia scandens* var. *mairei*. *Z. Naturforsch.* **1993** 48c, 563–569.

Kleene, R.; Berger, E.G. The molecular and cell biology of glycosyltransferases. *Biochim. Biophys. Act.* **1993** 1154, 283–325.

Kleinehollenhorst, G.; Behrens, H.; Pegels, G.; Srunk, N.; Wiermann, R. Formation of flavonol 3-*O*-diglycosides and flavonol 3-*O*-triglycosides by enzyme extracts from anthers of *Tulipa* cv. Apeldoorn. *Z. Naturforsch.* **1982** 37c, 587–599.

Kroon, J.; E., S.; de Graaff, A.; Xue, Y.; Mol, J.; Koes, R. Cloning and structural analysis of the anthocyanin pigmentation locus Rt of *Petunia hybrida* : characterization of insertion sequences in two mutant alleles. *Plant J.* **1994** 5(1), 69–80.

Lopez, L.C.; Youakim, A.; Evans, S.C.; Shur, B.D. Evidence for a molecular distinction between Golgi and cell surface forms of $\beta$ 1,4-galactosyltransferase. *J. Biol. Chem.* **1991** 266(24), 15984–15991.

Mascarenhas, J.P. Molecular mechanisms of pollen tube growth and differentiation. *Plant Cell.* **1993** 5, 1303–1314.

Mo, Y.; Nagel, C.; Taylor, L.P. Biochemical complementation of chalcone synthase mutants defines a role for flavonols in functional pollen. *Proc. Natl. Acad. Sci. USA.* **1992** 89, 7213–7217.

Nasrallah, J.B.; Nasrallah, M.E. The molecular genetics of self-incompatibility in *Brassica*. *Annu. Rev. Genet.* **1989** 23, 121–139.

Pollak, P.E.; Hansen, K.; Astwood, J.D.; Taylor, L.P. Conditional male fertility in maize. *Sex. Plant Reprod.* **1995** 8, 231–241.

Pollak, P.E.; Vogt, T.; Y., M.; Taylor, L.P. Chalcone synthase and flavonol accumulation in stigmas and anthers of *Petunia hybrida*. *Plant Physiol.* **1993** 102, 925–932.

Schlupmann, H.; Bacic, A.; Read, S.M. Uridine diphosphate glucose metabolism and callose synthesis in cultured pollen tubes of *Nicotiana alata* Link et Otto. *Plant Physiol.* **1994** 105, 659–670.

Sembdner, G.; Atzorn, R.; Schneider, G. Plant hormone conjugation. *Plant Mol. Biol.* **1994** 26, 1459–1481.

Sutter, A.; Grisbach, H. Free reversibility of the UDP-glucose:flavonol 3-*O* - glucosyltranferase reaction. *Arch. Biochem. Biophys.* **1975** 167, 444–447.

Taylor, L.P.; Hepler, P.K. Pollen germination and tube growth. *Ann. Rev. Plant Physiol. Plant Mol. Bio.* **1997** 48, 461–491.

Taylor, L.P.; Jorgensen, R. Conditional male fertility in chalcone synthase-deficient petunia. *J. Heredity.* **1992** 83, 11–17.

van der Meer, I.M.; Stam, M.E.; van Tunen, A.J.; Mol, J.N.M.; Stuitje, A.R. Antisense inhibition of flavonoid biosynthesis in petunia anthers results in male sterility. *Plant Cell.* **1992** 4, 253–262.

Vogt, T.; Pollak, P.; Tarlyn, N.; Taylor, L.P. Pollination- or wound-induced kaempferol accumulation in petunia stigmas enhances seed production. *Plant Cell.* **1994** 6, 11–23.

Vogt, T.; Taylor, L.P. Flavonol 3-*O*-glycosyltransferases associated with petunia pollen produce gametophyte-specific flavonol diglycosides. *Plant Physiol.* **1995** 108, 903–911.

Vogt, T.; Wollenweber, E.; Taylor, L.P. The structural requirements of flavonols that induce pollen germination of conditionally male fertile *Petunia*. *Phytochem.* **1995** 38, 589–592.

Xu, P.; Vogt, T.; Taylor, L.P. Uptake and metabolism of flavonols during *in vitro* germination and tube growth of petunia pollen. *Planta.* **1997** In press.

Ylstra, B.; Busscher, J.; Franken, J.; Hollman, P.C.H.; Mol, J.N.M.; van Tunen, A.J. Flavonols and fertilization in Petunia hybrida: Localization and mode of action during pollen tube growth. *Plant J.* **1994** 6, 201–212.

Zerback, R.; Bokel, M.; Geiger, H.; Hess, D. A kaempferol 3-glucosylgalactoside and further flavonoids from pollen of *Petunia hybrida*. *Phytochemistry.* **1989** 28, 897–899.

# EXPRESSION OF GENES FOR ENZYMES OF THE FLAVONOID BIOSYNTHETIC PATHWAY IN THE EARLY STAGES OF THE *RHIZOBIUM*-LEGUME SYMBIOSIS

H. I. McKhann,[1*] N. L. Paiva,[2] R. A. Dixon,[2] and A. M. Hirsch[1]

[1]Department of Molecular, Cell, and Developmental Biology
405 Hilgard Avenue
University of California
Los Angeles, California 90095-1606
[2]Plant Biology Division
The Samuel Roberts Noble Foundation
Ardmore, Oklahoma 73402

## 1. INTRODUCTION

Isoflavonoids/flavonoids have been found to be involved in a number of critical plant activities, such as plant defense, pollen development, pigmentation, and the *Rhizobium*-legume symbiosis. The biosynthetic pathways for isoflavonoids/flavonoids are well established, and the genes for the majority of enzymes involved in these pathways have been isolated and cloned. For alfalfa (*Medicago sativa* L.), genes encoding phenylalanine ammonia lyase (PAL) (Gowri et al., 1991), chalcone synthase (CHS) (Esnault et al., 1993; Junghans et al., 1993; McKhann and Hirsch, 1994a), chalcone reductase (CHR) (Ballance and Dixon, 1995; Sallaud et al., 1995), chalcone isomerase (CHI) (McKhann and Hirsch, 1994a), chalcone 4'-*O*-methyltransferase (Maxwell et al., 1993), isoflavone reductase (IFR) (Paiva et al., 1991), and vestitone reductase (Guo and Paiva, 1995) have been cloned and sequenced. Earlier, we utilized two *CHS* cDNAs and *CHI* and *IFR* clones to study the temporal and spatial expression patterns of these genes during the establishment of the *Rhizobium meliloti*-alfalfa symbiosis (McKhann et al., 1997). Chalcone synthase is a key enzyme in the phenypropanoid biosynthetic pathway, catalyzing the condensation of malonyl-CoA with 4-coumaroyl CoA to yield naringenin chalcone, whereas chalcone

---

* Present Address: Laboratoire de Biologie Cellulaire, Route de Saint Cyr, 78026 Versailles, Cedex, France.

*Flavonoids in the Living System*, edited by Manthey and Buslig
Plenum Press, New York, 1998.

isomerase catalyzes the isomerization of chalcones to their corresponding (-)- flavanones. Isoflavone reductase catalyzes the penultimate step in the biosynthesis of medicarpin, an isoflavonoid phytoalexin.

In the *Rhizobium*-legume symbiosis, specific flavonoids have been shown to attract rhizobia chemotactically to plant roots (Caetano-Anollés et al., 1988), to enhance rhizobial growth (Hartwig et al, 1991), and to induce rhizobial *nod* genes (see review by Fisher and Long, 1992). In addition, flavonoids are produced as part of the Ini response (Increase in *nod* gene-inducing flavonoids) that takes place in a number of legume roots within a few days following inoculation by *Rhizobium*. The Ini response was first described for the *Vicia sativa-R. leguminosarum* bv. *viciae* symbiosis (van Brussel et al., 1990). Either *R. leguminosarum* bv. *viciae* or Nod factor induced the Ini response, but not a strain cured of its symbiotic plasmid (van Brussel et al., 1990; Recourt et al., 1991). The increase in flavonoids is thought to be due to *de novo* synthesis; *CHS* transcript accumulation is enhanced following inoculation with *R. leguminosarum* bv. *viciae* (Recourt et al., 1992). An Ini response has also been reported for alfalfa (*Medicago sativa* L.) by Dakora et al. (1993) who found that inoculation with wild-type *R. meliloti* led to a 200% increase in the *nod* gene-inducing activity of root exudates. Inoculation with the heterologous *R. leguminosarum* bv. *phaseoli* induced the Ini response in alfalfa, but at a lower level than when the plants were inoculated with the homologous *R. meliloti*. White clover (*Trifolium repens*) also exhibited an Ini response, but in contrast to vetch, the Ini response occurred following inoculation with either the homologous *R. leguminosarum* bv. *trifolii* or with the heterologous *R. leguminosarum* bv. *viciae* (Rolfe et al., 1989). Lawson et al. (1994, 1996) later found that the *CHS5* gene is up-regulated in white clover roots upon inoculation, and also that 7,4'-dihydroxyflavone is detected as early as 3 days post-inoculation. We have confirmed for alfalfa that *nod*-gene-inducing flavonoids, specifically 7,4'-dihydroxyflavone and 4,4'-dihydroxy-2'-methoxychalcone, are synthesized within 5 days post-inoculation and also that two different *CHS* genes, *CHS6–4* and *CHS4–1*, are up-regulated several days after inoculation (McKhann et al., 1997).

Lawson et al. (1994) detected very early *CHS5* expression (6 h post-inoculation with *R. leguminosarum* bv. *trifolii*), before infection thread formation and cortical cell divisions were visible. This early increase in steady-state transcript levels of *CHS* gene(s) can be considered a pre-infection event. We predict that such an early increase should be common to other legume-*Rhizobium* symbioses. To address this question, we examined the expression of *CHS* and *CHI* genes in alfalfa 6 to 48 hours after inoculation with wild-type and mutant *R. meliloti*.

## 2. MATERIALS AND METHODS

### 2.1. Growth Conditions

*Medicago sativa* L. cv. Iroquois seeds were sterilized as described in Löbler and Hirsch (1993). The seedlings were either grown in plastic dish pans containing vermiculite and perlite (Hirsch et al., 1989), in dish pans on wire mesh covered with cheesecloth, suspended over Jensen's medium, or in Magenta jars (Magenta Corp., Chicago, IL) (Löbler and Hirsch, 1993). Three days after germination, the seedlings were inoculated with wild-type (Rm1021; Meade et al., 1982), or with mutant *R. meliloti* strains, either Rm5610 (*nodA*::Tn*5*) (Klein et al., 1988) or Rm7094 (*exoB*::Tn*5*) (Finan, 1988). Wild-type *R. meliloti* were heat-killed by autoclaving them. Total root systems were harvested at specified time intervals and frozen immediately in liquid nitrogen. *R. meliloti* cultures, grown in RDM (*Rhizobium* Defined Medium; Vincent, 1970) to an O.D.$_{600}$ of 0.2 to 0.5, were used to inoculate plants.

## 2.2. RNase Protection Studies

RNase protection was performed as described previously (McKhann and Hirsch, 1994a). Riboprobes were generated from *CHS4–1*P (175 nt), *CHS6–4*P (148 nt), and *Msc27*P (105 nt) (McKhann and Hirsch, 1994a). Approximately $1.2 \times 10^6$ cpm of probe RNA were added to each sample. The three probes, *CHS4–1*P, *CHS6–4*P, and *Msc27*P were in the same reaction mixture. Hybridizations were done at 42°C and in 50% formamide. The transcript levels were corrected for loading differences by expressing the transcript level as the ratio of *CHS4–1* or *CHS6–4* to *Msc27* (relative transcript level). All RNase protections were done with RNA samples from at least two, and in some cases three, independent experiments.

## 2.3. RNA Isolation and Northern Analysis

Total RNA was isolated from roots as described previously (McKhann and Hirsch, 1994a). RNA was subjected to electrophoresis on a formaldehyde gel (Sambrook et al., 1989) and blotted onto Nytran following the instructions of Schleicher & Schuell. Each lane was loaded with 10 µg RNA. Restriction fragments used as probes were labeled by random priming using $\alpha$-$^{32}$P-dCTP (NEN). For the *CHI* probe, the entire EcoRI insert of *CHI-1* was used (McKhann and Hirsch, 1994a). The *Msc27* clone (Kapros et al., 1992) used to monitor RNA loading, is a 694 bp PstI fragment. The autoradiograms were quantified using an UltroScan XL laser densitometer (Pharmacia) or with a PhosphorImager (Molecular Dynamics). Samples of RNA from two, and in some cases three, different experiments were used for northern blot analysis.

## 2.4. *In Situ* Hybridizations

*In situ* hybridizations were performed as described by McKhann and Hirsch (1993) except that $^{33}$P-UTP was used. Hybridizations were performed at 45°C in 50% formamide using *CHS6–4*P antisense and sense probes. For the radiolabeled *in situ* hybridizations, the photographic emulsion was developed after 4 weeks. Slides were stained with 1% toluidine blue in borate buffer. Photographs were taken on a Zeiss Axiophot microscope using Ektachrome Tungsten 160 film.

## 2.5. Spot-Inoculations

For the *in situ* hybridizations, spot-inoculation was performed. Seeds were sterilized as described above and placed on top of 1.5% nitrogen-free Jensen's agar in square Petri dishes (LabTek). Seeds were allowed to germinate in the dark and after 72 h, the seedlings were inoculated with 1 to 2 nL of Rm1021 as described by Dudley et al. (1987). Control plants were mock-inoculated with RDM alone, and India ink was used to mark the point of inoculation. Root hair deformation and nodulation were monitored by light microscopy. At various times after the start of the treatments, approximately 5 mm of the root adjacent to the spot of inoculation were harvested and fixed in formaldehyde-alcohol-acetic acid (FAA) and subsequently prepared for *in situ* hybridization (McKhann and Hirsch, 1993).

## 3. RESULTS

### 3.1. Temporal and Spatial Patterns of *CHS* and *CHI* Transcript Accumulation from 0 to 48 h Post-Inoculation

*CHS6–4* and *CHS4–1* are 94.2% identical to each other and both are highly expressed in the root tips of uninoculated plants (McKhann and Hirsch, 1994a). Because of the high sequence identity, we used RNase protection analysis to distinguish between the two *CHS* genes. Clones *CHS4–1*P and *CHS6–4*P were determined to give specific protected fragments in RNase protection analysis (McKhann and Hirsch, 1994a). We found that the amount of *CHS4–1* transcripts significantly increased in alfalfa root tissue 48 h after inoculation with *R. meliloti* (Fig. 1). An increase in *CHS6–4* transcript accumulation was not apparent at either 24 or 48 h using RNase protection assays (Fig. 1). Thus, *in situ* hybridization was used to localize *CHS6–4* transcripts in individual alfalfa roots spot-inoculated with wild-type *R. meliloti*. *CHS6–4* transcripts were detected 2 days post-inoculation in root hairs and no transcripts were found in uninfected roots even 10 days after the start of the experiment (Fig. 2). We concluded from these experiments that both *CHS6–4* and *CHS4–1* expression were enhanced in response to wild-type *R. meliloti* within 48 h after inoculation, although *CHS6–4* could only be detected in individual roots and not in total RNA isolated from a population of roots.

We used northern analysis to examine the temporal expression patterns of transcripts corresponding to *CHI*. In contrast to *CHS*, where there are likely to be at least 12 different

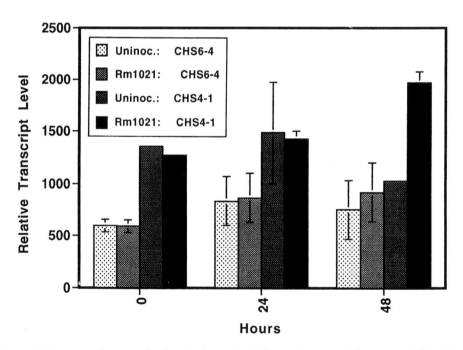

**Figure 1.** RNase protection assay showing the changes in *CHS* transcript accumulation in roots 0, 24, 48 h post-inoculation with wild-type *R. meliloti*. The transcript levels were normalized for the amount of *Msc27*-hybridizing mRNA present in each sample. The line above and below the average values indicates the standard deviation. If absent, the standard deviations were too small to be displayed. For *CHS6–4*, three different samples of RNA were utilized for the RNase protection assays and for *CHS4–1*, two different experiments are represented.

# Expression of Genes for Enzymes of the Flavonoid Biosynthetic Pathway

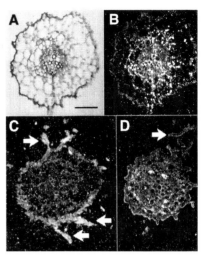

**Figure 2.** *In situ* hybridizations of uninfected alfalfa roots (A, B, and D) and an inoculated alfalfa root (C). All magnifications are the same. (A, B) Transverse-section of an uninoculated alfalfa root 72 h after the start of the experiment. A) Bright-field photograph, Bar, 50 μm. B) Dark field. There is no accumulation of *CHS6–4* mRNA, only the amyloplasts are refractile. (C) Transverse-section of an alfalfa root harvested 48 h post-inoculation The arrows point to the silver grains (white dots) all over the root hairs indicating the expression of *CHS6–4*. (D) An uninoculated root harvested 10 days after the start of the experiment. No silver grains are evident in the epidermis or the root hairs.

genes in alfalfa, there are only one or two *CHI* genes in alfalfa. As found previously (McKhann and Hirsch, 1994a) and as was also observed for *CHS*, *CHI* transcripts accumulated in uninoculated roots only very slightly 6 to 24 h after the start of the experiment (Fig. 3A). The level of *CHI* transcript accumulation remained low for the next 48 h (Fig. 3A). In contrast, inoculation with wild-type *R. meliloti* led to a significant increase in the level of *CHI* transcripts 6 h after the start of the experiment compared to the uninoculated controls (Fig. 3B). A densitometric analysis of the two northern blots showing the increase in *CHI* transcript levels over time is presented in Fig. 3C.

Alfalfa roots were then inoculated with two different *R. meliloti* mutants for the purpose of testing their effect on *CHS* transcript accumulation (Fig. 4). Increased *CHS6–4* and *CHS4–1* mRNA levels were detected in alfalfa root RNA 6 h after inoculation with either Rm5610, a *nodA*::Tn5 mutant that is Nod⁻ on alfalfa because it does not produce Nod factor (Fisher and Long, 1992), or with Rm7094, an *exoB*::Tn5 (exopolysaccharide-minus) mutant that forms Fix⁻, bacteria-free nodules on alfalfa (Finan et al., 1985). To examine the level of *CHI* gene expression, alfalfa roots were inoculated either with the Nod⁻ mutant Rm5610 or with heat-killed wild-type *R. meliloti*. An increase in *CHI* mRNA levels was detected 6 to 12 h after inoculation (Fig. 5). However, the effect of inoculating with the Nod⁻ mutant was not as great as inoculating with the heat-killed bacteria (cf. Fig. 5A and 5B).

## 4. DISCUSSION

Higher levels of gene expression for the phenylpropanoid pathway as well as accumulation of flavonoids/isoflavonoids in ineffective (Fix⁻) nodules has been interpreted as a defense response on the part of the plant to rhizobia. The invasion by rhizobia in some ways resembles that of pathogenic bacteria, and some antimicrobial flavonoid derivatives (phytoalexins) are synthesized in response to pathogen attack (see review by McKhann and Hirsch, 1994b; Buffard et al., 1996). Nevertheless, for a defense reaction to occur, it must occur soon after inoculation and preferably with enough magnitude to elicit a hypersensitive response (HR), defined as rapidly induced tissue necrosis in response to an incompatible pathogen (Klement, 1982). However, as shown by Lawson et al. (1994), increased *CHS5* mRNA levels were not observed in response to inoculation with a Sym⁻ *R. leguminosarum* bv. *trifolii* strain or with the heterologous *R. meliloti*, arguing against the concept of an HR because a failed symbiosis should result in increased *CHS* transcript levels. Similarly, we found that an *exoB* mutant of *R. meliloti* did not elicit increased *CHS*, *CHI*, or *IFR* gene expression early in nodule development (McKhann et al., 1997).

We find that there are at least two time points in the interaction between *Rhizobium* and its host in which flavonoid biosynthetic gene transcripts accumulate compared to un-

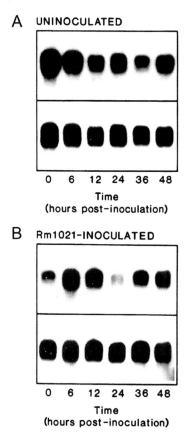

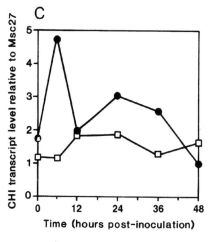

Figure 3. (A) A representative northern blot showing *CHI* transcript accumulation (top panel) in uninoculated alfalfa roots collected at the same times as the inoculated roots. The bottom panel shows the amount of RNA loaded per lane as indicated by *Msc27* transcript levels. (B) A representative northern blot showing *CHI* transcript accumulation (top panel) in Rm1021-inoculated alfalfa roots from 0 to 48 h. The bottom panel shows the amount of RNA loaded per lane as indicated by *Msc27* transcript levels. (C) A densitometric analysis of the amount of RNA in Rm1021-inoculated roots (filled circles) and the uninoculated roots (open squares) over time. The RNA levels were normalized to the amount of *Msc27* mRNA present at each time point.

# Expression of Genes for Enzymes of the Flavonoid Biosynthetic Pathway

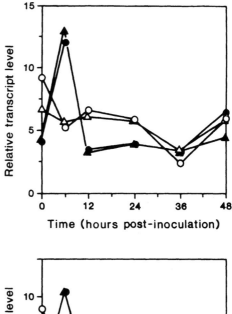

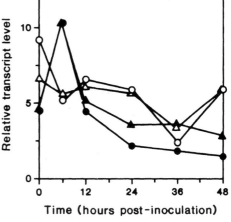

**Figure 4.** RNase protection assays comparing relative transcript levels of *CHS4–1* (circles) and *CHS6–4* (triangles) in uninoculated roots (open circles or triangles) and roots inoculated with mutant *R. meliloti* (closed circles or triangles) over a 48 h-time period. (A) RNA isolated from roots inoculated with Rm7094 (Fix⁻). (B) RNA isolated from roots inoculated with Rm5610 (Nod⁻). A transient increase in *CHS* transcript accumulation is observed 6 h post-inoculation.

inoculated controls. The first increase in *CHS* and *CHI* steady-state transcript accumulation occurs approximately 6 h after inoculation, whereas a second increase, however, only in *CHS* gene expression, takes place 2 to 4 days after inoculation (see McKhann et al., 1997). We believe that the second increase in *CHS* transcript accumulation is correlated with the Ini response because *nod*-gene inducing flavonoid production as well as *CHS* gene expression is enhanced during this time period (McKhann et al., 1997). The question is what does the first increase represent?

Lawson et al. (1994) reported that *CHS* gene expression was induced in *Trifolium subterraneum* within 6 h after inoculation with a wild-type strain of *R. leguminosarum* bv. *trifolii*, but not by a Sym⁻ strain or by the heterologous *R. meliloti* strain Rm1021. In addition, there was a steady accumulation of *CHS* transcripts over the 36-h experiment (Lawson et al., 1994). In contrast, we find that the initial 6-h increase in *CHS* transcript levels is transient and induced by both heat-killed wild-type and *nodA* mutant *R. meliloti*; none of these rhizobia elicits nodulation. Moreover, the *exoB R. meliloti* mutant also triggered a transient increase in *CHS* and *CHI* transcript levels. This suggests that the accumulation of *CHS* and *CHI* mRNAs at 6 h is non-specific because it is not dependent on

## A

**Rm5610-INOCULATED**

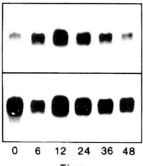

0   6   12   24   36   48
Time
(hours post-inoculation)

## B

**HEAT-KILLED
Rm1021-INOCULATED**

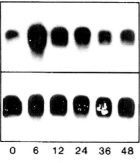

0   6   12   24   36   48
Time
(hours post-inoculation)

**Figure 5.** (A) A representative northern blot showing *CHI* transcript accumulation (top panel) in alfalfa roots inoculated with Rm5610. The bottom panel shows the amount of RNA loaded per lane as indicated by *Msc27* transcript levels. Relative to *Msc27*, there is a slight increase in *CHI* transcript accumulation 6 to 12 h post-inoculation. (B) A representative northern blot showing *CHI* transcript accumulation (top panel) in alfalfa roots inoculated with heat-killed Rm1021. The bottom panel shows the amount of RNA loaded per lane as indicated by *Msc27* transcript levels. Relative to *Msc27*, there is a significant increase in *CHI* transcript accumulation 6 h post-inoculation.

rhizobial *nod* genes or even on the presence of living bacteria. Jakobek and Lindgren (1993), who investigated induction of *PAL*, *CHS*, and *CHI* expression in bean in response to *hrp* mutants of *Pseudomonas syringae* pv. *tabaci*, suggested that early accumulation of these defense transcripts is a non-specific plant response which occurs independently of a hypersensitive response. These researchers found that inoculation with heat-killed *P. syringae*, non-pathogenic *P. fluorescens*, or even *E. coli* induced the accumulation of defense transcripts, without eliciting a hypersensitive response (Lindgren et al., 1992). In addition, Junghans et al. (1993) reported that extracts from yeast, *Erwinia chrysanthemi*, and *R. meliloti* all triggered increased *CHS* transcript levels in alfalfa 2 h after treatment, after which transcript levels decayed as a function of time.

We do not know the stimulus for induction of *CHS* and *CHI* transcripts early in the interaction between rhizobia and alfalfa roots. Ethylene accumulation elicited by inoculation may be a possibility. Environmental conditions such as light play a role in the timing and types of flavonoids secreted (Lawson et al., 1996). Moreover, the possibility also exists that the initial enhancement of *CHS* and *CHI* gene expression is related to Nod-factor independent stages in nodule organogenesis, i.e., flavonoids resulting from *de novo CHS* gene expression may modify endogenous hormone levels, which results in the triggering of cell divisions (Hirsch, 1992; Yang et al., 1992). Flavonoids have been reported as endogenous auxin transport inhibitors (Jacobs and Rubery, 1988). We are pursuing additional experiments to answer these questions.

## ACKNOWLEDGMENTS

We thank D. Dudits (Szeged, Hungary) for *Msc27* and Agway Inc. (Syracuse, NY) for supplying us with seeds of *M. sativa* cv. Iroquois. Yiwen Fang and James Shieh helped with quantifying the northern blots. Jacob S. Seeler and Richard Gaynor (Dallas, TX) are thanked for their help with the RNase protections. We are grateful to the members of our laboratory for their helpful comments on the manuscript. Special thanks to Margaret Kowalczyk and Rita Widjaja for help with the illustrations.

A grant from the USDA (CRGO 91–37307–6603) to AMH funded this research in part. HIM was supported by a USDA training fellowship and a California Biotechnology training fellowship.

## REFERENCES

Ballance, G.M.; Dixon, R.A. *Medicago sativa* cDNAs encoding chalcone reductase. *Plant Physiol.* **1995**, *107*, 1027–1028.

Buffard, D.; Esnault, R.; Kondorosi, A. Role of plant defence in alfalfa during symbiosis. *World J. Microbiol. Biotechnol.* **1996**, *12*, 175–188.

Caetano-Anollés, G.; Wall, L.G.; DeMicheli, A.T.; Macchi, E.M.; Bauer, W.D.; Favelukes, G. Role of motility and chemotaxis in efficiency of nodulation by *Rhizobium meliloti*. *Plant Physiol.* **1988**, *86*, 1228–1235.

Dakora, F.D.; Joseph, C.M.; Phillips, D.A. Alfalfa (*Medicago sativa* L.) root exudates contain isoflavonoids in the presence of *Rhizobium meliloti*. *Plant Physiol.* **1993**, *101*, 819–824.

Dudley, M.E.; Jacobs, T.W.; Long, S.R. Microscopic studies of cell divisions induced in alfalfa roots by *Rhizobium meliloti*. *Planta (Berl.)* **1987**, *171*, 289–301.

Esnault, R.; Buffard, D.; Breda, C.; Sallaud, C.; El-Turk, J.; Kondorosi, A. Pathological and molecular characterizations of alfalfa interactions with compatible and incompatible bacteria, *Xanthomonas campestris* pv. *alfalfae* and *Pseudomonas syringae* pv. *pisi*. *Mol. Plant-Microbe Inter.* **1993**, *6*, 655–664.

Finan, T. Genetic and physical analysis of Group E *exo* mutants of *Rhizobium meliloti*. *J. Bacteriol.* **1988**, *170*, 474–477.

Finan, T.; Hirsch, A.M.; Leigh, J.A.; Johansen, E.; Kuldau, G.A.; Deegan, S.; Walker, G.C.; and Signer, E.R. Symbiotic mutants of *Rhizobium meliloti* that uncouple plant from bacterial differentiation. *Cell.* **1985**, *40*, 869–877.

Fisher, R.F.; Long, S.R. *Rhizobium*-plant signal exchange. *Nature.* **1992**, *387*, 655–660.

Gowri, G.; Paiva, N.L.; Dixon, R.A. Stress responses in alfalfa (*Medicago sativa* L.) 12. Sequence analysis of phenylalanine ammonia-lyase (PAL) cDNA clones and appearance of PAL transcripts in elicitor-treated cell cultures and developing plants. *Plant Mol. Biol.* **1991**, *17*, 415–429.

Guo, L.N.; Paiva, N.L. Molecular cloning and expression of alfalfa (*Medicago sativa* L.) vestitone reductase, the penultimate enzyme in medicarpin biosynthesis. *Arch. Biochem. Biophys.* **1995**, *320*, 353–360.

Hartwig, U.A.; Joseph, C.M.; Phillips, D.A. Flavonoids released naturally from alfalfa seeds enhance growth rate of *Rhizobium meliloti*. *Plant Physiol.* **1991**, *95*, 797–803.

Hirsch, A.M. Developmental biology of legume nodulation. *New Phytol.* **1992**, *122*, 211–237.

Hirsch, A.M.; Bhuvaneswari, T.V.; Torrey, J.G.; Bisseling, T. Early nodulin genes are induced in alfalfa root outgrowths elicited by auxin transport inhibitors. *Proc. Natl. Acad. Sci. USA* **1989**, *86*, 1244–1248.

Jacobs, M.; Rubery, P.H. Naturally occurring auxin transport regulators. *Science.* **1988**, *241*, 346–349.

Jakobek, J.L.; Lindgren, P.B. Generalized induction of defense responses in bean is not correlated with the induction of the hypersensitive reaction. *Plant Cell.* **1993**, *5*, 49–56.

Junghans, H.; Dalkin, K.; Dixon, R.A. Stress responses in alfalfa (*Medicago sativa* L.). 15. Characterization and expression patterns of a subset of the chalcone synthase multigene family. *Plant Mol. Biol.* **1993**, *22*, 239–253.

Kapros, T.; Bogre, L.; Nemeth, K.; Bako, L.; Györgyey, J.; Wu, S.C.; Dudits, D. Differential expression of histone H3 gene variants during cell cycle and somatic embryogenesis in alfalfa. *Plant Physiol.* **1992**, *98*, 621–625.

Klein, S.; Hirsch, A.M.; Smith, C.A.; Signer, E.R. Interaction of *nod* and *exo Rhizobium meliloti* in alfalfa nodulation. *Mol. Plant-Microbe Inter.* **1988**, *1*, 94–100.

Klement, Z. *Hypersensitivity*. In *Phytopathogenic Procaryotes*. Mount, M.S., Lacey, G.S., Eds.; Academic Press: New York, 1982; pp 150–178.

Lawson, C.G.R.; Djordjevic, M.A.; Weinman, J.J.; Rolfe, B.G. *Rhizobium* inoculation and physical wounding result in the rapid induction of the same chalcone synthase copy in *Trifolium subterraneum*. *Mol. Plant-Microbe Inter.* **1994**, *7*, 498–507.

Lawson, C.G.R.; Rolfe, B.G.; Djordjevic, M.A. *Rhizobium* inoculation induces condition-dependent changes in the flavonoid composition of root exudates from *Trifolium subterraneum*. *Aust. J. Plant Physiol.* **1996**, *23*, 913–101.

Lindgren, P.B.; Jakobek, J.; Smith, J.A. Molecular analysis of plant defense responses to plant pathogens. *J. Nematol.* **1992**, *24*, 330–337.

Löbler, M.; Hirsch, A.M. A gene that encodes a proline-rich nodulin with limited homology to *PsENOD12* is expressed in the invasion zone of *Rhizobium meliloti*-induced alfalfa root nodules. *Plant Physiol.* **1993**, *103*, 21–30.

Maxwell, C.A.; Harrison, M.J.; Dixon, R.A. Molecular characterization and expression of alfalfa isoliquiritigenin 2'-*O*-methyltransferase, an enzyme specifically involved in the biosynthesis of an inducer of *Rhizobium meliloti* nodulation genes. *Plant J.* **1993**, 6, 971–981.

McKhann, H.I; Hirsch, A.M. *In situ* localization of specific mRNAs in plant tissues. In *Methods in Plant Molecular Biology and Biotechnology*. Glick, B.R., Thompson, J.E., Eds.; CRC Press, Inc.: Boca Raton, 1993; pp 179–205.

McKhann, H.I; Hirsch, A.M. Isolation of chalcone synthase and chalcone isomerase cDNAs from alfalfa (*Medicago sativa* L.): highest transcript levels occur in young roots and root tips. *Plant Mol. Biol.* **1994a**, *24*, 767–777.

McKhann, H.I; Hirsch, A.M. Does *Rhizobium* avoid the host response? In *Current Topics in Microbiology and Immunology*, Vol. 192, Bacterial Pathogenesis of Plants and Animals. Dangl, J., Ed.; Springer Verlag: Berlin, 1994b; pp 139–162.

McKhann, H.I; Paiva, N.L.; Dixon, R.A.; Hirsch, A.M. Chalcone synthase transcripts are detected in alfalfa root hairs following inoculation with wild-type *Rhizobium meliloti*. *Mol. Plant-Microbe Inter.* **1997**, *10*, 50–58.

Meade, H.M.; Long, S.R.; Ruvkun, G.B.; Brown, S.E.; Ausubel, F.M. Physical and genetic characterization of symbiotic and auxotrophic mutants of *Rhizobium meliloti* induced by transposon Tn*5* mutagenesis. *J. Bacteriol.* **1982**, *149*, 114–122.

Paiva, N.L.; Edwards, R.; Sun, Y.; Hrazdina, G.; Dixon, R.A. Stress responses in alfalfa (*Medicago sativa* L.). 11. Molecular cloning and expression of alfalfa isoflavone reductase, a key enzyme of isoflavonoid phytoalexin biosynthesis. *Plant Mol. Biol.* **1991**, *17*, 653–667.

Recourt, K.; Schripsema, J.; Kijne, J.W.; van Brussel, A.A.N.; Lugtenberg, B.J.J. Inoculation of *Vicia sativa* subsp. *nigra* roots with *Rhizobium leguminosarum* biovar *viciae* results in release of *nod* gene activating flavanones and chalcones. *Plant Mol. Biol.* **1991**, *16*, 841–852.

Recourt, K.; van Tunen, A.J.; Mur, L.A.; van Brussel, A.A.N.; Lugtenberg, B.J.J.; Kijne, J.W. Activation of flavonoid biosynthesis in roots of *Vicia sativa* subsp. *nigra* plants by inoculation with *Rhizobium leguminosarum* biovar *viciae*. *Plant Mol. Biol.* **1992**, *19*, 411–420.

Rolfe, B.G.; Sargent, C.L.; Weinmann, J.J.; Djordjevic, M.A.; McIver, J.; Redmond, J.W.; Batley, M.; Yuan, D.C.; Sutherland, M.W. Signal exchange between *R. trifolii* and clovers. In *Signal Molecules in Plants and Plant-Microbe Interactions*; Lugtenberg, B.J.J., Ed.; Springer-Verlag: Berlin, 1989; pp 303–310.

Sallaud, C.; El-Turk, J.; Bigarré, L.; Sevin, H.; Welle, R.; Esnault, R. Nucleotide sequences of three chalcone reductase genes from alfalfa. *Plant Physiol.* **1995**, *108*, 869–870.

Sambrook J.; Fritsch E.F.; Maniatis, T. *Molecular Cloning: A Laboratory Manual*. Cold Spring Harbor Press, Cold Spring Harbor, 1988.

van Brussel, A.A.N.; Recourt, K.; Pees, E.; Spaink, H.P.; Tak, T.; Wijffelman, C.A.; Kijne, J.W.; Lugtenberg, B.J.J. A biovar-specific signal of *Rhizobium leguminosarum* bv. *viciae* induces increased nodulation gene-inducing activity in root exudate of *Vicia sativa* subsp. *nigra*. *J. Bacteriol.* **1990**, *172*, 5394–5401.

Vincent, J.M. *A Manual for the Practical Study of Root-Nodule Bacteria*, Blackwell Scientific Publications, London, 1970.

Yang, W.-C.; Cremers, H.C.J.C.; Hogendijk, P.; Katinakis, P.; Wijffelman, C.A.; Franssen, H.; van Kammen, A.; Bisseling, T. *In-situ* localization of chalcone synthase mRNA in pea root nodule development. *Plant J.* **1992**, *2*, 143–151.

# PROSPECTS FOR THE METABOLIC ENGINEERING OF BIOACTIVE FLAVONOIDS AND RELATED PHENYLPROPANOID COMPOUNDS

Richard A. Dixon,[1] Paul A. Howles,[1,2] Chris Lamb,[2] Xian-Zhi He,[1] and J. Thirupathi Reddy[1]

[1]Plant Biology Division
Samuel Roberts Noble Foundation
P. O. Box 2180
Ardmore, Oklahoma 73402
[2]Plant Biology Laboratory
Salk Institute for Biological Studies
10010 N. Torrey Pines Road
La Jolla, California 92037

## 1. ABSTRACT

The successful engineering of complex metabolic pathways will require, in addition to availability of cloned genes and promoters, knowledge of the regulatory mechanisms that control metabolic flux into the pathway including post-translational phenomena such as metabolite channeling. We are interested in modifying pathways for the synthesis of isoflavonoids and other bioactive phenylpropanoid compounds in transgenic plants. We describe studies on flux control utilizing transgenic tobacco plants that under- and over-express key biosynthetic enzymes, and outline experimental approaches for the molecular dissection of potential metabolic channels in the synthesis of antimicrobial flavonoid derivatives in alfalfa and other species.

## 2. THE FLAVONOID/ISOFLAVONOID PATHWAY IN PLANT-MICROBE INTERACTIONS

Plants synthesize an enormous variety of secondary metabolites via often complex biosynthetic pathways, and many of these metabolites play important functions in the plant's interactions with its environment. There is considerable research interest on understanding how the synthesis of secondary metabolites is regulated during development and in response to environmental factors such as light, drought, and pathogen infection. Flavonoid derivatives have been shown to exhibit a number of biological activities, functioning as insect attractants, UV protectants, antimicrobial phytoalexins, insect and herbivore feeding deterrents, allelochemicals, signals for initiation of interactions with symbiotic bacteria, regulators of auxin transport, and stimulators of pollen germination (Dixon & Paiva 1995). Although considerable progress has been made in identifying the enzymes of flavonoid biosynthesis, cloning their genes, and dissecting the signal transduction pathways linking transcriptional activation to initial signal perception (Dangl 1992; Dixon et al. 1995; Hahlbrock & Scheel 1989), much less is known about the post-transcriptional regulation of flavonoid synthesis, or indeed of any pathway for biosynthesis of plant secondary metabolites.

Much of our recent work has centered on the identification and expression of genes encoding enzymes for the synthesis of elicitor-induced flavonoid derivatives and the associated pathways of primary metabolism in alfalfa (Dixon et al., 1995; Fahrendorf et al., 1995; Paiva et al., 1994) in relation to the response of this plant to fungal infection, and the establishment of beneficial interactions with symbiotic bacteria. The compounds of most interest to us include the antimicrobial pterocarpan phytoalexin medicarpin (a product of the isoflavonoid branch pathway), and the potent nodulation gene inducer 2'-methoxy,4,4'-dihydroxychalcone. Medicarpin, as its malonyl glycoside MGM, is constitutively produced at low levels in root cortical cells and stored in the cell vacuoles (Mackenbrock et al., 1992). 2'-Methoxy,4,4'-dihydroxychalcone is released from the root cortex to the rhizosphere along with other less potent nodulation gene inducing flavonoids (Maxwell et al., 1989).

A long term goal of our program is to improve plant performance by genetic manipulation of pathways for the biosynthesis of biologically active phenylpropanoid compounds such as antimicrobial isoflavonoids and flavonoid nodulation signals. Strategies for such improvements have been reviewed elsewhere (Dixon et al., 1994; Dixon et al., 1993; Lamb et al., 1992) and will not be discussed further here. It is the purpose of the present chapter to highlight important but relatively little studied aspects of pathway regulation that might impact the success of secondary metabolite pathway engineering.

## 3. THE FLAVONOID/ISOFLAVONOID PATHWAY IS UNDER TIGHTLY CO-ORDINATED TRANSCRIPTIONAL CONTROL IN ELICITED CELLS

All flavonoids contain a phenylpropanoid nucleus derived from phenylalanine, and a second aromatic ring (the A-ring) derived from three molecules of malonyl CoA, originating via the acetyl CoA carboxylase reaction (Figure 1). The rapid deployment of isoflavonoid phytoalexin synthesis *de novo* as a defensive chemical barrier in leguminous plants requires the co-ordinated induction of all the enzymes of the basic flavonoid pathway, the isoflavonoid branch pathway, and a number of enzymes of primary metabolism to provide NADPH (pentose phosphate pathway) and activated methyl groups (S-adenosyl methionine synthase). Thus, as a result of corresponding increases in the steady state transcript levels, the elicitation of medicarpin is preceded by the increase in the activities of over 12 enzymes required for its synthesis from L-phenylalanine (Dixon et al. 1995; Fahrendorf et

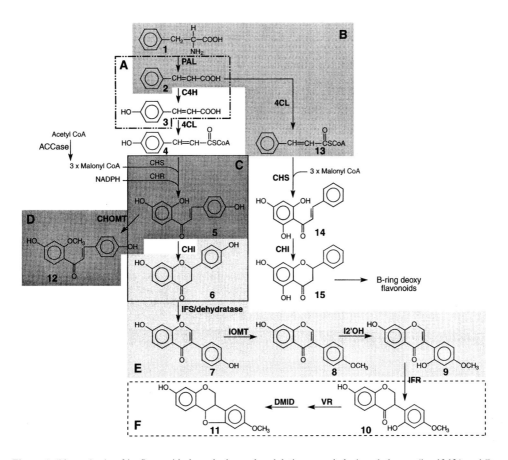

Figure 1. Biosynthesis of isoflavonoid phytoalexins and nodulation-gene inducing chalcones (in alfalfa) and B-ring deoxy flavonoids (in *Cephalocereus senilis*), indicating potential metabolic channeling. The enzymes are: PAL, L-phenylalanine ammonia-lyase; C4H, cinnamate 4-hydroxylase; 4CL, 4-coumarate CoA ligase; ACCase, acetyl CoA carboxylase; CHS, chalcone synthase; CHR, chalcone reductase; CHI, chalcone isomerase; CHOMT, chalcone 2'-O-methyltransferase; IFS, isoflavone synthase; IOMT, isoflavone O-methyltransferase; I2'-OH, isoflavone 2'-hydroxylase; IFR, isoflavone reductase; VR, vestitone reductase; DMID, 7,2'-dihydroxy-4'-methoxy-isoflavanol dehydratase. The individual compounds are: **1**, L-phenylalanine; **2**, *trans*-cinnamic acid; **3**, 4-coumaric acid; **4**, 4-coumaroyl CoA; **5**, 2',4,4'-trihydroxychalcone; **6**, liquiritigenin; **7**, daidzein; **8**, formononetin; **9**, 2'-hydroxy-formononetin; **10**, vestitone; **11**, medicarpin; **12**, 2'-methoxy, 4,4'-dihydroxychalcone; **13**, cinnamoyl CoA; **14**, 2',4',6'-trihydroxychalcone; **15**, pinocembrin. Each shaded or outlined box (A-F) represents a series of reactions through which metabolic channeling may occur (see text). All the reactions shown occur in roots and cell cultures of alfalfa, except for the pathway from cinnamate to pinocembrin, which has been studied in elicited cell cultures of *C. senilis*.

al. 1995). Such increases in enzyme activities have been shown to involve *de novo* transcriptional activation (Lawton & Lamb 1987; Somssich et al., 1989).

In bacteria and fungi, secondary metabolic pathway genes are clustered in operons (Desjardins et al., 1993; Plater & Strohl 1994) whereas genes encoding enzymes of plant secondary metabolic pathways are generally distributed throughout the plant genome, with the exception of clustering of some members of multigene families such as that encoding chalcone synthase (CHS) (Ryder et al., 1987; Akada & Dube 1995). The mechanisms underlying the rapid co-ordinated activation of these sets of genes remain to be identified.

However, much attention has been given to the identification of *cis*-elements and *trans*-acting factors for elicitor- and infection-mediated activation of individual defense response genes (Da Costa e Silva et al., 1993; Loake et al., 1992; Raventós et al. 1995; Yu et al. 1993), and it is hoped that the complete signal transduction pathway(s) leading from elicitor/pathogen recognition to defense gene activation will soon be elucidated.

Towards this goal, we have used nuclear run-on transcription assays to measure transcriptional activation kinetics for genes encoding enzymes of isoflavonoid phytoalexin biosynthesis and related pathways in elicited alfalfa cells. Every gene that we have studied in this pathway undergoes rapid but transient transcriptional activation in response to elicitor (Ni et al., 1996a). Genes encoding L-phenylalanine ammonia-lyase (PAL), CHS and chalcone reductase (CHR) are the most rapidly activated (increased transcription measurable within 10–20 minutes post-elicitation), whereas cinnamic acid 4-hydroxylase (C4H), chalcone isomerase (CHI), isoflavone reductase (IFR) and the lignin biosynthetic enzyme caffeic acid 3-*O*-methyltransferase (COMT) genes are activated at a slower initial rate. In contrast, initiation of transcription of chalcone 2'-*O*-methyltransferase (CHOMT) exhibits a lag period of 60 min post-elicitation. Figure 1 shows the reactions catalyzed by these various enzymes. Although lignin biosynthetic genes are transcriptionally activated along with the phytoalexin pathway genes in elicited alfalfa cells, their steady state transcript levels and corresponding enzymatic activities do not increase, and lignin does not accumulate (Ni et al., 1996b). Clearly, the final pattern of phenylpropanoid synthesis is determined by a combination of transcriptional and post-transcriptional events.

## 4. POTENTIAL LIMITATIONS TO THE SUCCESSFUL ENGINEERING OF COMPLEX PATHWAYS SUCH AS FLAVONOID/ISOFLAVONOID SYNTHESIS IN TRANSGENIC PLANTS

For several reasons, availability of a complete set of cloned genes, such as those encoding enzymes for the synthesis of biologically active flavonoids or isoflavonoids, and the knowledge of the mechanisms underlying their transcriptional regulation, are in themselves insufficient to guarantee success in the genetic manipulation of a complex biosynthetic pathway. The incorporation of a new gene or a new set of genes in transgenic plants under the appropriate transcriptional promoter sequences, even if leading to high expression of the desired protein(s), does not necessarily mean that significant metabolic flux will occur into the desired end product(s). It is important to know which steps in the pathway are rate limiting, and this may be far from obvious in complex pathways. In the following section, we discuss transgenic approaches for determining rate limiting steps in the phenylpropanoid pathway. A second complication that may arise is that the secondary metabolic pathways, or portions of them, may be organized in "metabolic compartments," through which pathway intermediates are channeled without equilibrating with cytoplasmic pools (Hrazdina & Jensen 1992; Hrazdina & Wagner 1985 a,b; Srere 1987). These compartments may arise as a consequence of close physical associations between consecutive biosynthetic enzymes (Srere 1987). Metabolic compartmentation adds an extra level of flux control over that determined by the activity levels and kinetic properties of the individual enzymes of the pathway. It is possible that individual members of multigene families, such as those encoding PAL and CHS in the phenylpropanoid/flavonoid pathway, function in different metabolic compartments, even though the expression patterns and kinetic properties of the individual isoenzymes may be very similar. This is an important area for future study, and specific examples of potential metabolic channeling in flavonoid/isoflavonoid synthesis will be discussed later in this chapter.

## 5. TRANSGENIC APPROACHES FOR DETERMINING THE CONTROL POINTS FOR FLUX INTO THE BRANCH PATHWAYS OF PHENYLPROPANOID/FLAVONOID SYNTHESIS

We have generated, as a result of epigenetic sense-suppression resulting from transformation with a heterologous (bean) PAL transgene, a series of transgenic tobacco plants (a convenient model system) with severely reduced PAL activity (Elkind et al., 1990). The subsequent phenotypes of the progeny of two of these individual sense-suppressed primary transformants differed in PAL activity retention. In line YE6-16, the reduced PAL enzyme activity of the primary transformant was gradually lost in successive selfed generations, reaching near wild-type levels by the T5 generation (Bate et al., 1994). In contrast, PAL-suppressed primary transformant YE10–6 changed to a PAL overexpressing line in a single selfed generation (Howles et al., 1996). Thus, we produced near isogenic tobacco lines that exhibit a wide range of PAL activities, from approximately 5% to 200% of the wild-type activity (measured in stem tissue) or up to 600% of wild-type activity in leaf tissue. These plants provide a means of assessing the role of PAL in controlling flux into downstream pathways of phenylpropanoid synthesis.

Analysis of the levels of various phenylpropanoid compounds by HPLC in the above lines revealed that PAL is the rate determining step for synthesis of chlorogenic acid (CGA) in tobacco leaves, with a flux control coefficient of 1.0 (Kacser et al., 1995). Decreasing PAL activity below wild-type levels also results in decreased flavonoid (rutin) levels in leaves, although PAL appears to have a flux control coefficient of less than 1.0 for rutin synthesis in tobacco (Bate et al. 1994), indicating the operation of further sites of flux control. PAL also becomes a rate limiting step for lignin synthesis, but only when reduced below a threshold of approximately 20% of the wild-type level (Bate et al., 1994; Sewalt et al., 1997).

Transgenic tobacco plants overexpressing PAL were obtained from seed of primary transformant YE 10–6. We used transcript-specific RNA probes and antibodies that discriminate between endogenous and transgene-encoded PAL proteins to analyze PAL expression in the primary transformant ($T_0$) and first generation ($T_1$) overexpressor plants. The results indicated that the transgene-encoded PAL was the cause of the higher PAL activity (up to five- and two-fold greater activities over that of the wild-type in leaf and stem tissue, respectively) in the $T_1$ as compared to $T_0$ plants (Howles et al. 1996). Leaves of PAL overexpressing plants contained higher levels of CGA, but not of rutin, again indicating that although PAL is the key control point for flux into CGA, there is an extra level of flux control for flavonoid synthesis, possibly at the level of the first enzyme of the flavonoid branch pathway, chalcone synthase. However, levels of the glucoside of 4-coumaric acid were also increased in the PAL overexpressing plants, suggesting that the 4-coumarate:CoA ligase reaction might become a limiting step for flavonoid synthesis under these conditions. These results indicate the feasibility of engineering selective changes in the phenylpropanoid pathway by the over-expression of a single early pathway gene. Clearly, however, engineering increases in flavonoid synthesis will require manipulation of further downstream enzymes.

We have generated, by transformation with gene constructs containing a full length alfalfa C4H cDNA sequence in sense or antisense orientations, another series of transgenic tobacco plants with altered levels of expression of the second enzyme of the phenylpropanoid pathway, the C4H cytochrome P450 (Sewalt et al., 1997). Transformation with sense constructs led to either the over-expression or sense-suppression of C4H activity. Low levels of C4H activity resulted in a reduction of the levels of CGA and rutin in leaves and of lignin in stems. Higher C4H activity resulted in increased CGA and rutin accumulation, but did not significantly affect lignin levels. A 50% reduction in C4H activity from wild-type was associated with a 50–60% decrease in CGA levels, and a 72% reduction in C4H activity was associated with a 75% decrease in lignin, suggesting that C4H has a high flux control coefficient for synthesis of both CGA and lignin in tobacco. This at first appears paradoxical, since the sum of all flux control coefficients in a linear pathway can not exceed 1.0 (Kacser et al., 1995), and we have shown that PAL has a flux control coefficient of near unity for synthesis of CGA in tobacco. However, reduction in C4H activity in transgenic plants leads to a similar-fold reduction in PAL activity, probably due to complex feedback control mechanisms involving cinnamic acid as a regulatory molecule (Mavandad et al., 1990). Thus, a direct relation between C4H and PAL activities was observed in stems and young leaves of plants in which C4H activity had been down-regulated by C4H transgene expression. However, decreasing PAL activity by genetic manipulation did not appear to affect C4H activity, at least in leaves (S.A. Masoud, J.W. Blount and R.A. Dixon, unpublished results).

## 5.1. Examples of Potential Metabolic Channeling in the Phenylpropanoid/Flavonoid/Isoflavonoid Pathway

*5.1.1. Chalcone Isomerization versus Chalcone 2'-O-Methylation.* CHS, the first enzyme specific for flavonoid biosynthesis, catalyzes the condensation of three molecules of malonyl CoA with one molecule of 4-coumaroyl CoA to yield a polyketide intermediate which is cyclized on the enzyme to form 2',4,4',6'-tetrahydroxychalcone. The cDNAs encoding at least 7 different CHS isoforms have been characterized from alfalfa (Esnault et al., 1993; Junghans et al., 1993; McKhann & Hirsch 1994), and we have shown that at least 5 of them are expressed in roots (Junghans et al., 1993). It is not known whether functional differences exist between the different CHS isoforms in legumes. However, as mentioned above, co-expression of multiple forms in a single cell type provides potential for metabolic channeling.

In many legume species, the flavonoids and isoflavonoids found constitutively in roots, and the elicitor/infection induced isoflavonoid phytoalexins, lack the hydroxyl group at the 6' position (chalcone numbering; 5-position flavanone numbering) (Figure 1). This OH group is removed by the action of a polyketide reductase enzyme (chalcone reductase, CHR) which reduces a carbonyl group on the CHS enzyme-linked polyketide prior to cyclization of the aromatic A-ring, with subsequent loss of the oxygen function as water (Welle et al., 1991). This results in the formation of 2',4,4'-trihydroxychalcone. CHR contains a leucine zipper motif region which may be involved in its physical interaction with CHS (Welle et al., 1991). We have recently isolated and characterized cDNA clones encoding two distinct isoforms of CHR from alfalfa (Ballance & Dixon 1994). It is not yet known whether these interact differentially with different forms of CHS.

The second reaction specific for flavonoid/isoflavonoid synthesis is the closure of the central heterocyclic ring catalyzed by CHI, which is probably encoded by a single gene in alfalfa (McKhann & Hirsch 1994). CHI from legumes is generally active with both 2',4,4',6'-tetrahydroxy- and 2',4,4'-trihydroxychalcones, whereas the trihydroxychalcone is not utilized by the enzyme from those plant species that do not make 5-deoxy flavonoid derivatives (Dixon et al., 1988).

2',4,4'-Trihydroxychalcone is the branch point metabolite for synthesis of isoflavonoid phytoalexins and nodulation gene-inducing methoxychalcone in alfalfa. Thus, CHI converts the chalcone to 4',7-dihydroxyflavanone (liquiritigenin), a precursor of medicarpin, whereas a highly regiospecific chalcone 2'-O-methyltransferase (CHOMT) catalyzes the formation of 2'-methoxy,4,4'-dihydroxychalcone, utilizing S-adenosyl methionine as methyl group donor. CHOMT has been purified and characterized from alfalfa, and corresponding cDNAs cloned (Maxwell et al., 1992, 1993) as a prelude to attempts to up-regulate nodulation gene inducer production by genetic engineering.

Alfalfa CHI and CHOMT have very similar $K_m$ values for the trihydroxychalcone substrate (Dixon et al., 1988; Maxwell et al., 1992), and, in the absence of compartmentalization or other forms of regulation, these enzymes would be expected to effectively compete with each other for the substrate. Furthermore, both enzymes appear to be operationally "soluble" in the cytosol, co-induced in elicitor-treated cell suspension cultures, and co-expressed temporally and spatially in root cortical cells which appear to synthesize both isoflavonoids and methoxychalcone (Maxwell et al., 1992). Chalcone 2'-O-methylation blocks the cyclization reaction catalyzed by CHI. Thus, at first glance, the CHI and CHOMT reactions are mutually exclusive. The question therefore arises as to how the flux into the two pathways is controlled. One possibility is that specific complexes exist, perhaps involving different isoforms of CHS, whereby chalcone precursor can be channeled directly to the isomerase or O-methyltransferase reactions. To investigate the possibility that two or more of the enzymes CHS, CHI, CHR, and CHOMT have the potential to form a multienzyme complex we are currently using the yeast two-hybrid system of cloning genes based on physical interactions between their protein products (Fields & Song 1989). We have constructed a large cDNA library (more than 10 million initial unamplified clones) from RNA isolated from alfalfa seedling roots in a yeast expression vector with the cloned sequences fused to the yeast GAL4 transactivation domain. Mobilization of the library into yeast cells harboring the flavonoid pathway genes as fusions to the GAL4 DNA binding domain should provide information on whether the above enzymes can physically interact among themselves, or with other alfalfa root proteins.

*5.1.2. The O-Methylation of Daidzein.* The isoflavonoid phytoalexins of alfalfa are methylated at the 4'-position of the B-ring (Figure 1). Although an isoflavone 4'-O-methyltransferase was described from chickpea in 1974, the regiospecificity of the enzyme was not unequivocally determined (Wengenmayer et al., 1974). Previous work in our laboratory had demonstrated that the major isoflavone O-methyltransferase (IOMT) activity purified from elicited alfalfa cell cultures methylates the A-ring rather than the B-ring (Edwards & Dixon 1991), and this has been confirmed for the chickpea enzyme (W. Barz, personal communication). Remarkably, the induced isoflavonoids of alfalfa and chickpea are not A-ring methylated, although the activity of the alfalfa isoflavone A-ring OMT is induced approximately 200-fold by elicitor (Edwards & Dixon 1991). A low level of 4'-OMT (B-ring) activity was observed in alfalfa cell cultures, but this was only very weakly induced by elicitor, and it has not proven possible to "follow" this activity through an enzyme purification scheme He and Dixon, 1996).

Labeled precursor feeding experiments in chemically elicited alfalfa seedlings suggest the possibility that the IOMT reaction is part of a metabolic channel, and the association of this enzyme with adjacent isoflavonoid biosynthetic enzymes could account for the different enzyme specificities observed *in vivo* and *in vitro*. Thus, although labeled 2',4,4'-trihydroxychalcone and formononetin are precursors of medicarpin, daidzein (the presumed substrate of the 4'-*O*-methyltransferase) was not incorporated (Dewick & Martin 1979) (Figure 1). The substrate specificities of the enzymes preceeding and following daidzein in the isoflavonoid pathway indicate that daidzein is the substrate for 4'-*O*-methylation. Also, a mutant of subterranean clover that can not make methylated isoflavonoids accumulates high levels of daidzein (Wong & Francis 1968). These observations are consistent with this compound being a precursor for 4'-*O*-methylation. The enzymes preceding and following IOMT in the medicarpin pathway, isoflavone synthase and isoflavone 2'-hydroxylase, are both microsomal cytochrome P45Os (Gunia et al., 1991; Kessmann et al., 1990; Kochs & Grisebach 1986), which potentially could be physically associated with the 4'-OMT as a complex weakly associated with the exterior of the endoplasmic reticulum. The most likely interpretation of the precursor feeding experiments is, therefore, the existence of a metabolic channel through which transient levels of daidzein, formed *in situ* by microsomal isoflavone synthase, are used in the synthesis of medicarpin. However, exogenously supplied daidzein can not be a substrate for IOMT, as it does not gain access to the metabolic channel.

We have recently purified the elicitor-induced isoflavone OMT to homogeneity by affinity chromatography on an immobilized daidzein matrix, using elution with the co-substrates genistein and S-adenosyl L-methionine (He and Dixon, 1996). Internal peptide sequences have been obtained, and an 800 bp partial IOMT cDNA clone isolated. The alfalfa IOMT shows extensive sequence identity to a 6a-hydroxypterocarpan OMT involved in synthesis of the pea phytoalexin pisatin (H.D. VanEtten, personal communication). The next step is to utilize reverse genetics to address the product specificity of the IOMT *in vivo*. Utilizing the yeast two-hybrid system it may also be possible to use the cloned IOMT to isolate cDNA clones encoding isoflavone synthase and isoflavone 2'-hydroxylase (Fields & Song 1989). Cloning these genes would enable us to start reconstructing expression of the complete isoflavonoid phytoalexin pathway in plants that do not possess it. This is a long term goal of our research program, and is of both fundamental interest (e.g for understanding how to co-transform multiple genes encoding sequential enzymatic reactions and maintain correct regulation) and applied relevance (for increasing a plant's spectrum of antimicrobial agents, or incorporating compounds (such as the isoflavone genistein) with human health promoting benefits. In the shorter term, cloning these genes might give insights into how the components of this potential metabolic compartment are physically associated.

*5.1.3. Synthesis of B-Ring Deoxy Flavonoids.* Elicited cell suspension cultures of the cactus *Cephalocereus senilis* accumulate flavonoids lacking the hydroxyl group at the 4'-position on the B-ring (Liu et al., 1993a,b). This hydroxyl group originates from the action of C4H prior to formation of the 15 carbon flavonoid skeleton. In contrast, flavonoids found in healthy intact plants possess the normal B-ring 4'-hydroxylation pattern. In collaboration with the group of Dr Tom Mabry, we have shown that the cactus cultures contain the necessary enzymatic activities for the conversion of phenylalanine to the B-ring deoxy flavanone pinocembrin, i.e. a single form of CoA ligase which is (unusually) highly active with cinnamic acid, a CHS which is active with cinnamoyl CoA, and a CHI which is active with B-ring deoxy chalcones (Liu et al. 1995; Paré et al. 1992) (Figure 1). The simplest explanation for formation of the unusual flavonoids would be a lack of induction of C4H activity, leading to a pool of cinnamic acid that can be converted to the CoA derivative. However, although not used, C4H activity is strongly induced in the cultures (Liu et al. 1995). This suggests a novel form of metabolic channeling from PAL to the CoA ligase, thereby by-passing the microsomal C4H (Figure 1). We do not know how many different forms of PAL exist in *C. senilis*. Elicitation may induce a specific PAL isoform that is not present at significant levels in healthy tissues of the intact plant or in unelicited cell cultures. This would then allow for a mechanism to by-pass C4H in which this PAL isoform physically interacted with the CoA ligase, rather than with C4H. A further possibility is stimulation of associations or dissociations as a result of post-translational modification of PAL, C4H or CoA ligase. Further studies are in progress to provide kinetic evidence for or against these various models, and to elucidate the physical nature of the apparent interactions.

## 6. FUTURE PROSPECTS

Rapid progress is being made towards elucidating the details of the biosynthetic pathways leading to the synthesis of the major classes of plant secondary metabolites, and towards cloning the genes which encode the various biosynthetic enzymes. This Herculean task is being made easier by emerging molecular techniques including PCR-based strategies for generation of cDNA sequences from gene family members with conserved sequence elements, such as the cytochrome P450s. Systems for expressing cloned genes in *E. coli*, yeast, insect cells and transgenic plants are facilitating the functional identification of cloned genes (Dixon & Paiva 1995). Methods for transferring single genes into plant cells are now routine for a wide range of species. A major challenge for the future will be to develop gene delivery sytems for plants that will enable simultaneous transfer of whole pathways rather than single genes. Preliminary progress has been made in this area using vectors harboring a potyviral (potato virus Y)-based polyprotein processing system (Marcos & Beachy 1994). Although our knowledge of transcriptional control of plant secondary metabolic pathway enzymes is increasing, it is now important to complement the biochemical and molecular approaches currently being pursued with genetic approaches. There is a wealth of genetic information on the regulatory genes controlling the synthesis of colored anthocyanin pigments (Almeida et al., 1989), but very little for other compounds such as the isoflavonoids. Development of model legumes such as *Medicago truncatula* and *Lotus corniculatus* as tractable genetic systems will provide a critical resource for such studies.

Our knowledge of how metabolic flux is channeled through the various branch pathways of plant secondary metabolism is woefully incomplete. It is hoped that experiments with model systems such as those discussed in this chapter will lead to a better understanding of metabolite channeling and the regulatory architecture of complex pathways.

## ACKNOWLEDGMENTS

We thank Cuc Ly for the artwork. Work in the authors' laboratories was supported by the Samuel Roberts Noble Foundation.

## REFERENCES

Akada, S.; Dube, S. K. Organization of soybean chalcone synthase gene clusters and characterization of a new member of the family. *Plant Mol. Biol.* **1995**, 29, 189–199.

Almeida, J.; Carpenter, R.; Robbins, T. P.; Martin, C.; Coen, E. S. Genetic interactions underlying flower color patterns in *Antirrhinum majus*. *Genes Devel.* **1989**, 3, 1758–1767.

Ballance, G. M.; Dixon, R. A. Medicago sativa cDNAs encoding chalcone reductase. *Plant Physiol.* **1994**, 107, 1027–1028.

Bate, N. J.; Orr, J.; Ni, W.; Meroni, A.; Nadler-Hassar, T.; Sowenwe, P. W.; Dixon, R. A.; Lamb, C. J.; Elkind, Y. Quantitative relationship between phenylalanine ammonia-lyase levels and phenylpropanoid accumulation in transgenic tobacco identifies a rate determining step in natural product synthesis. *Proc. Natl. Acad. Sci., U.S.A.* **1994**, 91, 7608–7612.

Da Costa e Silva, O.; Klein, L.; Schmelzer, E.; Trezzini, G. F.; Hahlbrock, K. BPF-1, a pathogen-induced DNA-binding protein involved in the plant defense response. *Plant J.* **1993**, 4, 125–135.

Dangl, J. L. Regulatory elements controlling developmental and stress induced expression of phenylpropanoid genes. *Plant Gene Res.* **1992**, 8, 303–326.

Desjardins, A. E.; Hohn, T. M.; McCormick, S. P. Trichothecene biosynthesis in *Fusarium* species: chemistry, genetics, and significance. *Microbiol. Rev.* **1993**, 57, 595–604.

Dewick, P. M.; Martin, M. Biosynthesis of pterocarpan, isoflavan and coumestan metabolites of *Medicago sativa*: chalcone, isoflavone and isoflavanone precursors. *Phytochemistry* **1979**, 18, 597–602.

Dixon, R. A.; Bhattacharyya, M. K.; Paiva, N. L. Engineering disease resistance in plants: an overview. In *Advanced Methods in Plant Pathology*; Singh, R. P.; Singh U. S., Eds. CRC Press:Boca Raton, **1994**; pp 249–270.

Dixon, R. A.; Blyden, E. R.; Robbins, M. P.; van Tunen, A. J.; Mol, J. N. M. Comparative biochemistry of chalcone isomerases. *Phytochemistry* **1988**, 27, 2801–2808.

Dixon, R. A.; Harrison, M. J.; Paiva, N. L. The isoflavonoid phytoalexin pathway: from enzymes to genes to transcription factors. *Physiol. Plant.* **1995**, 93, 385–392.

Dixon, R. A.; Maxwell, C. A.; Ni, W.; Oommen, A.; Paiva, N. L. . Genetic manipulation of lignin and phenylpropanoid compounds involved in interactions with microorganisms. In *Recent Advances in Phytochemistry*; Stafford, H. A. Ed.; Plenum: New York, 1993; pp 153–178.

Dixon, R. A.; Paiva, N. L. Stress-induced phenylpropanoid metabolism. *Plant Cell* **1995**, 7, 1085–1097.

Edwards, R.; Dixon, R. A. Isoflavone O-methyltransferase activities in elicitor-treated cell suspension cultures of *Medicago sativa*. *Phytochemistry* **1991**, 30, 2597–2606.

Elkind, Y.; Edwards, R.; Mavandad, M.; Hedrick, S. A.; Ribak, O.; Dixon, R. A.; Lamb, C. J. Abnormal plant development and down regulation of phenylpropanoid biosynthesis in transgenic tobacco containing a heterologous phenylalanine ammonia-lyase gene. *Proc. Natl. Acad. Sci. U.S.A.* **1990**, 87, 9057–9061.

Esnault, R.; Buffard, D.; Breda, C.; Sallaud, C.; Turk, J. E.; Kondorosi, A. Pathological and molecular characterizations of alfalfa interactions with compatible and incompatible bacteria, *Xanthomonas campestris* pv. *alfalfae* and *Pseudomonas syringae* pv. *pisi*. *Mol. Plant-Microbe Interact.* **1993**, 6, 655–664.

Fahrendorf, T.; Ni, W.; Shorrosh, B. S.; Dixon, R. A. Stress responses in alfalfa (*Medicago sativa* L.) XIX. Transcriptional activation of oxidative pentose phosphate pathway genes at the onset of the isoflavonoid phytoalexin response. *Plant Mol. Biol.* **1995**, 28, 885–900.

Fields, S.; Song, O.-K. A novel genetic system to detect protein-protein interactions. *Nature* **1989**, 340, 245–246.

Gunia, W.; Hinderer, W.; Wittkampf, U.; Barz, W. Elicitor induction of cytochrome P-450 monooxygenases in cell suspension cultures of chickpea (*Cicer arietinum* L.) and their involvement in pterocarpan phytoalexin biosynthesis. *Z. Naturforsch* **1991**, 46c, 58–66.

Hahlbrock, K.; Scheel, D. Physiology and molecular biology of phenylpropanoid metabolism. *Annu. Rev. Plant Physiol. Plant Mol. Biol.* **1989**, 40, 347–369.

He, X.-Z.; Dixon, R. A. Affinity chromatography, substrate/product specificity and amino acid sequence analysis of an isoflavone O-methyltransferase from alfalfa (*Medicago sativa* L.). *Arch. Biochem. Biophys.* **1996**, 336, 121–129.

Howles, P. A.; Paiva, N. L.; Sewalt, V. J. H.; Elkind, N. L.; Bate, Y.; Lamb, C. J.; Dixon, R. A. Overexpression of L-phenylalanine ammonia-lyase in transgenic tobacco plants reveals control points for flux into phenylpropanoid biosynthesis. *Plant Physiol.* **1996**, 112, 1617–1624.

Hrazdina, G.; Jensen, R. A. Spatial organization of enzymes in plant metabolic pathways. *Annu. Rev. Plant Physiol. Plant Mol. Biol.* **1992**, 43, 241–267.

Hrazdina, G.; Wagner, G. J. Metabolic pathways as enzyme complexes: evidence for the synthesis of phenylpropanoids and flavonoids on membrane associated enzyme complexes. *Arch. Biochem. Biophys.* **1985a**, 237:88–100.

Hrazina, G.; Wagner, G. J. Compartmentation of plant phenolic compounds; sites of synthesis and accumulation. *Annu. Proc. Phytochem. Soc. Europe* **1985b**, 25, 119–133.

Junghans, H.; Dalkin, K.; Dixon, R. A. Stress responses in alfalfa (*Medicago sativa* L.) XV. Characterization and expression patterns of members of a subset of the chalcone synthase multigene family. *Plant Mol. Biol.* **1993**, 22, 239–253.

Kacser, H.; Burns, J. A.; Fell, D. A. The control of flux. *Biochem. Soc. Trans.* **1995**, 23:341–366.

Kessmann, H.; Choudhary, A. D.; Dixon, R. A. Stress responses in alfalfa (*Medicago sativa* L.) III. Induction of medicarpin and cytochrome P450 enzyme activities in elicitor-treated cell suspension cultures and protoplasts. *Plant Cell Rep.* **1990**, 9, 38–41.

Kochs, G.; Grisebach, H. Enzymic synthesis of isoflavones. *FEBS Lett.* **1986**, 155, 311–318.

Lamb, C. J.; Ryals, J. A.; Ward, E. R.; Dixon, R. A. Emerging strategies for enhancing crop resistance to microbial pathogens. *Bio/technology* **1992**, 10, 1436–1445.

Lawton, M. A.; Lamb, C. J. Transcriptional activation of plant defense genes by fungal elicitor, wounding, and infection. *Mol. Cell. Biol.* **1987**, 7, 335–341.

Liu, Q.; Dixon, R. A.; Mabry, T. J. Additional flavonoids from elicitor-treated cell cultures of *Cephalocereus senilis*. *Phytochemistry* **1993a**, 34, 167–170.

Liu, Q.; Markham, K. R.; Paré, P. W.; Dixon, R. A.; Mabry, T. J. Flavonoids from elicitor-treated cell suspension cultures of *Cephalocereus senilis*. *Phytochemistry* **1993b**, 32, 925–928.

Liu, Q.; Seradge, E.; Bonness, M. S.; Liu, M.; Mabry, T. J.; Dixon, R. A. Enzymes of B-ring-deoxy flavonoid biosynthesis in elicited cell cultures of "old man" cactus (*Cephalocereus senilis*). *Arch. Biochem. Biophys.* **1995**, 321, 397–404.

Loake, G. J.; Faktor, O.; Lamb, C. J.; Dixon, R. A. Combination of H-box (CCTACC($N_7$)CT) and G-box (CACGTG)cis-elements is necessary for feedforward stimulation of a chalcone synthase promoter by the phenylpropanoid pathway intermediate *p*-coumaric acid. *Proc. Natl. Acad. Sci. U.S.A.* **1992**, 89, 9230–9234.

Mackenbrock, U.; Vogelsang, R.; Barz, W. Isoflavone and pterocarpan malonylglucosides and ß-1,3-glucan-and chitin-hydrolases are vacuolar constituents in chickpea (*Cicer arietinum* L.). *Z. Naturforsch.* **1992**, 47, 815–822.

Marcos, J. F.; Beachy, R. N. *In vitro* characterization of a cassette to accumulate multiple proteins through synthesis of a self-processing polypeptide. *Plant Mol. Biol.* **1994**, 24, 495–503.

Mavandad, M.; Edwards, R.; Liang, X.; Lamb, C. J.; Dixon, R. A. Effects of *trans*-cinnamic acid on expression of the bean phenylalanine ammonia-lyase gene family. *Plant Physiol.* **1990**, 94, 671–680.

Maxwell, C. A.; Edwards, R.; Dixon, R. A. Identification, purification and characterization of S-adenosyl-L-methionine: isoliquiritigenin 2'-O-methyltransferase from alfalfa (*Medicago sativa* L.). *Arch. Biochem. Biophys.* **1992**, 293, 158–166.

Maxwell, C. A.; Harrison, M. J.; Dixon, R. A. Molecular characterization and expression of alfalfa isoliquiritigenin 2'-O-methyltransferase, an enzyme specifically involved in the biosynthesis of an inducer of *Rhizobium meliloti* nodulation genes. *Plant J.* **1993**, 4, 971–981.

Maxwell, C. A.; Hartwig, U. A.; Joseph, C. M.; Phillips, D. A. A chalcone and two related flavonoids released from alfalfa roots induce nod genes of *Rhizobium meliloti*. *Plant Physiol.* **1989**, 91, 842–847.

McKhann, H. I.; Hirsch, A. M. Isolation of chalcone synthase and chalcone isomerase cDNAs from alfalfa (*Medicago sativa* L.): highest transcript levels occur in young roots and root tips. *Plant Mol. Biol.* **1994**, 24, 767–777.

Ni, W.; Fahrendorf, T.; Ballance, G. M.; Lamb, C. J.; Dixon, R. A. Stress responses in alfalfa (*Medicago sativa* L.). XX. Transcriptional activation of phenylpropanoid pathway genes in elicitor-treated cell suspension cultures. *Plant Mol. Biol.* **1996a**, 30, 427–438.

Ni, W.; Sewalt, V. J. H.; Korth, K. L.; Blount, J. W.; Ballance, G. M.; Dixon, R. A. Stress responses in alfalfa (*Medicago sativa* L.) XXI. Activation of caffeic acid 3-O-methyltransferase and caffeoyl CoA 3-O-methyltransferase genes does not contribute to changes in metabolite accumulation in elicitor-treated cell suspension cultures. *Plant Physiol.* **1996b**, 112, 717–726.

Paiva, N. L.; Oommen, A.; Harrison, M. J.; Dixon, R. A. Regulation of isoflavonoid metabolism in alfalfa. *Plant Cell, Tissue Organ Cult.* **1994**, 38, 213–220.

Paré, P.; Mischke, C. F.; Edwards, R.; Dixon, R. A.; Norman, H.; Mabry, T. J. Induction of phenylpropanoid pathway enzymes in elicitor-treated cultures of *Cephalocereus senilis*. *Phytochemistry* **1992**, 31, 149–153.

Plater, R.; Strohl, W. R. Polyketide biosynthesis: antibiotics in *Streptomyces*. In *Genetic Engineering of Plant Secondary Metabolism*, Ellis, B. E., Kuroki, G. W., Stafford, H. A. Eds.; Plenum Press: New York, **1994**; pp. 61–91.

Raventós, D.; Jensen, A. B.; Rask, M. B.; Casacuberta, J. M.; Mindy, J.; Segundo, B. S. A 20 bp *cis*-acting element is both necessary and sufficient to mediate elicitor response of a maize *PRms* gene. *Plant J.* **1995**, 7, 147–155.

Ryder, T. B.; Hedrick, S. A.; Bell, J. N.; Liang, X.; Clouse, S. D.; Lamb, C. J. Organization and differential activation of a gene family encoding the plant defense enzyme chalcone synthase in *Phaseolus vulgaris*. *Mol. Gen. Genet.* **1987**, 210, 219–233.

Sewalt, V. J. H.; Ni, W.; Blount, J. W.; Jung, H. G.; Howles, P. A.; Masoud, S. A.; Lamb, C.; Dixon, R. A. Reduced lignin content and altered lignin composition in transgenic tobacco down-regulated in expression of phenylalanine ammonia-lyase or cinnamate 4-hydroxylase. *Plant Physiol.* **1997**, 115, 41–50.

Somssich, I. E.; Bollmann, J.; Hahlbrock, K.; Kombrink, E.; Schulz, W. Differential early activation of defense-related genes in elicitor-treated parsley cells. *Plant Mol. Biol.* **1989**, 12, 227–234.

Srere, P. A. Complexes of sequential metabolic enzymes. *Annu. Rev. Biochem.* **1987**, 56, 89–124.

Welle, R.; Schröder, G.; Schiltz, E.; Grisebach, H.; Schröder, J. Induced plant responses to pathogen attack. Analysis and hererologous expression of the key enzyme in the biosynthesis of phytoalexins in soybean (*Glycine max* L. Merr. cv. Harosoy 63). *Eur. J. Biochem.* **1991**, 196, 423–430.

Wengenmayer, H.; Ebel, J.; Grisebach, H. Purification and properties of a S-adenosylmethionine: isoflavone 4'-O-methyltransferase from cell suspension cultures of *Cicer arietinum* L. *Eur. J. Biochem.* **1974**, 50, 135–143.

Wong, E.; Francis, C. M. Flavonoids in genotypes of *Trifolium subterraneum*-II. Mutants of the Geraldton variety. *Phytochemistry* **1968**, 7, 2131–2137.

Yu, L. M.; Lamb, C. J.; Dixon, R. A. Purification and biochemical characterization of two proteins which bind to the H-box *cis*-element implicated in transcriptional activation of plant defense genes. *Plant J.* **1993**, 3, 805–816.

# 6

# FLAVONOID ACCUMULATION IN TISSUE AND CELL CULTURE*

## Studies in *Citrus* and Other Plant Species

Mark A. Berhow

United States Department of Agriculture
Agricultural Research Service
National Center for Agricultural Utilization Research
1815 N. University Street
Peoria, Illinois 61604

## 1. INTRODUCTION

The genus *Citrus* provides a number of economically important horticultural crops, including the orange, the lemon and the grapefruit. The flavonoid composition of these species are of interest biosynthetically, taxonomically, and because they affect the market quality of the fruit and its products. The work on flavonoid metabolism and function in citrus has been done with cell free extracts derived from whole plant tissues and some work using citrus cell cultures. This chapter will summarize some of the more important contributions of cell culture studies to the flavonoid research literature, summarizing the general work on flavonoid accumulation in plants, with a detailed overview of the work done in *Citrus* species.

The diversity of "secondary metabolites" produced by plants has served man with both fascination and utility for centuries. Secondary metabolites have been defined historically as naturally-occurring substances that do not seem to be vital to the immediate survival of the organism that produces them, and are not an essential part of the process of building and maintaining living cells. However, recent research indicates that secondary metabolites play pivotal roles in the chemical functionality of the plant, determining the role a particular species will play in the environment. In addition, research is beginning to show roles for these compounds at the cellular level as plant growth regulators and modu-

---

* Original research described in this chapter was conducted at the USDA, ARS, Fruit and Vegetable Chemistry Laboratory, Pasadena, California, which was closed by the ARS in 1994.

*Flavonoids in the Living System*, edited by Manthey and Buslig
Plenum Press, New York, 1998.

lators of gene expression. Secondary compounds are crucial in the plant's response to stresses, such as changes in light or temperature, competition, herbivore pressure, and pathogenic attack. They appear to play critical roles in the plant's ability to survive and reproduce. Plants have also incorporated these secondary metabolites into specialized physiological functions such as seed production and intracellular signaling. These secondary metabolites are often functionally unique at the species level, and taxonomically discrete individual compounds are responsible for identical functions in separate species. One class of secondary metabolites, the flavonoids, which are derivatives of the phenylpropanoid pathway, has been implicated in all these roles and is an ideal system for study.

Much of this "emerging picture" for secondary metabolite biosynthesis and function has come about from studies initiated by the development and use of some key techniques in the years that followed the Second World War: 1) the widespread availability of radioactive isotopes of carbon, hydrogen, sulfur and phosphorus for research in the 1950s, 2) the development of new chromatography techniques for the isolation and quantitation of plant natural products in the 1950s and 60s, especially thin-layer chromatography (TLC), gas chromatography (GC), and later, in the 1970s, high pressure (or performance) liquid column chromatography (HPLC), 3) the development of protocols for the extraction of active enzymatic preparations involved in secondary metabolite biosynthesis from plant tissues in the 1960s, and 4) the development of culturing techniques for growing undifferentiated plant cells in the laboratory in the 1970s. With the advent of these tools, and the recent interest in the field of plant natural products, research in this field has exploded in recent years (Cordell, 1995).

This is especially true of the plant flavonoids. The flavonoids are derived from phenylalanine via the general phenylpropanoid biosynthetic pathway. This pathway is highly regulated and controlled by both normal growth and development as well as being induced by wounding and attack by pathogens. Flavonoids have attracted the attention of man for centuries. Certain plant flavonoids were among the first dyes used by man. While they have been systematically studied since around the turn of the century, only recently have their biochemical roles in plants been fully appreciated (Dakora, 1995; Shirley, 1996).

With the development of modern scientific techniques, these compounds have been identified and purified from their biosynthetic sources, and their modes of action, especially in mammalian health, are being examined and commercially exploited. By 1975, the study of flavonoid accumulation and function was well advanced, with careful biosynthetic studies underway. This ubiquitous class of phenolic compounds is found in nearly all vascular plants. Among the phenolics, flavonoids are of particular interest, since they appear to function in all roles in which plant secondary metabolites have been implicated (Dakora, 1995).They, and the other phenolics derived from the phenylpropanoid biosynthetic pathway, have been reported as having a wide variety of physiological effects in both plants and animals, serving as enzyme activators and inhibitors, metal chelators, antioxidants, free radical scavengers, transcription regulators, phytohormones, and as mutagenic, antimutagenic, carcinogenic, anticarcinogenic, cytotoxic, antineoplastic, anti-inflammatory and anti-allergenic substances. In plants they have been shown to function in protection from UV radiation damage; in mineral nutrition; in temperature and water stress; in pollination and seed dispersal (by their color properties); and as constitutive chemoprotective agents against other plants, microbial pathogens, fungi, insects, and herbivores. Flavonoids have also been shown to be naturally occurring auxin transport regulators (Jacobs and Rubery, 1988). Much of the early work was done in plant species other than *Citrus*.

As a specific result of the plant's interaction with changes in its the environment, specific flavonoids and other phenolic metabolites are produced and accumulated (Snyder and Nicholson, 1990; Laks and Pruner, 1989; Hahlbrock and Scheel, 1989; Dixon and Paiva, 1995). These compounds have been postulated as being one of the earliest chemical classes evolved by plants for protection from the stronger UV radiation that struck the earth before the development of the ozone layer (Kubitzki, 1987). The biosynthesis of new flavonoids is constitutive in plants, being internally controlled during normal growth and development, such as new vegetative leaf growth and reproductive organ development (Heller and Forkmann, 1988). Flavonoid biosynthesis may also be induced by exogenous stimuli, such as changes in light and temperature (Hahlbrock and Scheel, 1989). Biosynthesis and/or modification of constituative flavonoids can be triggered, or elicited, by damage to the plant caused by physical agents (wind, freezing, water stress, ozone, heavy metal ions, certain herbicides), herbivore attack (insects, grazing animals) and microbial invasion (bacteria, fungi) (Nahrstedt, 1990; Nicholson and Hammerschmidt, 1992; Ebel, 1986; Dixon and Paiva, 1995; Dixon et al., 1992; Chappell and Hahlbrock, 1984).

They are, as one reviewer has pointed out, "biological molecules for useful exploitation" (Dakora, 1995). An excellent series of books have been published on plant flavonoids. These include the first comprehensive work on the subject by Geisman (1962), a series of books edited by Harborne (Harborne et al., 1975; Harborne and Mabry, 1982; Harborne, 1994) and one by Stafford (Stafford, 1990). For general reviews of flavonoid biosynthesis in plants see Heller and Forkmann (1988); Hahlbrock and Scheel (1989); and Stafford (1990).

## 2. CELL CULTURE AND FLAVONOID RESEARCH

Cell culture provides a uniform mass of undifferentiated cells free from microbial contamination, which is available for study regardless of the season. The technique also provides unparalleled control of the environment through the manipulation of the cell media. And, most importantly, cell culture provides a means to induce the formation of secondary metabolites at higher levels and over shorter periods of time than occurs in differentiated plants. Through refinement of the technique, it was found that sterile cultures of undifferentiated cells could be procured from almost all species of plants.

Since the activity levels of many of the enzymes in the flavonoid biosynthetic pathway are low and/or transient in nature, much of the early work on this pathway was conjectural in nature. With the advent of plant cell culture, however, increased levels of some secondary enzyme activities could be attained by altering the media conditions. It is now well established that many cell cultures produce relatively large quantities of one or more of the secondary metabolites produced by the whole plant, but other cultures from the same species or even the same plant will produce trace amounts of secondary metabolites, different compounds or sometimes none at all. Some cultures will produce compounds not normally found in the whole plant. The reasons for the differences in metabolite production in the cell cultures are unknown. They may be due to the nature of cell culture, as they are undifferentiated cells which were initiated from an injury to a part of the whole plant (Becker et al., 1987; Charlwood and Rhodes, 1990). There are also disadvantages to the use of cell culture for these types of studies (Ellis, 1984). The possibility of genomic instability, the loss of morphological differentiation, the loss of the ability to accumulate all the compounds found in the tissues of the intact plant, and the loss of regulatory control.

In spite of the technique's limitations, many of the biosynthetic steps in plants have been elucidated mostly through the use of cultured cells. Flavonoids are synthesized from phenylalanine via a biosynthetic route in plants termed the "general phenylpropanoid pathway." This important pathway generates a large number of compounds which have in common a phenyl structural element. Secondary metabolites derived from this pathway include (in addition to flavonoids) tannins, phenols, benzoic acids, stilbenes, cinnamate esters, and coumarins. A sequence of reactions converts phenylalanine into co-enzyme A (CoA) derivatives of substituted cinnamic acids. These are further converted to various other metabolites. Flavonoids are derived from the conjugation of *p*-coumaryl-CoA with three malonyl-CoA molecules to form naringenin chalcone, which is considered to be the precursor of all the flavonoids. Naringenin chalcone is rapidly converted to the flavanone form, naringenin. Naringenin can then be further modified enzymatically by reactions including reduction, oxidation, hydroxylation, *O*-methylation, *O*-glycosylation, *C*-glycosylation, acylation, sulfonation, rearrangement, and polymerization to form other flavonoids.

Flavonoids have generally been classified into 12 different subclasses by the state of oxidation and the substitution pattern at the C-2—C-3 unit (Harborne et al., 1975). These include flavanones, flavones, flavonols, chalcones, dihydrochalcones, anthocyanidins, aurones, flavanols, dihydroflavonols, proanthocyanidins (flavan-3,4-diols), isoflavones, and neoflavones. More than 10,000 flavonoids have been identified from natural sources and more continue to be identified at a rate of more than 10 per month (Dakora, 1995). Of particular interest is the use of flavonoids as taxonomic markers, as individual species within plant families often vary widely in flavonoid type and content (Seigler, 1981). In addition, individual plants within a species will produce and accumulate different flavonoids depending on several factors, such as plant growth stage, reproductive stage, the particular plant tissue involved, and the type of environmental stress or pathogenic attack involved (Dixon and Paiva, 1995).

The control mechanisms for complex modifications in flavonoid biosynthesis (such as B-ring hydroxylation, methylation, or glycosylation) have been studied in a small number of cultured cell suspensions and whole plants, including parsley (acylated flavone and flavonol glycosides) (Ebel and Hahlbrock, 1977), soybean (isoflavones) (Dixon et al., 1983), green bean (isoflavones) (Banks and Dewick, 1983; Ebel et al., 1984; Ebel et al., 1985; Ebel, 1986), alfalfa and chickpea systems (Heller and Forkmann, 1988; Hahlbrock and Scheel, 1989; Hahlbrock, 1977; Hahlbrock, 1976), and carrot (anthocyanins) (Cheng et al., 1985; Hinderer et al., 1984; Ozeki and Komamine, 1982; Ozeki and Komamine, 1985a; Ozeki and Komamine, 1985b)). Much of the recent work on the control of these pathways has been determined using new molecular biological techniques. Specific modification of plant genomes will allow researchers to delve further into the controls operating in this complex system. These pathways appear to be very tightly regulated and controlled by other metabolic, developmental, and stress-related factors. The enzymes in these pathways are often present as several different isozymes, indicating groups of tightly controlled metabolic chains which are driven by the formation of very specific end products (Douglas et al., 1992; Dixon et al., 1992; Stafford, 1990).

Induction has been studied extensively in cell culture systems. Phenylproponoid pathway genes are activated in cells treated with either microbes or elicitors, resulting in the formation of newly synthesized flavonoids that inhibit microbial growth. In some cases, these same biosynthetic pathways can be induced by UV light or physical damage to the cells (Logemann et al., 1995; Nojiri et al., 1996; Negrel and Javelle, 1995; Liu et al., 1993a; Liu et al., 1993b). Modern genetic engineering techniques may be used to modify plant cell cultures to produce large amounts of flavonoids (Logemann et al., 1995; Nojiri et al., 1996; Negrel and Javelle, 1995; Liu et al., 1993a; Liu et al., 1993b; Kuc, 1995; Zenk, 1991). Several cell lines have already been shown to produce and accumulate flavonoids that are different than those normally found in mature plants, some of which have interesting biocidal properties (Charlwood and Rhodes, 1990; Becker, 1987; Yamamoto et al., 1993; Yamamoto et al., 1992).

## 3. CITRUS FLAVONOIDS

The study of the flavonoids in *Citrus* species is of interest because they accumulate high concentrations of flavonoid glycosides. The most common glycosidic group attached to the flavonoids in citrus is the rhamnose-glucose diglycoside. This sugar group is present in two isomeric forms—neohesperidose and rutinose. The two sugars differ only in the position of the attachment of rhamnose to the glucose. The main flavonoids found in most cultivated citrus species are the flavanone glycosides. The flavanone rutinosides, such as hesperidin and narirutin, are tasteless, while the flavanone neohesperidosides, such as naringin and neohesperidin, are bitter, with the bitterness intensity being relative to the composition of the flavanone aglycone structure. These compounds can account for up to 5 % of the dry wieght of the leaf and fruit tissue. The accumulation of the naringin and neohesperidin in citrus species related to the pummelo causes these fruits to have a bitter taste. This includes the economically important sour orange and grapefruit varieties. Pummelo accumulates the neohesperidosides exclusively, while sour orange and grapefruit accumulate both the neohesperidosides and the rutinosides found in other citrus species related to the citron and the mandarin (Albach and Redman, 1969; Horowitz and Gentili, 1977).

The biosynthesis of flavonoids in citrus is similar to other plant species (Hasegawa and Maier, 1981). The flavonoid glycosides are accumulated in young leaves and fruits during the cell division stage (Fisher, 1968; Berhow and Vandercook, 1989). The woody portions of the plant do not biosynthesize the characteristic flavonoids of the leaves and fruits, though these compounds are found in stem and root tissues (Berhow and Vandercook, 1991; Jourdan et al., 1985). The major flavanone glycosides are probably accumulated in the cell vacuoles. During the cell elongation and subsequent maturation of the leaves and fruit, there is little further biosynthesis. As the fruit reaches maturity, flavonoid concentrations in the fruit decrease due to dilution effects (Jourdan et al., 1985; Hagen et al., 1966; Castillo et al., 1992; Shaw et al., 1991; Vandercook and Tisserat, 1989).

Interestingly, however, the metabolic fate of the flavanone glycosides in the citrus plant tissues is not known, and there is as yet no clue as to what functional role these compounds play in the plant themselves. It may be that these compounds are accumulated constitutively in the developing leaves and fruits, glycosylated for storage purposes, and saved for utilization in response to stress, either as the flavanone aglycone or for further modification in response to the specific stress. Hence, citrus species may be accumulating large quantities of an "inactive" form of phenolics stored in a non-toxic form for future use. Several of the citrus flavanone aglycones, including naringenin and hesperetin, have been shown to have biological activities in other plant and animal systems, yet their function in citrus has yet to be demonstrated.

Flavonoid composition in citrus has been extensively examined. The most comprehensive reviews on the composition of the flavonoids found in citrus species are by Horowitz and Gentili (1977) and Kefford and Chandler (1970). Published flavonoid composition surveys in citrus have been published by Kanes et al. (1993) and Nishiura et al. (1969, 1971a, 1971b). Other aspects of flavonoid composition and accumulation in citrus are more fully covered in other chapters in this book.

## 4. CITRUS CELL CULTURE

The techniques to culture cells derived from citrus species have been around for a number of years (Tisserat et al., 1989; Tisserat and Galletta, 1987). Extensive research was done with explanted tissues from citrus in culture, such as leaf and flower buds, embryo, shoot apex, juice vesicles, stem and root cultures. The development of a defined media by Murashige and Skoog (1962) for culturing of plant cells, refined by Murashige and Tucker (1969) for citrus, promoted further interest in this area. Most of these cultures either produced dedifferentiated cells (callus) or degenerated into callus soon after explanting.

With the emergence of new molecular biology techniques, especially the ability to transform cells derived from woody plant species and regenerate whole plants from these cultures, interest in plant cell culture has expanded greatly in the last 15 years. Many laboratories around the world have established callus and cell culture lines from embryonic tissues of many different citrus species, varieties and relatives (Vardi and Galun, 1988). Techniques have been developed to regenerate whole plants from cell lines that maintain their embryogenic capability (Kobayashi et al., 1985; Moore, 1985; Hidaka and Omura, 1989; Beloualy et al., 1991). These cultures have been mostly used for the production of protoplasts for genetic alteration with the aim of introducing desirable characteristics as either new fruit or rootstock cultivars. Protoplast culture for citrus is discussed by Vardi and Galun (1988), and has been used for the production of somatic hybrids using protoplast fusion (Grosser and Gmitter, 1990; Kobayashi et al., 1991a; Kobayashi et al., 1991b; Ohgawara and Kobayashi, 1991; Tusa et al., 1992; Hidaka and Omura, 1992). These techniques have been used to create interspecies hybrids between sexually incompatible species such as *Citrus* and *Severinia* (Grosser et al., 1988), and *Citrus* and *Atalantia* (Louzada et al., 1993). Foreign DNA fragments have been inserted into cultured citrus cells and whole plants successfully regenerated via *Agrobacterium*-mediated transformation (Hidaka et al., 1990; Feng et al., 1991; Moore et al., 1992; Peña et al., 1995), or direct DNA transfer (Kobayashi and Uchimiya, 1989; Vardi et al., 1990).

Actual published studies of flavonoid biosynthesis and accumulation in citrus species has been rather limited. This is in part due to the observation noted above about secondary metabolism in cell cultures, that cell cultures established from plants are inconsistent and variable in the production of these metabolites. Also, the fairly recent development of techniques to consistently extract active enzymatic preparations from intact plant tissues that have high levels of phenolics and other oxidizing species has only been applied to *Citrus* species since the early 1970s. These two techniques applied to cells and tissues of *Citrus* species would provide the biochemical evidence for the mechanisms of the accumulation of the citrus flavonoids. Though not completely understood even to this day, much of the early part of the *Citrus* phenylpropanoid biosynthetic pathway has been worked out and confirmed by the use of radioactive isotopes and the isolation and characterization of the enzymes involved.

## 5. CITRUS FLAVONOID BIOSYNTHESIS

### 5.1. Quantitation of Flavonoids

Early citrus flavonoid biochemistry research was driven by the need to understand the mechanisms of the accumulation of the bitter principles in citrus juices. Hence much of the biosynthetic research on citrus flavonoids has been in grapefruit, pummelo and sour orange. Though the main bitter principle in grapefruit was known to be the flavanone glycoside naringin, its complete chemical structure was not confirmed until 1963 (Horowitz and Gentili, 1963). Early speculation on the biosynthetic pathway was based on the characterization of the various flavonoids found in the grapefruit, especially in immature tissues (Maier and Metzler, 1967; Maier, 1969).

Definitive biochemical work on flavonoid biosynthesis in citrus species began with the work of Fisher who showed that $^{14}$C-labeled phenylalanine fed to detached immature grapefruit leaves resulted in the formation of $^{14}$C-labeled naringin, but this conversion did not occur in fully expanded leaves (Fisher, 1968). TLC evaluation of developing grapefruit showed high concentrations of naringin in young, developing fruit (Albach et al., 1969). This demonstrated that flavanone glycoside accumulation was most active in young fruit. Grapefruit tissues at various stages of development were later examined with radioimmunoassays specific for flavanone glycosides (Jourdan et al., 1985). Naringin, which comprises 80% of all the flavanone glycosides found in grapefruit, was found in all tissues examined: seeds, germinating seedlings, 5 month old plants, 1 year old plants, (leaves, stems and roots) flushes from mature trees, developing flowers and fruit, and mature fruit. The highest concentrations were found in young developing leaves, flowers and fruit. The lowest concentrations were found in the roots, stems, and older leaves. Feeding either $^{14}$C-phenylalanine or $^{14}$C-acetate to detached immature grapefruit resulted in the formation of labeled naringin and prunin (naringenin 7-$O$-glucoside) (Berhow and Vandercook, 1989). Radiolabeled naringin was found in all parts of intact grapefruit seedlings fed labeled acetate—roots, stems and leaves (Berhow and Vandercook, 1991). The highest levels were found in the young developing leaves. If the seedling was taken apart and labeled acetate fed to each part—leaves, stem and roots, labeled naringin was found only in the young expanding leaves. In lemon, hesperidin, the major flavanone glycoside present, is accumulated in a pattern similar to that of the grapefruit. A rapid increase occurred in concentration of hesperidin in developing lemons, followed by a steady decrease in concentration as the fruit expanded to full size and matured (Vandercook and Tisserat, 1989). Interestingly, two other lemon flavonoids, eriocitrin and diosmin, are accumulated later in the course of fruit development and continued to increase in concentration throughout the maturation process. Similar results were found in naringin, neohesperidin and neodiosmin accumulation in developing leaves and fruit of sour orange (Benavente-Garcia et al., 1993; Castillo et al., 1992). Both hesperetin 7-$O$-glucoside and naringenin 7-$O$-glucoside were found in very early stages of the developing sour orange fruit (Castillo et al., 1993) indicating a stepwise addition of sugar molecules as the final steps of biosynthesis. A unique flavanone glycoside, naringin-6"-malonate, was found in immature grapefruit fruit, but little or none could be found in mature fruit (Berhow, 1991). A similar compound, the flavone apigenin 7-$O$-(6-$O$-malonyl) glucoside, was shown to be selectively taken up by isolated vacuoles of parsley, while apigenin 7-$O$-glucoside, was not (Matern et al., 1986). This may be true for citrus flavonoids as well.

These quantitation studies showed that flavonoids are accumulated constitutively in young developing leaves and reproductive tissues of citrus (i.e. during periods of rapid cell division). Citrus species can dedicate a relatively large amount of carbon resource to this synthesis. Immature grapefruit, just after the abscission of the ancillary flower structures, can have up to 40% on a dry weight basis of a single compound, the flavonoid naringin (Maier, 1969). This constitutive accumulation appears to be similar to the pathways in other more studied dicot species, the conversion of phenylalanine into the flavanone aglycone in the cytoplasm, followed by the stepwise addition of sugars, mainly glucose and rhamnose in the case of citrus species, to give the major flavanone glycosides. These are probably earmarked by acylation for transport to the cell vacuole where they are deacylated and stored. The concentration of the major flavanone glycoside, naringin, hesperidin or neohesperidin in most species, decreases in the fruit tissues as the fruit matures (Albach et al., 1981; Albach and Wutscher, 1988; Jourdan et al., 1985; Vandercook and Tisserat, 1989), but some other minor flavonoids are accumulated throughout the maturation process (Vandercook and Tisserat, 1989).

## 5.2. Enzymes in Cell-Free Extracts

Enzymatic characterization of flavonoid biosynthesis began in the late 1950s, with the demonstration of the activities of the enzymes chalcone flavanone isomerase (Shimokoriyama, 1957), phenylalanine ammonia lyase (PAL) (Maier and Hasegawa, 1970), and cinnamate 4-hydroxylase (Hasegawa and Maier, 1972) in cell free extracts prepared from young citrus fruits. The presence of PAL in citrus fruit tissue cultured *in vitro* has also been reported (Thorpe et al., 1971). *O*-methylation of flavonoids was demonstrated in cell free extracts prepared from calamondin (Brunet and Ibrahim, 1980). The activities of chalcone synthase, flavanone 7-*O*-glucosyltransferase and flavanone 7-*O*-glucoside-2"-*O*- rhamnotransferase were detected in cell free extracts prepared from several citrus species (Lewinsohn et al., 1989a & 1989b). Flavanone glucosyltransferase activities were reported in cell free extracts of young leaves of grapefruit seedlings (McIntosh and Mansell, 1990) and lemon seedling leaves (Berhow and Smolensky, 1995). A methyltransferase was detected in extracts prepared from young leaves and fruits of sour orange (Benavente-Garcia et al., 1995), and a cytochrome P450-mediated 3'-hydroxylase activity was detected in an extract prepared from sweet orange cell cultures (Doostdar et al., 1995). Purified or partially purified enzymes from citrus sources include PAL from grapefruit (Hasegawa and Maier, 1970), chalcone-flavanone isomerase from sweet orange (Fouche and Dubery, 1994), flavanone 7-*O*-glucoside-2"-*O*-rhamnotransferase from pummelo (Bar-Peled et al., 1991) and a flavanone 7-*O*-glucosyltransferase from grapefruit leaves (McIntosh et al., 1990).

Again, with a few exceptions noted below, all of these activities are found in either young developing leaves or fruit and are part of the constitutive pathway. The rhamnotransferase in pummelo was shown by both activity assays and immunodetection techniques to be specifically located in young leaves, and all flower parts except the anthers and young fruit (Bar-Peled et al., 1993). Both the enzyme activity and the protein molecule itself rapidly disappear as leaves reach full length, and as the fruit approaches about a quarter of its mature weight. Interestingly, there continues to be a net increase of naringin up to the point that the fruit reaches about half of its mature weight. This would seem to point to an import of naringin from the leaves of the plant.

Very little research has been done on induced biosynthetic pathways in citrus. There are two reports on the induction of PAL activity in citrus. It was observed that wounding Valencia orange peel will induce an increase in the activity of PAL (Ismail and Brown, 1979), while fungal infection will suppress the ethylene-induced PAL activity in grapefruit (Lisker et al., 1983). The lack of work on the induced pathway may be due to the fact that there is not a good cell culture model system to study. As will be discussed below, citrus cell cultures generally do not produce and accumulate flavonoids. This is important, as much of the work on the induced pathways have been done with cell cultures in other species of plants.

## 5.3. Flavonoid Metabolism in Cell Cultures

As noted above, much of the research work in citrus cell culture has been focused on either the culture of fruit tissues or genetic alteration of citrus species. With few exceptions, citrus cell cultures generally do not accumulate flavonoids. When they were accumulated, it was usually in either differentiated cultures like fruit and juice vesicle cultures or very young callus cultures recently initiated from fruit tissue. The first report of flavonoid accumulation in cultured citrus tissues came from the laboratory of Kordan (Kordan and Morganstern, 1962; Kordan, 1965). He observed that phenolic-like substances were produced and excreted into the medium from callus cultures derived from juice vesicles of lemon proliferating *in vitro*. Kordan and Morganstern considered these substances to be flavonoids based on thin-layer chromatography evaluation, though they could not confirm this structurally.

In a cultured juice vesicle study, the concentrations of accumulated flavonoids were much lower than those of fruit developing on the tree. Also, the phenolic patterns of callus derived from the cultured vesicles were significantly different than those found in the cultured vesicles or vesicles in fruit from trees (Tisserat et al., 1989; Vandercook and Tisserat, 1989). Del Rio measured the accumulation of naringin and neohesperidin in callus cultures derived from young fruit of sour orange (del Rio et al., 1992). These cultures were eight months old, having been subcultured at four week intervals. One could speculate that these cultures would have continued to produce flavonoids similar to that found in the mature fruit. Low levels of methoxylated flavones occurred in callus cultures of orange and lemon flavedo (Brunet and Ibrahim, 1973), however, these compounds could have been simply be carried over from the original flavedo cells from which the calli were derived.

In general, callus cultures derived from citrus fruit tissues do not accumulate flavonoids, but I have found an exception. In 1991, I examined callus cultures derived from the embryonic fruit tissues of 9 mandarin cultivars, 1 grapefruit cultivar, 1 pummelo hybrid cultivar, 1 sour orange cultivar, 2 tangelo cultivars, 3 papeda hybrids cultivars, 1 kumquat cultivar, 10 sweet orange cultivars, and 2 citrus rootstock cultivars that were maintained by the Japanese Fruit Tree Research Station in Okitsu, Japan. With one exception, which will be discussed in detail below, none of these callus cultures produced measurable levels of methanol-extractable phenolic compounds. Most of these cultures were over 2 years old, and some had been maintained in culture for over 10 years and were still capable of regenerating embryos and ultimately whole plants (Hidaka and Omura, 1989). Other reports have commented on the lack of flavonoid production in callus cultures. Callus derived from sweet orange and grapefruit rapidly loose the ability to produce hesperidin and naringin respectively (Barthe et al., 1987). Regeneration of shoots from cultured cells restored the production and accumulation of these flavonoids. Callus cultures derived from grapefruit were able to glucosylate endogenously added naringenin at the 7 position, but did not accumulate any flavonoids on their own (Lewinsohn et al., 1986). Examination of other cultures by the same research group found that a culture derived from sour orange ovules was able to glucosylate naringenin to prunin and rhamnosylate prunin to form narirutin (Lewinsohn et al., 1989a). Cultures derived from lemon and grapefruit were able to glucosylate exogenously added hesperetin. However, the ability to do this was inconsistant among the cultures examined. The rhamnotransferase protein could not be detected in cell free extracts of grapefruit callus cultures by the antibody prepared against purified protein isolated from young leaves (Bar-Peled et al., 1993). As callus cells were induced to form embryos, the enzymatic activities of both the glucosyltransferase and the rhamnotransferase increased markedly (Gavish et al., 1989). Cytochrome P-450 mediated naringenin 3'-hydroxylase enzyme activity was detected in cell free extracts prepared from sweet orange cultures (Doostdar et al., 1995). This activity is required for the conversion of eriodictyol (5, 7, 3', 4'-tetrahydroxyflavanone) to hesperetin (5, 7, 4'-trihydroxy, 3'-methoxyflavanone) in oranges and lemons. This activity was also found in young developing leaves and fruit, but not in mature fully expanded tissues. While the presence of flavonoids in the cell cultures was not discussed in this paper, it was noted that the hydroxylase activity was lost when the cells were transferred to a medium that induced the formation of embryos. This is in contrast to other observations noted above.

## 5.4. Flavonoid Accumulation in Callus Cultures Derived from Mexican Lime

One of the recurring themes in this review is that callus cultures derived from citrus species generally do not accumulate flavonoids. Indeed, the few reported occurrences of this are in either very young callus lines or those in which embryogenesis has been induced. Most callus cultures I have examined are white in color and have no detectable levels of UV-absorbing phenolics.

However, during a 1991 visit to the Okitsu Fruit Tree Research Station citrus genetic research group in Japan, I found one callus line derived from *Citrus aurantifolia* 'Mexican lime' that formed bright yellow cultures. These cells had been in continuous culture for over 8 years. Extraction of freeze-dried cells from these cultures revealed the presence of low levels of UV-absorbing flavonoids. The UV spectra obtained from a diode array detector showed them to be flavonols. We were able to purify the three major peaks and identify them as being kaempferol 3-*O*-ß-rutinoside (K-R), kaempferol 3-*O*-ß-D-glucopyranoside-6"-(3-hydroxy-3-

Table 1. Major flavonoids found in Mexican lime callus and in Mexican lime leaves

| HPLC elut. order | Spectral pattern | Identification | Conc. mg/g.d.w. |
|---|---|---|---|
| | | *Callus* | |
| 1 | flavone | apigenin triglycoside | |
| 2 | flavone | diosmetin triglycoside | |
| 3 | flavone | diosmetin triglycoside | |
| 4 | flavonol | kaempferol diglucoside-HMG ester | |
| 5 | cinnamic acid | unknown | |
| 6 | cinnamic acid | unknown | |
| 7 | flavone | unknown | |
| 8 | flavonol | rutin | |
| 9 | flavone | isorhoifolin | |
| 10 | flavone | diosmin | |
| 11 | flavonol | kaempferol 7-*O*-rutinoside | |
| 12 | flavonol | kaempferol 3-*O*-glucoside-HMG ester | |
| | | *Leaves* | |
| 1 | flavanone | Narirutin 4'-glucoside | ~ 0.1 |
| 2 | flavone | Apigenin triglycoside | 0.1 |
| 3 | flavanone | Eriocitrin | 0.1 |
| 4 | flavanone | Hesperidin | 1.8 |
| 5 | flavonol | Rutin | 0.6 |
| 6 | flavone | Diosmin | 0.5 |

methylglutarate) (K-HMG-G) and kaempferol 3-$O$-ß-D-glucopyranoside-6"-(3-hydroxy-3-methylglutarate)-7-$O$-ß-D-glucopyranoside (K-HMG-diG) (Berhow et al., 1994) as shown in Table 1.

Acylated flavonoids have not been routinely reported as being present in citrus species. Sugiyama reported an HMG acylated flavonoid found in methanol extracts prepared from the fruit of mandarin oranges (Sugiyama et al., 1993). There are probably other acylated flavonoids in citrus, but their levels are probably normally very low in healthy mature tissues.

These kaempferol compounds are not found in the differentiated tissues of the Mexican lime plants. HPLC examination of extracts from leaf and fruit tissues from mature Mexican lime plants did not detect any of these compounds as indicated in Table 1. Indeed, kaempferol glycosides have not been reported in lime.

It is interesting that for other plant species there are many reports on the accumulation of phenolic compounds in plant cell culture. Three major observations can be made. 1) Production is highly variable and may be different even among cultures prepared from the same plant. Medium composition has an effect on how much of these compounds are accumulated. 2) Higher levels and additional phenolics can be induced by the addition of certain chemicals to the medium, by altering the osmotic potential of the media, by light, etc., these changes in the medium often mimic a "stress," such as adding an elicitor like chitin, a phytohormone such as 2,4-D, heavy metals, and others. 3) Often, the phenolic compounds accumulated in the cell cultures under these conditions are very different from the compounds that are predominately found in mature healthy tissues of the original plant.

A few examples from the literature of the study of phenolic accumulation are given here to compare with what may be going on in the Mexican lime cell cultures. Cell cultures derived from *Vancouveria hexandra* produced large amounts of kaempferol glycosides that were not found in the original plant. The production was influenced by the culture media and required the addition of 2,4-D (Yamamoto et al., 1993). Some cell cultures derived from various *Hypericum* species produced quercetin glycosides. The cultures produced were extremely variable in their ability to produce these flavonoids and other secondary metabolites. A cell culture derived from *H. balearicum* accumulated high levels of quercetin, isoquercitrin and rutin (Kartnig et al., 1996). Several kaempferol and quercetin glycosides were produced in cultures derived from *Dionaea muscipula* plants. Some of the kaempferol glycosides were acylated (Pakulski and Budzianowski, 1996). Regulation studies have shown that light, specifically UV light, is required for anthocyanin accumulation in carrot cell cultures (Takeda, 1990), flavonoid accumulation in parsley cell cultures (Logemann et al., 1995), and that anthocyanin accumulation in grape cell culture (Kakegawa et al., 1995) and flavonoid accumulation in parsley cell culture (Logemann et al., 1995) is related to the cessation of cell division. Addition of elicitors to cell cultures induces the formation of phenolics, including flavonoids. Fungal cells added to tomato cell cultures induce the formation of cinnamic acid esters (Keller et al., 1996; Bernards and Ellis, 1989; Bernards et al., 1991). Addition of chitin to cactus cell cultures induces the formation of several flavonoids. This can also be induced by the addition of yeast extract or autoclaved fungal mycelia (Liu et al., 1993a, b). Flavonoids are also induced by fungal elicitors added to parsley cell cultures (Logemann et al., 1995). These studies generated some ideas as to which chemicals could be added to the medium to affect the flavonoid accumulation in the Mexican lime cultures. Some of the results of studies in my laboratory are shown in Table 2.

Growing the cultures in the dark stops the accumulation, and returning the cultures to light reinitiates the accumulation. The addition of phytohormones has a profound effect

Table 2. Effect of various compounds added to the culture media on flavonoid accumulation in Mexican lime callus cultures

| Compound/Classification | Effect |
| --- | --- |
| No light | — — |
| Benzyladenine/PGR-cytokinin | — |
| Gibberellic acid/PGR | + |
| Abscisic acid/PGR | + |
| Napthaleneacetic acid/PGR-auxin | + |
| 2,4 Dichlorophenoxyacetic acid/PGR-auxin | NE |
| Zeatin/PGR-cytokinin | + |
| Zeatin riboside/PGR-cytokinin | + |
| Indoleacetic acid/PGR-auxin | NE |
| Indolebutyric acid/PGR-auxin | NE |
| Salicylic acid/signal | NE |
| Acetylsalicylic acid/signal | NE |
| Arachadonic acid/elicitor | NE |
| Chitin/elicitor | NE |
| Fungal extracts/elicitor | NE |
| Leaf extracts/elicitor | + |
| Pectin digests/elicitor | + |

+: Increase in flavonoid production
—: Decrease in flavonoid production
NE: No effect

on the accumulation of the acylated kaempferol glycosides. The auxins 2,4-D and NAA, gibberellic acid, the cytokinin zeatin, and abscisic acid have a stimulatory effect, while the cytokinin BA had an inhibitory effect. Abscisic acid was stimulatory at all levels examined (Table 2 and Figure 1).

Classical elicitors, such as chitin, arachadonic acid, and autoclaved fungal mycelial preparations, did not seem to have an effect, and neither did the plant "secondary messenger" salicylic acid. Interestingly, carbohydrate fractions prepared from macerated citrus leaves stimulated the formation of additional flavonoids in the Mexican lime cell cultures. This could be mimicked to a lesser degree by the addition of partially digested fractions prepared from polygalacturonic acid and pectin. Addition of these oligosaccharide fractions to Marsh grapefruit callus cultures, which do not normally produce any phenolic compounds, induces them to form low levels of flavonoids including hesperidin. This work will be more fully discussed in an forthcoming paper.

## 6. CONCLUSION

The use of cell cultures to study secondary metabolism, such as the accumulation of flavonoids, in plants has led to the an understanding of how these pathways are regulated and controlled. Without a doubt, cell cultures are indispensable tools for secondary metabolism research and are, as one reviewer has put it, a pot of gold (Zenk, 1991). Cell cultures are being exploited not only as tools, but as sources for the generation of flavor and colors (Dörnenburg and Knorr, 1996). This approach is being eyed for many other uses including the production of important pharmaceuticals. Though some researchers caution that cells in culture may never live up to its promise as "living cell factories" for pharma-

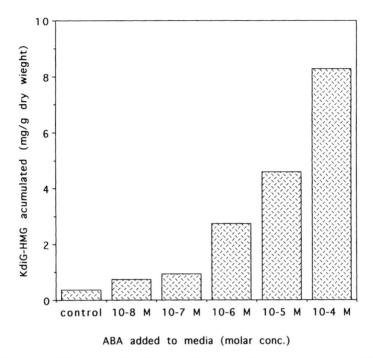

**Figure 1.** Effect of the addition of abscisic acid to the cell culture medium on the accumulation of kampferol-4,7-O-diglucoside HMG ester in Mexican lime callus cultures grown on solid media.

ceuticals and other useful chemicals (Kuc, 1995), new techniques may well overcome some of these obstacles. Flavonoids appear to be universal chemical tools which plants use to deal with their environment. They have been implicated as defenses against other plants, fungi, insects and bacteria, and as regulators of interactions between beneficial fungi, herbivores, and insects, as plant hormones, as important constituents of animal diets, both nutritionally and medicinally. As such, plant flavonoids are the phytochemicals of the future (Dakora, 1995; Shirley, 1996). Citrus flavonoids are unique in that they have qualities, such as taste, that contribute to the overall quality of the juice. Can citrus cell cultures be exploited commercially? The possibility that citrus cell cultures could be used to produce neohesperidin, which can be easily converted into the sweetener neohesperidin dihydrochalcone, has already been broached. The abundance of flavonoids in many agricultural commodities and processing streams allows for the possible exploitation of chemicals like the flavonoids from the processing wastes for agricultural, pesticidal, nutritional, and medicinal purposes looms ahead. To that end, citrus cell culture studies will provide useful information for the bioregulation of the accumulation of these compounds in citrus species. The future can only grow brighter for research in this important field.

## REFERENCES

Albach, R. F.; Juarez, A. T.; Lime, B. J. Time of naringin production in grapefruit. *Journal of the American Society of Horticultural Science* **1969**, *94*, 605–609.

Albach, R. F.; Redman, G. H. Composition and Inheritance of Flavanones in Citrus Fruit. *Phytochemistry* **1969**, *8*, 127–143.

Albach, R. F.; Redman, G. H.; Cruse, R. R. Annual and seasonal changes in naringin concentration of Ruby Red grapefruit juice. *Journal of Agriculture and Food Chemistry* **1981**, *29*, 808–811.

Albach, R. F.; Wutscher, H. K. Flavanone content of whole grapefruit and juice as influenced by fruit development. *Journal of the Rio Grande Valley Horticultural Society* **1988**, *41*, 89–95.

Banks, S. W.; Dewick, P. M. Biosynthesis of glyceolins I, II, and III in soybean. *Phytochemistry* **1983**, *22*, 2729–2733.

Bar-Peled, M.; Fluhr, R.; Gressel, J. Juvenile-specific localization and accumulation of a rhamnosyltransferase and its bitter flavonoid in foliage, flowers, and young citrus fruits. *Plant Physiology* **1993**, *103*, 1377–1384.

Bar-Peled, M.; Lewinsohn, E.; Fluhr, R.; Gressel, J. UDP-rhamnose:flavanone-7-O-glucoside-2"-O-rhamnosyltransferase. Purification and characterization of an enzyme catalyzing the production of bitter compounds in citrus. *Journal of Biological Chemistry* **1991**, *266*, 20953–20959.

Barthe, G. A.; Jourdan, P. S.; McIntosh, C. A.; Mansell, R. L. Naringin and limonin production in callus cultures and regenerated shoots from Citrus sp. *Journal of Plant Physiology* **1987**, *127*, 55–65.

Becker, H. Regulation of Secondary Metabolism in plant cell cultures. In *Plant Tissue and Cell Culture*; C. E. Green; D. A. Somers; W. P. Hackett and D. D. Biesboer, Eds.; Alan R. Liss: New York, 1987; pp 199–220.

Benavente-Garcia, O.; Castillo, J.; del Rio, J. A. Changes in neodiosmin levels during the development of *Citrus aurantium* leaves and fruits. Postulation of a neodiosmin biosynthetic pathway. *Journal of Agricultural and Food Chemistry* **1993**, *41*, 1916–1919.

Benavente-Garcia, O.; Castillo, J.; Sabater, F.; Rio, J. A. D. Characterization of a S-adenosyl-L-methionine: eriodictyol 4'-O-methyltransferase from *Citrus aurantium*. *Plant Physiology and Biochemistry* **1995**, *33*, 263–271.

Berhow, M. A.; Bennett, R. D.; Poling, S. M.; Vannier, S.; Hidaka, T.; Omura, M. Acylated flavonoids in callus cultures of *Citrus aurantifolia*. *Phytochemistry* **1994**, *36*, 1225–1227.

Berhow, M. A.; Smolensky, D. Developmental and substrate specificity of hesperetin-7-O-glucosyltransferase activity in *Citrus limon* tissues using high-performance liquid chromatographic analysis. *Plant Science* **1995**, *112*, 139–147.

Berhow, M. A.; Vandercook, C. E. Biosynthesis of naringin and prunin in detached grapefruit. *Phytochemistry* **1989**, *28*, 1627–1630.

Berhow, M. A.; Vandercook, C. E. Sites of naringin biosynthesis in grapefruit seedlings. *Journal of Plant Physiology* **1991**, *138*, 176–179.

Bernards, M. A.; Ellis, B. E. Phenylpropanoid metabolism in tomato cell cultures co-cultivated with *Verticillium albo-atrum*. *Journal of Plant Physiology* **1989**, *135*, 21–26.

Bernards, M. A.; Strack, D.; Wray, V.; Ellis, B. E. Caffeoyl glucosides in fungal challenged tomato suspension cultures. *Phytochemistry* **1991**, *30*, 497–499.

Brunet, G.; Ibrahim, R. K. Tissue culture of citrus peel and its potential for flavonoid synthesis. *Zietschrift fur Pflanzenphysiologia* **1973**, *69*, 152–162.

Brunet, G.; Ibrahim, R. K. Methylation of flavonoids by cell-free extracts of calamondin orange *Citrus mitis*. *Phytochemistry* **1980**, 741–746.

Castillo, J.; Benavente, O.; del Rio, J. A. Naringin and neohesperidin levels during development of leaves, flower buds, and fruits of *Citrus aurantium*. *Plant Physiology* **1992**, *99*, 67–73.

Castillo, J.; Benavente, O.; del Rio, J. A. Hesperetin 7-O-glucoside and Prunin in *Citrus* Species *(C. aurantium* and *C. paradisi)*. A study of their quantitative distribution in immature fruits and as intermediate precursors of neohesperidin and naringin in *C. aurantium*. *Journal of Agricultural and Food Chemistry* **1993**, *41*, 1920–1924.

Chappell, J.; Hahlbrock, K. Transcription of plant defence genes in response to UV light or fungal elicitor. *Nature* **1984**, *311*, 76–78.

Charlwood, B. V.; Rhodes, M. J. C., Eds. *Secondary Products from Plant Tissue Culture*; Proceedings of the Phytochemical Society of Europe Vol. 30, Clarendon Press: Oxford, 1990.

Cheng, C. L.; Wetherell, D. F.; Dougall, D. K. 4-Coumarate: CoA ligase in wild carrot cell culture clones which accumulate different amounts of anthocyanin. In *Primary and Secondary Metabolism of Plant Cell Cultures*; K. H. Neumann; W. Barz and E. Reinhard, Eds.; Springer-Verlag: Berlin, 1985; pp 87–98.

Cordell, G. A. Changing stratagies in natural products chemistry. *Phytochemistry* **1995**, *40*, 1586–1612.

Dakora, F. D. Plant flavonoids: biological molecules for useful exploitation. *Australian Journal of Plant Physiology* **1995**, *22*, 87–99.

del Rio, J. A.; Ortuño, A.; Marin, F. R.; Garcia Puig, D.; Sabater, F. Bioproduction of neohesperidin and naringin in callus cultures of *Citrus aurantium*. *Plant Cell Reports* **1992**, *11*, 592–596.

Dixon, R. A.; Choudhary, A. D.; Dalkin, K.; Edwards, R.; Fahrendorf, T.; Gowri, G.; Harrison, M. J.; Lamb, C. J.; Loake, G. J.; Maxwell, C. A. Molecular biology of stress-induced phenylpropanoid and isoflavonoid biosynthesis in alfalfa. In *Phenolic Metabolism in Plants*; H. A. Stafford and R. K. Ibrahim, Eds.; Plenum Press.: New York, N.Y, 1992; pp 91–138.

Dixon, R. A.; Dey, P. M.; Lamb, C. J. Phytoalexins: enzymology and molecular biology. *Advances in Enzymology and Molecular Biology* **1983**, *55*, 1–136.

Dixon, R. A.; Paiva, N. L. Stress-induced phenylpropanoid metabolism. *Plant Cell* **1995**, *7*, 1085–1097.

Doostdar, H.; Shapiro, J. P.; Niedz, R.; Burke, M. D.; McCollum, T. G.; McDonald, R. E.; Mayer, R. T. A cytochrome P450 mediated naringenin 3'-hydroxylase from sweet orange cell cultures. *Plant and Cell Physiology* **1995**, *36*, 69–77.

Dörnenburg, H.; Knorr, D. Generation of colors and flavors in plant cell and tissue cultures. *Critical Reviews in Plant Sciences* **1996**, *15*, 141–168.

Douglas, C. J.; Ellard, M.; Hauffe, K. D.; Molitor, E.; Sá, M. M. d.; Reinold, S.; Subramaniam, R.; Williams, F. General phenylpropanoid metabolism: regulation by environmental and developmental signals. In *Phenolic Metabolism in Plants.*; H. A. Stafford and R. K. Ibrahim, Eds.; Plenum Press: New York, 1992; pp 63–89.

Ebel, J. Phytoalexin synthesis: the biochemical analysis of the induction process. *Annual Review of Phytopathology* **1986**, *24*, 235–264.

Ebel, J.; Hahlbrock, K. Enzymes of flavone and flavanone glycoside biosynthesis; coordinated and selective induction in cell-suspension cultures of *Petroselinum hortense*. *European Journal of Biochemistry* **1977**, *75*, 201–209.

Ebel, J.; Schmidt, W. E.; Loyal, R. Phytoalexin synthesis in soybean cells: elicitor induction of phenylalanine ammonia-lyase and chalcone synthase mRNAs and correlation with phytoalexin accumulation. *Archives of Biochemistry and Biophysics* **1984**, *232*, 240–248.

Ebel, J.; Stäb, M. R.; Schmidt, W. E. Induction of enzymes of phytoalexin synthesis in soybean cells by fungal elicitor. In *Primary and Secondary Metabolism of Plant Cell Cultures*; K. H. Newmann; W. Barz and E. Reinhard, Eds.; Springer-Verlag: New York, 1985; pp 247–254.

Ellis, B. E. Probing secondary metabolism in plant cell cultures. *Canadian Journal of Botany* **1984**, *62*, 2912–2917.

Fisher, J. F. A procedure for obtaining radioactive naringin from grapefruit leaves fed L-phenylalanine-14C. *Phytochemistry* **1968**, *7*, 769–771.

Fouche, S. D.; Dubery, I. A. Chalcone isomerase from *Citrus sinensis:* purification and characterization. *Phytochemistry* **1994**, *37*, 127–132.

Gavish, H.; Lewinsohn, E.; Vardi, A.; Fluhr, R. Production of flavanone-neohesperidosides in Citrus embryos. *Plant Cell Reports* **1989**, *8*, 391–394.

Geissman, T. A. *The Chemistry of Flavonoid Compounds.*; Pergamon Press: New York, 1962.

Grosser, J. W.; Gmitter, F. J.; Chandler, J. L. Intergeneric somatic hybrid plants from sexually incompatible woody species: *Citrus sinensis* and *Severinia disticha*. *Theoretical and Applied Genetics* **1988**, *75*, 397–401.

Grosser, J.W.; Gmitter, F.J. Somatic hybridization of Citrus with wild relatives for germplasm enhancement and cultivar development. *HortSci.* **1990**, *25*, 147–151.

Hagen, R. E.; Dunlap, W. J.; Wender, S. H. Seasonal variation of naringin and certain other flavanone glycosides in juice sacs of Texas Ruby Red grapefruit. *Journal of Food Science* **1966**, *31*, 542–547.

Hahlbrock, K. Plant cell suspension cultures as a model system for studying coordinated changes in the enzyme activities of flavonoid biosynthesis. *Nova Acta Leopold Suppl* **1976**, *7*, 311–318.

Hahlbrock, K. Regulatory aspects of phenylpropanoid biosynthesis in plant cell cultures. In *Plant Tissue Culture and its Bio-technological Application*; W. Barz; E. Reinhard and M. H. Zenk, Eds.; Springer-Verlag: New York, 1977; pp 95–111.

Hahlbrock, K.; Scheel, D. Physiology and molecular biology of phenylpropanoid metabolism. *Annual Review of Plant Physiology and Molecular Biology* **1989**, *40*, 347–369.

Harborne, J. B., Ed. *The Flavonoids: Advances in Research Since 1986*; Chapman and Hall: London, 1994.

Harborne, J. B.; Mabry, T. J., Eds. *The Flavonoids: Advances in Research*; Chapman and Hall: London, 1982.

Harborne, J. B.; Mabry, T. J.; Mabry, H., Eds. *The Flavonoids*; Academic Press: New York, 1975.

Hasegawa, S.; Maier, V. P. Biosynthesis of trans-cinnamate from phenylpyruvate and L glutamate by cell-free extracts of grapefruit. *Phytochemistry* **1970**, *9*, 2483–2487.

Hasegawa, S.; Maier, V. P. Cinnamate hydroxylation and the enzymes leading from phenylpyruvate to p-coumarate synthesis in grapefruit tissues. *Phytochemistry* **1972**, *11*, 1365–1370.

Hasegawa, S.; Maier, V. P. Some aspects of citrus biochemistry and juice quality. *Proceedings of the International Society of Citriculture* **1981**, *2*, 914–918.

Heller, W.; Forkmann, G. Biosynthesis. In *The Flavonoids*; J. Harborne, Ed.; Chapman and Hall: New York, N.Y., USA, 1988; pp 398–425.

Hidaka, T.; Omura, M. Control of embryogenesis in citrus cell culture: regeneration from protoplasts and attempts to callus bank. *Bulletin of the Fruit Tree Research Station, Series B, Okitsu* **1989**.

Hidaka, T.; Omura, M.; Ugaki, M.; Tomiyama, M.; Kato, A.; Ohshima, M.; Motoyoshi, F. Agrobacterium-mediated transformation and regeneration of Citrus spp. from suspension cells. *Japanese Journal of Breeding* **1990**, *40*, 199–207.

Hidaka, T.; Omura, M. Electrical fusion between Satsuma Mandarin *(Citrus unshiu)* and Rough Lemon *(C. jambhiri)* or Yuzu *(C. junos)*. *Japanese Journal of Breeding* **1992**, *40*, 79–89.

Hinderer, W.; Petersen, M.; Seitz, H. U. Inhibition of flavonoid biosynthesis by gibberellic acid in cell suspension cultures of *Daucus carota* L. *Planta* **1984**, *160*, 544–549.

Horowitz, R. M.; Gentili, B. Flavonoids of Citrus IV: the structure of neohesperidose. *Tetrahedron* **1963**, *19*, 773–782.

Horowitz, R. M.; Gentili, B. Flavonoid constituents of citrus. In: *Citrus Science and Technology*; P. E. S. M. K. V. S. Nagy, Ed.; AVI Publishing Company, Inc.: Westport, CT, 1977; pp 397–426.

Ismail, M. A.; Brown, G. E. Postharvest wound healing in citrus fruit: induction of phenylalanine ammonia-lyase in injured 'Valencia' orange flavedo. *Journal of the American Horticultiral Society* **1979**, *104*, 126–129.

Jacobs, M.; Rubery, P. H. Naturally occurring auxin transport regulators. *Science* **1988**, *241*, 346–349.

Jourdan, P. S.; McIntosh, C. A.; Mansell, R. L. Naringin levels in citrus tissues. II. Quantitative distribution of naringin in *Citrus paradisi* Macfad. *Plant Physiology* **1985**, *77*, 903–908.

Kakegawa, K.; Suda, J.; Sugiyama, M.; Komamine, A. Regulation of anthocyanin biosynthesis in cell suspension cultures of *Vitis* in relation to cell division. *Physiologia Plantarum* **1995**, *94*, 661–666.

Kanes, K.; Tisserat, B.; Berhow, M.; Vandercook, C. Phenolic composition of various tissues of Rutaceae species. *Phytochemistry* **1993**, *32*, 967–974.

Kartnig, T.; Göbel, I.; Heydel, B. Production of hypericin, pseudohypericin, and flavonoids in cell cultures of various *Hypericum* species and their chemotypes. *Planta Medica* **1996**, *62*, 51–53.

Kefford, J. F.; Chandler, B. V. *The Chemical Constituents of Citrus Fruits*; Academic Press: New York, 1970.

Keller, H.; Hohlfeld, H.; Wray, V.; Hahlbrock, K.; Scheel, D.; Strack, D. Changes in the accumulation of soluble and cell wall-bound phenolics in elicitor-treated cell suspension cultures and fungus-infected leaves of *Solanum tuberosum*. *Phytochemistry* **1996**, *42*, 389–396.

Kobayashi, S.; Ikeda, I.; Uchimiya, H. Conditions for high frequency embryogenesis from orange *(Citrus sinensis)* protoplasts. *Plant Cell Tissue and Organ Culture* **1985**, *4*, 249–259.

Kobayashi, S.; Oiyama, I.; Yoshinaga, K.; Ishii, S. Fertility in an intergeneric somatic hybrid of Rutaceae. *HortScience* **1991a**, *26*, 207.

Kobayashi, S.; Ohgawara, T., Fujiwara, K.; Oiyama, I. Analysis of cytoplasmic genomes in somatic hybrids between Navel Orange *(Citrus sinensis* Osb.) and "Murcott" tangor. *Theoretical and Applied Genetics* **1991b**, *82*, 6–10.

Kobayashi, S.; Uchimiya, H. Expression and integration of a foreign gene in orange *(Citrus sinensis* Osb.) protoplasts by direct DNA transfer. *Japanese Journal of Genetics* **1989**, *64*, 91–97.

Kordan, H. A.; Morganstern, L. Flavononoid production by mature citrus fruit tissue proliferating in vitro. *Nature* **1962**, *195*, 163–164.

Kordan, H.A. Some morphological and physiological relationships of lemon fruit tissue grown *in vivo* and *in vitro*. *Bulletin of the Torrey Botany Club* **1965**, *92*, 209–216.

Kubitzki, K. Phenylpropanoid metabolism in relation to land plant origin and diversification. *Journal of Plant Physiology* **1987**, *131*, 17–24.

Kuc, J. Phytoalexins, stress metabolism, and disease resistance in plants. *Annual Review of Phytopathology* **1995**, *33*, 275–297.

Laks, P. E.; Pruner, M. S. Flavonoid biocides: structure/activity relations of flavonoid phytoalexin analogues. *Phytochemistry* **1989**, *28*, 87–91.

Lewinsohn, E.; Berman, E.; Mazur, Y.; Gressel, J. Glucosylation of exogenous flavanones by grapefruit *(Citrus paradisi)* cell cultures. *Phytochemistry* **1986**, *25*, 2531–2535.

Lewinsohn, E.; Berman, E.; Mazur, Y.; Gressel, J. (7) Glucosylation and (1–6) rhamnosylation of exogenous flavanones by undifferentiated Citrus cell cultures. *Plant Science* **1989a**, *61*, 23–28.

Lewinsohn, E.; Britsch, L.; Mazur, Y.; Gressel, J. Flavanone glycoside biosynthesis in citrus. Chalcone synthase, UDP-glucose:flavanone-7-O-glucosyl-transferase and -rhamnosyl-transferase activities in cell-free extracts. *Plant Physiology* **1989b**, *91*, 1323–1328.

Lisker, N.; Cohen, L.; Chalutx, E.; Fuchs, Y. Fungal infections suppress ethylene-induced phenylalanine ammonialyase activity in grapefruits infected by *Penicillium digitatum*. *Physiology of Plant Pathology* **1983**, *22*, 331–338.

Liu, Q.; Dixon, R. A.; Mabry, T. J. Additional flavonoids from elicitor-treated cell cultures of *Cephalocereus senilis*. *Phytochemistry* **1993a**, *34*, 167–170.

Liu, Q.; Markham, K. R.; Paré, P. W.; Dixon, R. A.; Mabry, T. J. Flavonoids from elicitor-treated cell suspension cultures of *Cephalocereus senilis*. *Phytochemistry* **1993b**, *32*, 925–928.

Logemann, E.; Wu, S.-C.; Schroder, J.; Schmetzer, E.; Somssich, I. E.; Hahlbrock, K. Gene activation by UV light, fungal elicitor or fungal infection in *Petroselinum crispum* is correlated with repression of cell cycle-related genes. *Plant Journal* **1995**, *8*, 865–876.

Louzada, E. S.; Grosser, J. W.; Gmitter, J. Intergenic somatic hybridation of sexually incompatable patents: *Citrus sinensis* and *Atalantia ceylanica*. *Plant Cell Reports* **1993**, *12*, 687–690.

Maier, V. P. Compositional studies of citrus: significance in processing, identification, and flavor. *Proceedings of the First International Citrus Symposium* **1969**, *1*, 235–243.

Maier, V. P.; Hasegawa, S. L-Phenylalanine ammonia-lyase activity and naringenin glycoside accumulation in developing grapefruit. *Phytochemistry* **1970**, *9*, 139–144.

Maier, V. P.; Metzler, D. M. Grapefruit phenolics II: principal aglycones of endocarp and peel and their possible biosynthetic relationship. *Phytochemistry* **1967**, *6*, 1127–1135.

Matern, U.; Reichenbach, C.; Heller, W. Efficient uptake of flavonoids into parsley *(Petroselium hortense)* vacuoles requires acylated glycosides. *Planta* **1986**, *167*, 183–189.

McIntosh, C. A.; Latchinian, L.; Mansell, R. L. Flavanone-specific 7-O-glucosyltransferase activity in *Citrus paradisi* seedlings: purification and characterization. *Archives of Biochemistry and Biophysics* **1990**, *282*, 50–57.

McIntosh, C. A.; Mansell, R. L. Biosynthesis of naringin in *Citrus paradisi*: UDP-glucosyltransferase activity in grapefruit seedlings. *Phytochemistry* **1990**, *29*, 1533–1538.

Nahrstedt, A. The significance of secondary metabolites for interactions between plants and insects. *Planta Medica* **1990**, *55*, 333–338.

Negrel, J.; Javelle, F. Induction of phenylpropanoid and tyramine metabolism in pectinase- or pronase-elicited cell suspension cultures of tobacco *(Nictotiana tabacum)*. *Physiologia Plantarum* **1995**, *95*, 569–574.

Nicholson, R. L.; Hammerschmidt, R. Phenolic compounds and their role in disease resistance. *Annual Review of Phytopathology* **1992**, *30*, 369–389.

Nishiura, M.; Esaki, S.; Kamiya, S. Flavonoids in citrus and related genera. I. Distribution of flavonoid glycosides in *Citrus* and *Poncirus*. *Agricultural and Biological Chemistry* **1969**, *33*, 1109–1118.

Nishiura, M.; Kamiya, S.; Esaki, S. Flavonoids in citrus and related genera. III. flavonoid pattern and Citrus taxonomy. *Agricultural and Biological Chemistry* **1971a**, *35*, 1691–1706.

Nishiura, M.; Kamiya, S.; Esaki, S.; Ito, F. Flavonoids in Citrus and related genera. II. Isolation and identification of isonaringin and neoeriocitrin from Citrus. *Agricultural and Biological Chemistry* **1971b**, *35*, 1683–1690.

Nojiri, H.; Sugimori, M.; Yamane, H.; Nishimura, Y.; Yamada, A.; Shibuya, N.; Kodama, O.; Murofushi, N.; Omori, T. Involvement of jasmonic acid in elicitor-induced phytoalexin production in suspension-cultured rice cells. *Plant Physiology* **1996**, *110*, 387–392.

Ohgawara, T.; Kobayashi, S. Application of protoplast fusion to *Citrus* breeding. *Food Biotechnology* **1991**, *5*, 169–184.

Ozeki, Y.; Komamine, A. Induction of anthocyanin synthesis in a carrot suspension culture. Correlation of metabolic differentiation with morphological differentiation in *Daucus carota*. In: *Plant Cell and Tissue Culture* , Proc. 5th Int. Cong. of Plant and Cell Tissue Culture, Tokyo, 1982, A. Fujiwara, Ed., Jap. Assn. for Plant Tissue Cult., Tokyo, 1982; pp 355–356.

Ozeki, Y.; Komamine, A. Effects of inoculum density, zeatin and sucrose on anthocyanin accumulation in a carrot suspension culture. *Plant Cell, Tissue and Organ Culture* **1985a**, *5*, 45–53.

Ozeki, Y.; Komamine, A. Induction of anthocyanin synthesis in relation to embryogenesis in a carrot suspension culture--a model system for the study of expression and repression of secondary metabolism. In *Primary and Secondary Metabolism of Plant Cell Cultures*; K. H. Neumann; W. Barz and E. Reinhard, Eds.; Springer-Verlag: Berlin, 1985b; pp 99–106.

Pakulski, G.; Budzianowski, J. Quercetin and kaempferol glycosides of *Dionaea muscipula* from *in vitro* cultures. *Planta Medica* **1996**, *62*, 95–96.

Seigler, D. S. Secondary metabolites and plant systematics. In *Secondary Plant Products*; P. K. Stumpf and E. E. Conn, Eds.; Academic Press: New York, 1981; pp 139–175.

Shaw, P. E.; Calkins, C. O.; McDonald, R. E.; Greany, P. D.; Webb, J. C.; Nisperos, C. M. O.; Barros, S. M. Changes in limonin and naringin levels in grapefruit albedo with maturity and the effects of gibberellic acid on these changes. *Phytochemistry* **1991**, 3215–3219.

Shimokoriyama, M. Interconversion of chalcones and flavanones of a phloroglucinol-type structure. *Journal of the American Chemical Society* **1957**, *79*, 4199–4202.

Shirley, B. W. Flavonoid biosynthesis: 'new' functions for an 'old' pathway. *Trends in Plant Sciences* **1996**, *1*, 377–382.

Snyder, B. A.; Nicholson, R. L. Synthesis of phytoalexins in sorghum as a site-specific response to fungal ingress. *Science* **1990**, *248*, 1637–1639.

Stafford, H. A. *Flavonoid Metabolism*; CRC Press, Inc.: Boca Ratan, 1990.

Sugiyama, S.; Umehara, K.; Kuroyanagi, M.; Ueno, A.; Taki, T. Studies on the differentiation inducers of myeloid leukemic cells from *Citrus* species. *Chem. Pharm. Bull.* **1993**, *41*, 714–719.

Takeda, J. Light-induced synthesis of anthocyanin in carrot cells in suspension. II. Effects of light and 2,4-D on induction and reduction of enzyme activities related to anthocyanin synthesis. *Journal of Experimental Botany* **1990**, *41*, 749–755.

Thorpe, T. A.; Maier, V. P.; Hasegawa, S. Phenylalanine ammonia-lyase activity in citrus fruit tissue cultured *in Vitro*. *Phytochemistry* **1971**, *10*, 711–718.

Tisserat, B.; Galletta, P. D. In vitro culture of lemon juice vesicles. *Plant Cell, Tissue and Organ Culture* **1987**, *11*, 81–95.

Tisserat, B.; Vandercook, C. E.; Berhow, M. Citrus juice vesicle culture: a potential research tool for improving juice yield and quality. *Food Technology* **1989**, 95–100.

Vandercook, C. E.; Tisserat, B. Flavonoid changes in developing lemons grown in vivo and in vitro. *Phytochemistry* **1989**, *28*, 799–803.

Vardi, A.; Bleichman, S.; Aviv, D. Genetic transformation of citrus protoplasts and regeneration of transgenic plants. *Plant Science* **1990**, *69*, 199–206.

Vardi, A.; Galun, E. Recent advances in protoplast culture of horticultural crops: Citrus. *Scientia Horticulturae* **1988**, *37*, 217–230.

Yamamoto, H.; Ieda, K.; Tsuchiya, S.-I.; Yan, K.; Tanaka, T.; Iinuma, M.; Mizuno, M. Flavonol glycoside production in callus cultures of *Epimedium diphyllum*. *Phytochemistry* **1992**, *31*, 837–840.

Yamamoto, H.; Yan, K.; Ieda, K.; Tanaka, T.; Iinuma, M.; Mizuno, M. Flavonol glycosides production in cell suspension cultures of *Vancouveria hexandra*. *Phytochemistry* **1993**, *33*, 841–846.

Zenk, M. H. Chasing the enzymes of secondary metabolism: plant cell cultures as a pot of gold. *Phytochemistry* **1991**, *33*, 3861–3863.

# FLAVONOIDS OF THE ORANGE SUBFAMILY AURANTIOIDEAE

John A. Manthey and Karel Grohmann

U.S. Citrus and Subtropical Products Laboratory
Agricultural Research Service, USDA
600 Avenue S, NW
Winter Haven, Florida 33881

## 1. INTRODUCTION

The flavonoids and related compounds of the orange subfamily Aurantioideae have attracted the attention of generations of chemical researchers, beginning with the first description of hesperidin by Lebreton (1828) to the many current pharmacological studies of these compounds in living systems. For many reasons (medicinal, herbal, agricultural), citrus fruit have been collected and used by societies throughout the centuries (Webber, 1967). However, our modern focus on the impact of citrus flavonoids on human health was perhaps started by the work of Szent-Györgyi, who, in calling citrus flavonoids Vitamin P, first indicated the importance of flavonoids in capillary function (Armentano et al., 1936; Rusznyák and Szent-Györgyi, 1936; Bentsath et al., 1937). While the term Vitamin P fell into disuse, the importance of flavonoids and ascorbic acid in proper capillary function was firmly established. Without question, the importance of the capillaries in many different aspects of human health cannot be overstated, and aspects of this are discussed in the chapters by Middleton and Kandaswami (1998), Gerritsen (1998), and Attaway and Buslig (1998). Extending from this, many pharmacological studies now show the important antioxidant and anticancer activities that citrus flavonoids contribute to human health through the diet. Much of this research relies directly on the isolation and structural characterizations of these diverse citrus phenolics, much of which was done by chemists at the U.S. Department of Agriculture. Although many of the major citrus flavonoids have now been well characterized, much still remains unclear about the biological activities of these compounds in mammalian systems, and about the biosynthesis, transport, and physiological roles of these compounds in the plants in which these compounds occur. It has been noted that in developing citrus plant tissue tremendous amounts of metabolic energy are expended in the biosynthesis of these compounds. In fact, flavonoids can constitute well above 50 percent of the dry weight of immature citrus fruit and leaf tissue undergoing rapid cell division. Yet, very little is known why this occurs, or how the biosyntheses of the different groups of flavonoids in citrus are connected. As part of this chapter, the remarkable diversity and distribution of the flavonoids in the orange subfamily Aurantioideae are reviewed, and evidence pertinent to the biosynthetic pathways of citrus flavonoids is reported.

The orange subfamily Aurantioideae belongs to the plant family Rutaceae, and is composed of 33 genera, of which 14 genera have only 1 species. Among the cultivated genera are *Citrus, Fortunella,* and *Poncirus*, all in the subtribe Citrinae. A complete botanical description of the Aurantioideae was made by Swingle (1943,1967) and others (Engler, 1931; Tanaka, 1932). The botanical listing of the Aurantioideae reported by Swingle is shown in Table 1. Although cultivated extensively in many desert, tropical, and subtropical regions throughout the world, the large majority of citrus and their wild relatives are native to southeastern Asia, the East Indian Archipelago, New Guinea, Melanesia, New Caledonia, and Australia. A few species are also native to tropical Africa (Swingle 1967). With a few exceptions, the species of the Aurantioideae are trees or shrubs with persistent leaves, white, typically fragrant flowers, and with subglobose fruit with peel containing numerous oil glands (Schneider, 1968; Erickson, 1968; Grieve and Scora, 1980). In addition to these morphological similarities, it has been the widespread abundance of the flavonoids typically found among many species of the Aurantioideae that has attracted such a great deal of chemical study. These studies have shown that fairly unique to citrus are the abundance of several classes of flavonoids, including the flavanone glycosides, and a large number of polymethoxylated flavones.

## 2. FLAVONOID DIVERSITY AND DISTRIBUTION IN THE AURANTIOIDEAE

### 2.1. Structural Diversity

Extensive characterizations of the flavonoids in citrus (Horowitz, 1961; Horowitz and Gentili, 1977 and references therein; Kumamoto, et al., 1985a-d; Matsubara et al., 1985) have shown that these compounds (nearly 90 known compounds in the main citrus cultivars) occur in three broad classes, including the flavanone glycosides (mainly di- and tri-*O*-glycosides), the flavone glycosides (mainly di- and tri-*O*-glycosides and *C*-glycosides), and the polymethoxylated flavones. Much of what is known about the flavonoids in citrus has been derived from chemical analyses of the three main cultivated genera of the Citrinae subtribe (*Citrus, Poncirus*, and *Fortunella*) (Table 1). Recently, work has also been done on the other genera of the Citrinae and other subtribes of the Aurantioideae. The flavonoid distributions that have been found to occur among the 33 genera of the Aurantioideae are discussed in the following sections and are summarized in Table 1. The findings in Table 1 include mainly those results of chemical investigations in which the compounds were isolated, and whose structures were determined by chemical derivatization, and by various spectroscopic techniques. Also included in Table 1 are the results of an extensive chemical survey of the Aurantioideae by Grieve and Scora (1980). In contrast to the other cited results, these later findings were based mainly on the separation and identification of a large number of flavonoids by thin layer and paper chromatography, and by comparisons of the ultraviolet spectra and co-migration of these compounds and their hydrolyzed products with known standard compounds.

Table 1. Flavonoid distributions within the Aurantioideae

| | total flavonoids | flavanone glycosides | flavanone aglycones | methoxylated flavones (b1) | prenylated flavones | O-glycosyl flavones | C-glycosyl flavones | flavonol glycosides |
|---|---|---|---|---|---|---|---|---|
| **Tribe 1. Clauseneae**: (Very remote citroid fruit) | | | | | | | | |
| Subtribe 1. **Micromelinae**: (very remote citroid fruit trees) | | | | | | | | |
| *Micromelum*[a] | 1 | | | 1 | | | | |
| Subtribe 2. **Clauseninae**: (remote citroid fruit trees) | | | | | | | | |
| *Glycosmis*[b] | 2 (8) | (0) | 2 (a1)(0) | (0) | 2(a1) | (0) | (8) | (0) |
| *Clausena*[c] | 1 (6) | (0) | (0) | (0) | | 0(0) | (0) | 1(5) |
| *Murraya*[d-l] | 15 (15) | (0) | (0) | 15 (4) | (0) | (0) | (8) | (3) |
| Subtribe 3. **Merrilliinae**: (Large-Fruited Remote citroid trees) | | | | | | | | |
| *Merrillia*[m,n] | 3 (2) | 0 (0) | 0 (0) | 3 (0) | (0) | (0) | (2) | (0) |
| **Tribe 2. Citreae**: (Citrus and citroid trees) | | | | | | | | |
| Subtribe 1. **Triphasiinae**: (minor citroid fruit trees) | | | | | | | | |
| *Wenzelia* | NR | | | | | | | |
| *Monanthocitrus* | NR | | | | | | | |
| *Oxanthera* | NR | | | | | | | |
| *Merope* | NR | | | | | | | |
| *Triphasia* | NR (3) | (0) | (0) | (0) | (0) | (1) | (2) | (0) |
| *Pamburus*[o] | 1 (5) | (0) | (0) | (0) | 1(c1) | (1) | (2) | (2) |
| *Luvunga* | NR | | | | | | | |
| *Paramignya* | NR (0) | (0) | (0) | (0) | (0) | (0) | (0) | (0) |

Table 1. (continued)

| | total flavonoids | flavanone glycosides | flavanone aglycones | methoxylated flavones (b1) | prenylated flavones | O-glycosyl flavones | C-glycosyl flavones | flavonol glycosides |
|---|---|---|---|---|---|---|---|---|
| Subtribe 2. **Citrinae**: (citrus fruit trees) | | | | | | | | |
| Severinia | NR (2) | (0) | (0) | (0) | (0) | (0) | (1) | (1) |
| Pleiospermium | NR (10) | (1) | (0) | (0) | (0) | (6) | (3) | (0) |
| Burkillanthus | NR | | | | | | | |
| Limnocitrus | NR | | | | | | | |
| Hesperethusa | NR (4) | (0) | (0) | (2) | (0) | (0) | (0) | (2) |
| Citropsis | NR (6) | (0) | (0) | (2) | (0) | (0) | (6) | (0) |
| Atlantia[p-r] | 10 (10) | 3 (0) | (0) | (0) | 3(c1)(0) | 7(d1)(6) | (4) | (0) |
| Fortunella[s] | 13 | 6 | NR | 4 | NR | 1 | 2 | |
| Eremocitrus[t] | 1 (4) | 1 (3) | NR (0) | NR (0) | NR(0) | NR(0) | NR(0) | (1) |
| Clymenia[p] | 3 (8) | 1 (1) | NR (0) | NR (0) | NR(0) | 2(1)(d1) | (6) | (0) |
| Microcitrus[p,u] | 10 | 5 | NR | NR | NR | 5(d1) | | |
| Citrus[s] | 58 | 11 | 3 | 21 | 7 | 8 | 7 | (6) |
| Poncirus[s] | 4 | 3 | NR | 1 | NR | 0 | 0 | |

|  | total flavonoids | flavanone glycosides | flavanone aglycones | methoxylated flavones (b1) | prenylated flavones | O-glycosyl flavones | C-glycosyl flavones | flavonol glycosides |
|---|---|---|---|---|---|---|---|---|
| Subtribe 3. **Balsamocitrinae**: (hard-shelled citroid fruit trees) | | | | | | | | |
| *Swinglea* | NR (6) | (0) | (0) | (0) | (0) | (2) | (4) | (2) |
| *Aegle*[v,w] | 1 (5) | (0) | (0) | (0) | (0) | (1) | (2) | 1(0) |
| *Afaegle* | NR (5) | (0) | (0) | (0) | (1) | (0) | (1) | (3) |
| *Aeglopsis* | NR (4) | (0) | (0) | (0) | (0) | (1) | (0) | (3) |
| *Balsamocitrus* | NR (3) | (0) | (0) | (0) | (0) | (2) | (0) | (1) |
| *Feronia*[x,y] | 4 (8) | (0) | (0) | (1) | (0) | (0) | 4(6) | (1) |
| *Feroniella* | NR (8) | (0) | (0) | (0) | (0) | (1) | (7) | (0) |

a1. isoprenylated flavanone aglycones
b1. listed as flavone aglycones by Grieve and Scora (1980)
c1. isoprenylated flavone
g1. combined flavone and flavonol glycoside
a. Bowen and Perera (1982)
b. Chang and Wu (1995)
c. Huang et al. (1993)
d. Lin et al. (1975)
e. Dreyer (1968)
f. Wu et al. (1994)
g. Chowdhury and Chakraborty (1971)
h. Joshi and Kamat (1969)
I. Wu et al. (1980)
j. Kinoshita and Firman (1996)
k. Desoky (1992)
l. Bishay et al. (1987)
m. Fraser and Lewis (1974)
n. Kong et al. (1987)
o. Dreyer and Park (1975)
p. Mark Berhow (1997)
q. Fukai and Nomura (1993)
r. Banerji et al. (1988)
s. Horowitz and Gentili (1977)
t. Lastinger et al. (1979)
u. Kanes et al. (1993)
v. Karawya et al. (1982)
w. Sharma et al. (1980)
x. El Fishawy (1994)
y. Gupta et al. (1979)

It has been apparent from much of this research that one of the main distinguishing features of the citrus flavonoids is the diversity in the glycosidation of these compounds. Two of the most common glycosidation patterns are the flavone and flavanone 7-O-rutinosides (6-O-α-L-rhamnopyranosyl-β-D-glucopyranose) and 7-O-neohesperidosides (2-O-α-L-rhamnopyranosyl-β-D-glucopyranose) (Figure 1) (Horowitz and Gentili, 1963). Among the main flavanone glycosides in citrus, i.e. hesperidin, narirutin, and naringin (Figure 2), the 4'-O-glucosyl triglycosides are also found to occur. In addition to these, two other unusual triglycosides, naringenin 7-O-{[α-rhamnosyl(1–2)]-[α-rhamnosyl(1–6)]-β-glucose}, and hesperetin 7-O- {[α-rhamnosyl(1–2)]-[α-rhamnosyl(1–6)]-β-glucose}(Figure 3) have been reported (Kumamoto et al., 1985c). In citrus the flavone glycosides also typically occur as the 7-O-rutinosides and 7-O-neohesperidosides. For the flavonols, where the 3-hydroxy group is present, these compounds typically occur as the triglycosides, with O-glucose at the C-3 position. Rhamnosides are also common at the C-3 position, as found in rutin.

In many citrus species, there are also typically low concentrations of a number of C-glycosylflavones. However, in certain citrus relatives, i.e. species of *Glycosmis*, *Murraya*, and *Pamburus*, these compounds occur at much higher concentrations, and in fact, are the main flavonoids present in the leaf tissue (Table 1). The C-glycosylflavones occur mainly as C-glucosides (Figure 4), but in the genus *Fortunella*, such compounds also exist as C-neohesperidosides (Horowitz and Gentili, 1977; Kumamoto et al., 1985a).

In addition to the above mentioned flavone and flavanone glycosides, a small number of flavonoid esters have been reported in *Citrus* (Kumamoto et al., 1985b; Berhow et al., 1994; Sugiyama et al., 1993). In a number of species of the Aurantioideae, prenylated flavonoids (Figure 5) have also been reported (Table 1). These compounds appear in several of the different citrus subtribes, and have been recently reported in the root bark of several species of *Citrus* (Wu et al., 1988; Ito et al., 1989; Wu, 1989; Chang, 1990).

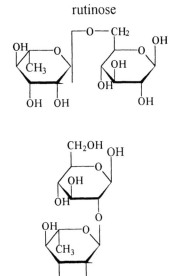

Figure 1. Structures of rutinose and neohesperidose.

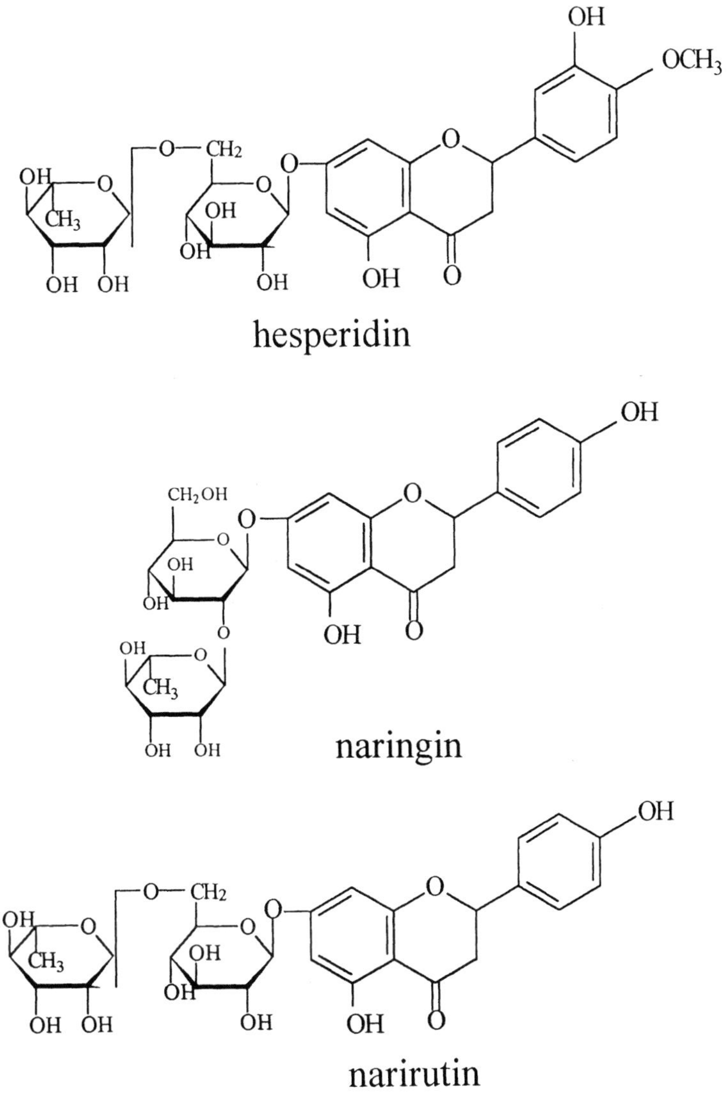

Figure 2. Structures of hesperidin, naringin, and narirutin.

Finally, while high concentrations of flavanone glycosides are typically found in citrus plant tissue, only very low levels of flavanone aglycones are ever detected. The three known flavanone aglycones thus far reported in citrus (citromitin, 5-$O$-desmethylcitromitin, and 6-$O$-desmethylcitromitin) occur in mandarin (*C. reticulata* Blanco), and are variably methoxylated at the 5,6,7,8,3' and 4' positions (Chahila et al., 1967; Chen et al., 1997). In contrast to the very low levels of these flavanone aglycones, the corresponding polymethoxylated flavone aglycones occur in much higher numbers and concentrations, and are one of the main distinctive classes of citrus flavonoids (Horowitz, 1961; Horowitz and Gentilli, 1977), especially in the Citrinae and Clauseneae subtribes. There are at least 53 known polymethoxylated compounds thus far characterized in citrus. These compounds typically demonstrate a number of interesting biological activities in living sys-

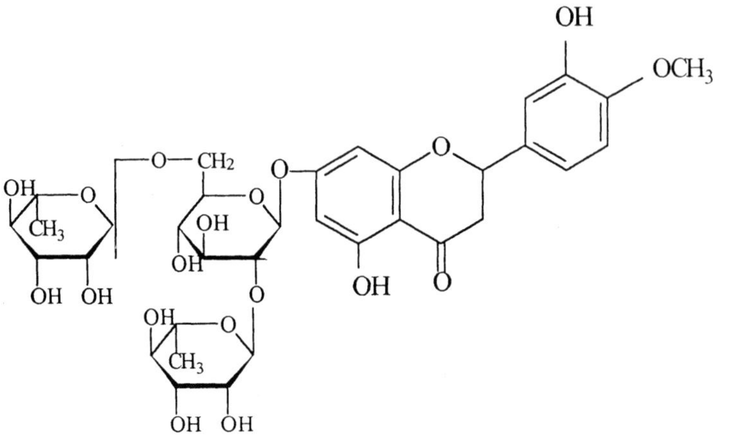

**hesperetin triglycoside**

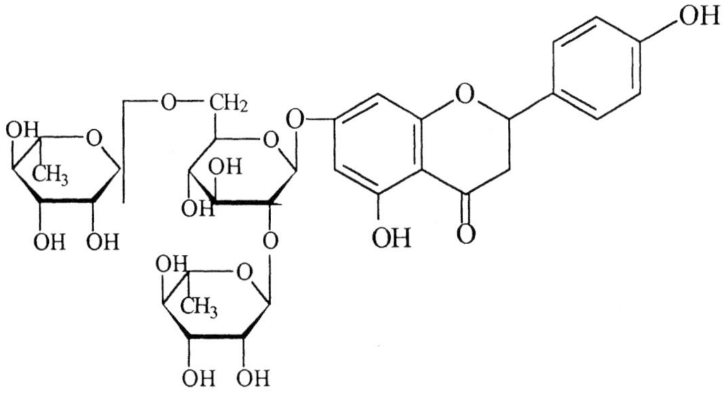

**naringenin triglycoside**

**Figure 3.** Structures of two hesperetin and naringenin triglycosides in *Citrus*.

tems, and are the focus of much of the current work on the pharmacological properties of citrus flavonoids. In lacking the glycosidic moities, these planar, less polar compounds are more likely to be able to interact with and penetrate cell membranes and thus have access to key enzyme systems. Recent reports show the uncoupling of mitochondrial respiration by a number of naturally-occurring flavonoids (Ravanel, 1986; Creuzet et al., 1988; Wagner and Brederode, 1996).

## 2.2. Chemotaxonomy

In an extensive survey of the flavonoid glycosides in citrus, primarily within the subtribe Citrinae, Kanes et al. (1993) observed that most *Citrus* species could be classified as either predominantly rutinosides or neohesperidosides at the C-7 position. Most notable was the predominance of the flavanone and flavone 7-*O*-neohesperidosides in pummelo (*C. grandis* (L.) Osb.), whereas in mandarin (*C. reticulata*) most flavonoids occurred as 7-*O*-ru-

**3,8-di-*C*-glucosyldiosmetin**

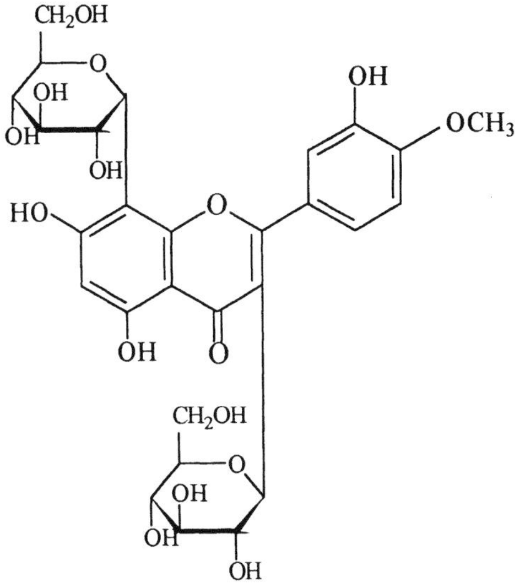

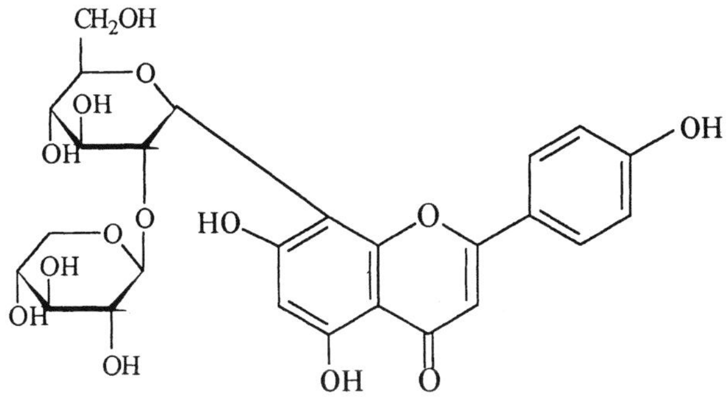

**2″-xylosylvitexin**

Figure 4. Examples of flavone *C*-glycosides in citrus.

tinosides. It was further observed that citron (*C. medica* L.) contained an extremely simple flavonoid pattern in comparison to many of the other citrus species. Hybrid species were shown to typically contain mixed flavonoid patterns of these 3 original *Citrus* species. These findings, along with other observations made in this work, supported the premise that the species, *C. medica*, *C. grandis*, and *C. reticulata* are likely the originial species responsible for many of the current *Citrus* species and cultivars (Swingle and Reece, 1967; Barrett and Rhodes, 1976). These findings were similar to those made in two previous studies of the flavonoid patterns in *Citrus* (Albach and Redman, 1969; Nishiura et al. 1971).

Due to the diversity and high concentrations of flavonoids typically found in *Citrus* species, and the current emphasis on new drug discovery, it seems that it would be of great taxonomic interest to consider the phylogenetic distributions of the different classes of flavonoids within the subtribe Citrinae and elsewhere within the Aurantioideae. Although flavonoids in the cultivated species of *Citrus* have been extensively characterized, relatively

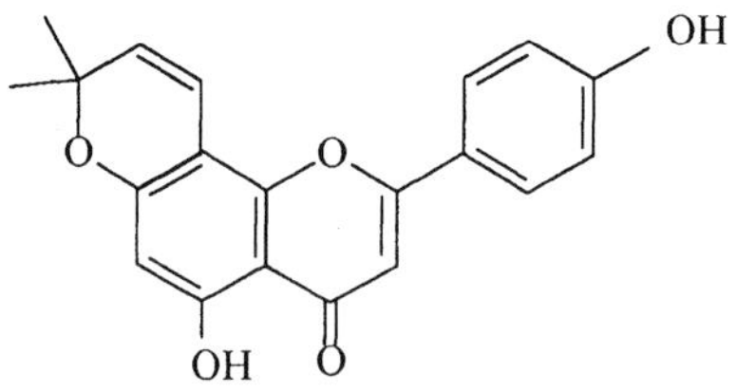

atalantoflavone

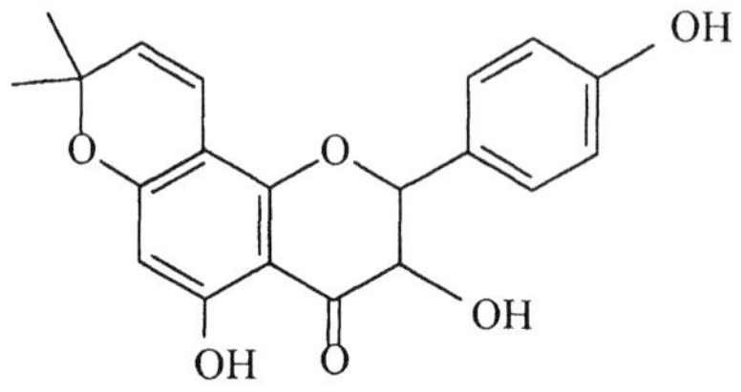

yukovanol

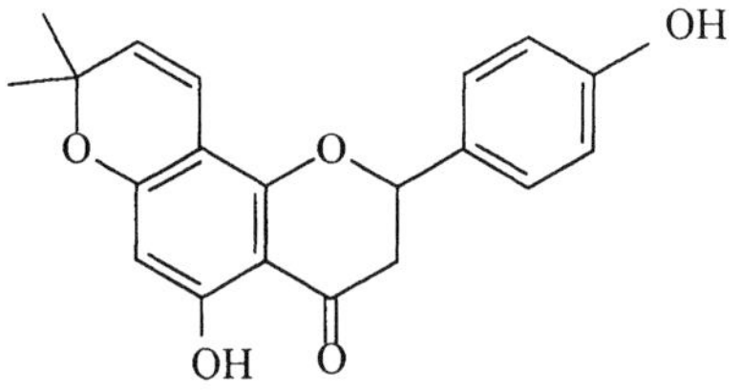

citflavanone

**Figure 5.** Examples of prenylated flavonoids in citrus.

few such studies have been done for other genera within the Aurantioideae (Table 1). The most notable exceptions to this are the genera of the 3 subtribes of the Clauseneae, in which a large number of flavonoids, mainly polymethoxylated flavones, have been reported. As shown in Table 1, very few flavonoid studies have been done on the genera in the Triphasiinae and Balsamocitrinae subtribes of the Citreae tribe. Nearly all of what we know of these citrus relatives comes from the initial screen of the Aurantioideae by Grieve and Scora (1980). From their analysis, it was generally observed that the genera within the Triphasiinae contain relatively few flavonoids, and specifically, in the species *Paramignya monophylla*, no flavonoids were detected. In the genera *Triphasia* and *Pamburus*, only apigenin-*O*- and *C*-glycosides were detected. In the subtribe Clauseninae there was found an abundance of *C*-glycosyl flavone derivatives of apigenin and luteolin, along with fewer numbers of flavonol glycosides. In spite of the large numbers of *C*-glycosylflavones, no corresponding *O*-glycosides were detected. Among the genera of the subtribe Balsamocitrinae, only the *Balsamocitrus* contained detectable levels of flavone glycosides, whereas among all the genera of the Balsamocitrinae the main flavonoids were various flavonol glycosides of kaempferol, quercetin, and isorhamnetin (Figure 6).

Interestingly, the flavanone glycosides were found in this study by Grieve and Scora to occur almost exclusively in the Citrinae subtribe, and among this subtribe, these com-

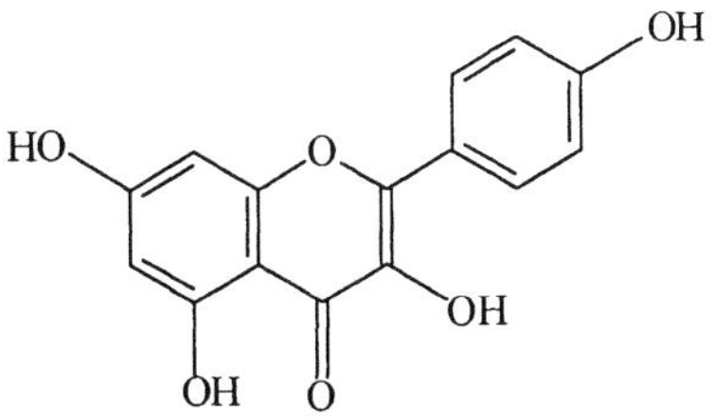


kaempferol

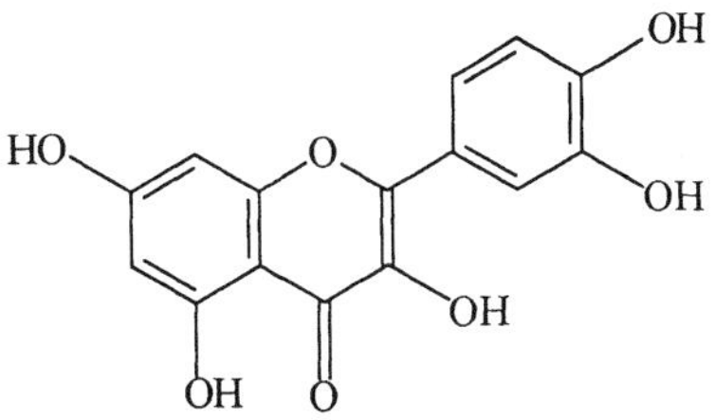


quercetin

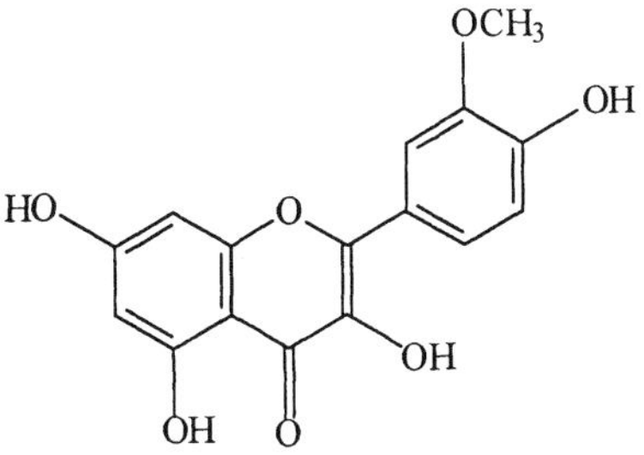


isorhamnetin

**Figure 6.** Structures of citrus flavonols.

pounds were primarily restricted to the subgroup of true citrus trees comprising of the genera: *Fortunella*, *Eremocitrus*, *Poncirus*, *Clymenia*, *Microcitrus*, and *Citrus*. As shown in Table 1, however, other flavanones, primarily prenylated flavanones, have been subquently detected in a few other species of remote citrus relatives.

In our own analysis of the occurrence of flavanone glycosides in several citrus relatives, we made use of the separation and detection capabilities of HPLC coupled with photodiode array detection. One result of this analysis was the detection of a number of compounds with retention times and UV spectra of flavanone glycosides in several species in which these compounds were not previously reported. In the leaf extracts of *Swinglea glutinosa* of the Balsamocitrinae, at least 5 compounds with very early elution times were detected with ultraviolet spectra closely similar to flavanones (Figure 7A). *Swinglea* was of further interest because of the presence of at least 10 compounds with unusual UV spectra suggestive of flavanone-like structures (Figure 7B). Nothing is known of these compounds, and thus far, these compounds do not appear to have been detected in any of the other citrus relatives. Similarly, in analyzing young immature fruit of *Afraegle paniculata*, also of the Balsamocitrinae, several very early eluting compounds with flavanone-

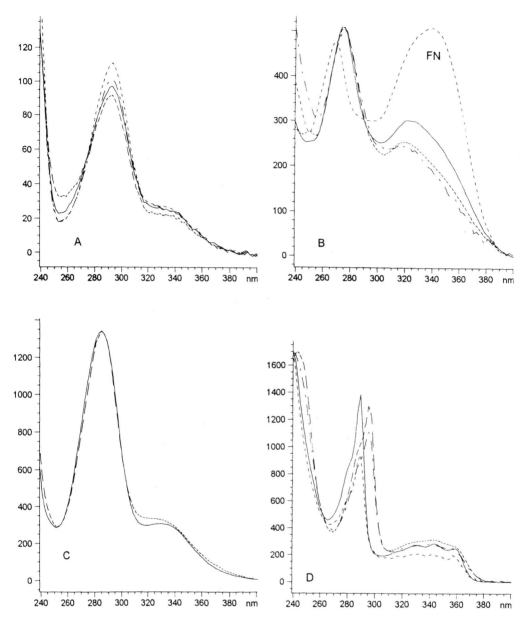

**Figure 7.** UV spectra of flavonoids in citrus relatives. A: Four spectra of early-eluting compounds in *Swinglea glutinosa* with flavanone-like UV spectra. B: Unusual spectra of three late-eluting compounds in *Swinglea glutinosa*. The spectrum labeled FN contrasts these spectra with the spectrum of a typical flavone(glycoside) in citrus leaf tissue. C: Flavanone-like UV spectra of compounds in immature *Afraegle paniculata* fruit. D: Unusual UV spectra of four late-eluting compounds in leaves of *Murraya koenigii*.

like UV spectra were detected (Figure 7C). These compounds were not detected in the mature leaf tissue. Two species in the Citrinae, *Atalantia* sp. and *Severinia buxifolia* were also observed to contain flavanone glycosides. This suggests a wider occurrence of flavanones in the Citrinae than previously considered. Finally, a set of compounds with novel UV spectra were shown to occur in the leaf extracts of *Murraya koenigii* (Figure

7D). Like the unusual compounds mentioned for *Swinglea glutinosa*, these compounds appear to represent a second new class of putative phenolic compounds that have yet to be characterized within the Aurantioideae. These compounds serve as excellent examples of the remarkable structural diversity that we continue to find associated with the flavonoids within the Aurantioideae orange subfamily.

## 3. FLAVONOID BIOSYNTHESIS

Because of this diversity of the several classes of flavonoids in citrus, there has been considerable interest in the study of the biosynthesis as well as the pharmacological properties of the flavonoids found in citrus leaf and fruit tissue. Much of the research on the biosynthesis of these compounds has centered around the characterization of the glycosidation pathways (Lewinsohn et al., 1986, 1989; Berhow and Vandercook, 1989; McIntosh and Mansell, 1990; McIntosh et al., 1990; Bar-Peled et al, 1991, 1993; Castillo et al., 1993), that lead to the high concentrations of flavanone neohesperidosides and rutinosides in the cultivated *Citrus* species (Jourdan et al., 1985; Castillo et al., 1992). Far less is known about the biosynthesis of the other classes of flavonoids, especially those in the citrus relatives. Characterization of the glycosidation reactions is of particular importance in ultimately understanding the regulation of not only the biosynthesis, but also the transport and storage of these compounds in citrus plant tissue.

In citrus, as in all plants, the initial steps in flavonoid biosynthesis involve the biosynthesis of naringenin (Figure 8) (Maier and Hasegawa, 1970). The failure to detect in citrus plant tissue the free hesperetin or naringenin aglycones, or the intermediate 7-$O$-glucosides, led early researchers to suggest that the hydroxylation and methoxylation of naringenin, leading to the production of hesperetin, occurs after the glycosidation reactions (Raymond and Maier, 1977). More recent studies, however, provide evidence supporting the pathways shown in Figure 8, where glycosidation occurs after the hydroxylation and methoxylation reactions, suggesting that hesperetin 7-$O$-glucoside and naringenin 7-$O$-glucoside are direct precursors of hesperidin and naringin (Lewinsohn et al., 1989; Bar-Peled et al., 1991; Castillo et al., 1993). The glycosyltransferases responsible for these reactions are highly specific for the flavanones (i.e. hesperetin and naringenin) and show negligible activities towards the corresponding flavones (diosmetin and apigenin) (McIntosh et al., 1990; McIntosh and Mansell, 1990; Benavente-Garcia et al., 1993). These findings, along with the fact that the highest rate of neodiosmin accumulation in sour orange (*C. aurantium*) occurs after the accumulation of naringin (Benavente-Garcia et al., 1993), support the direct biosynthesis of flavone glycosides from the flavanone glycosides (i.e. diosmin from hesperidin as shown in Figure 8).

Recent evidence further supports the role of hesperidin as the substrate in subsequent diosmin biosynthesis, and similarly, the role of naringin as the substrate in rhoifolin biosynthesis. In studies of zinc-deficient citrus leaf tissue, it was noted that significantly higher flavonoid levels occurred in zinc-deficient orange and grapefruit leaf tissue. These findings were particularly pertinent in understanding the physiological changes that accompany the zinc-deficiency induced in citrus trees with blight (Wutscher et al. 1977). In zinc-deficient 'Valencia' orange leaves parallel increases occurred in the concentrations of hesperidin, diosmin, and an unknown flavone with a diosmin-like ultraviolet spectrum (Table 2). Interestingly, the apigenin-derived flavones, (i.e. isorhoifolin, and 5 unknown flavones with ultraviolet spectra similar to apigenin) showed no increases. One other diosmin-like flavone also showed no increase. These observations suggest that

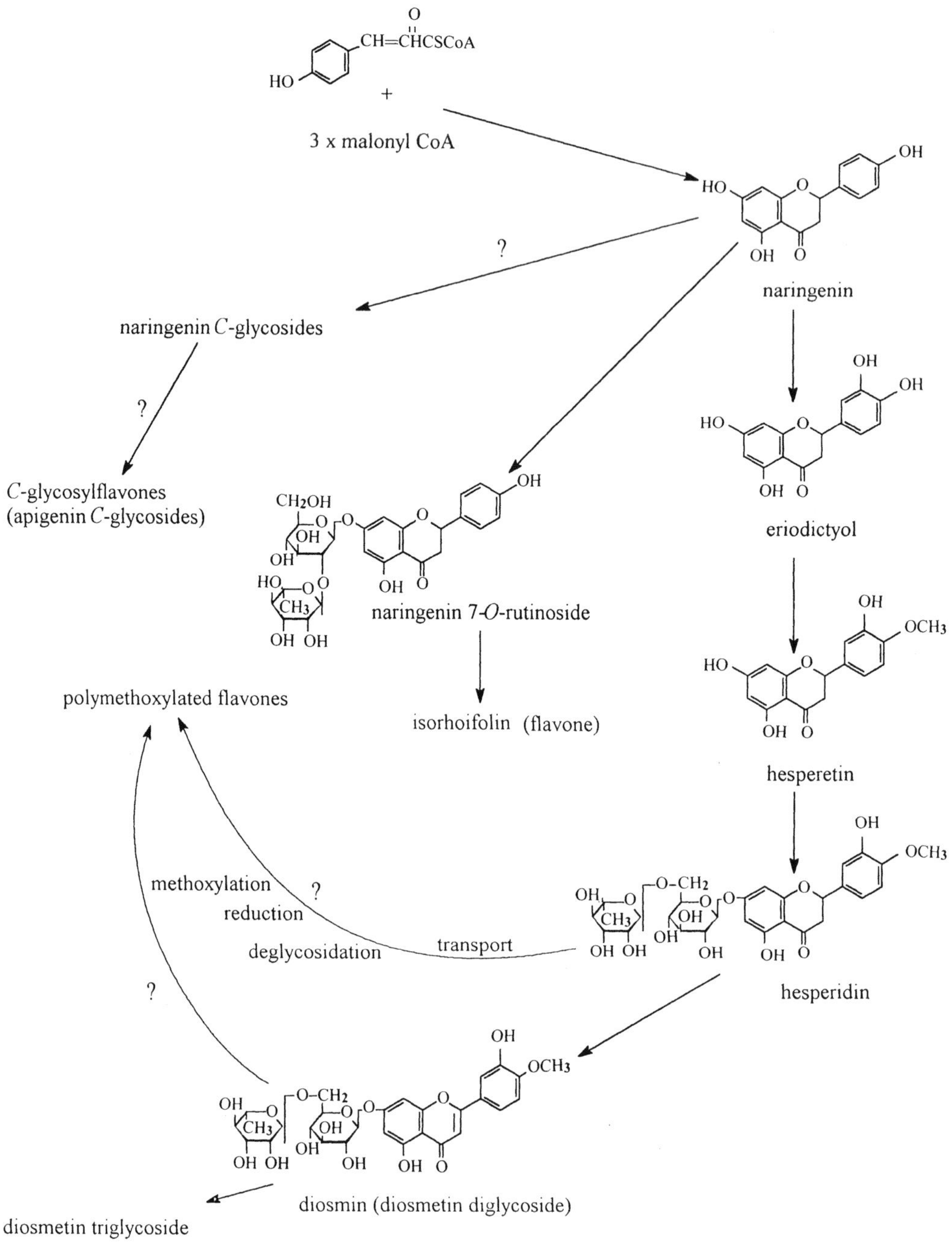

**Figure 8.** Proposed flavonoid biosynthetic pathways in sweet orange.

in this instance, the regulation of the biosynthesis of the flavones with the apigenin structure occurs distinct from the pathway leading to the biosynthesis of the diosmetin glycosides. It is also very likely that at least several of the apigenin compounds are C-glycosyl derivatives known to occur in orange (Horowitz and Gentili, 1977). The presence of these compounds raises other very interesting questions concerning the biosynthe-

Table 2. Concentrations (ppm) of flavonoids in orange leaves affected by zinc-deficiency. Leaf samples (20 leaves per group) were collected with no (control), moderate, and severe visual zinc-deficiency symptoms. Leaves were acid washed, dried, and analyzed for zinc content. Flavonoids were extracted from dried leaf samples as described by Kanes et al. (1993). Analyses were run in triplicate. Unknown compounds UK1 and UK2 both have UV spectra similar to apigenin, and are assumed to represent apigenin glycosides. The designation N/HEPTA signifies nobiletin and heptamethoxyflavone

| Flavonoid | Control 11 ppm Zn | Moderate 4 ppm Zn | Severe 3 ppm Zn |
|---|---|---|---|
| Hesperidin | 8256±225 | 15541±264 | 18775±1012 |
| UK1 flavone | 492±54 | 1024±194 | 1759±92 |
| Isorhoifolin | 786±33 | 1293±47 | 1248±93 |
| UK2 flavone | 815±61 | 1117±31 | 1043±65 |
| Diosmin | 1324±53 | 3013±54 | 3591±67 |
| Sinensetin | 286±12 | 237±86 | 167±38 |
| N/hepta | 512±14 | 590±79 | 470±19 |
| Tangeritin | 202±12 | 175±19 | 148±16 |

sis of such $C$-glycosylflavones, of which nothing has been studied in citrus. Glycosidation is considered to be important in the storage, transport, and solubility of these flavonoids within the plant tissue. It would be interesting to know what roles the two sets of glycosylflavones (i.e. $O$-glycosides and $C$-glycosides) play in citrus tissue, and how the sites of synthesis and storage of these compounds compare.

Similar studies of blight-induced zinc-deficiency in grapefruit leaves showed sharp increases in both naringin and rhoifolin levels (data not shown). In blighted, zinc-deficient grapefruit leaves, both compounds occurred at slightly over twice the levels found in healthy leaves, and these parallel increases support the direct synthesis of rhoifolin from naringin. With the exception of a few of the additional phenols measured, the concentrations of most of the other flavones and related coumarins and psoralens were also increased in the zinc-deficient leaf tissue. In this way, the changes in the flavonoids in grapefruit differ from what was found to occur in the orange leaf tissue.

Finally, the last set of flavonoids in citrus, the polymethoxylated flavones, remain the least understood relative to their biosynthesis. Very little is known of their sites of biosynthesis and storage. While it is speculated that in citrus fruit tissue, the polymethoxylated flavones are components of the oil glands, little else is known of their sites of biosynthesis and storage. Nothing is known of the link in the biosynthesis of the polymethoxylated flavones and the extremely rapid flavanone glycoside biosynthesis that occurs during early tissue development. In light of the rapidly growing interest in the pharmacological properties of citrus polymethoxylated flavones, the biosynthesis of these, as well as the other classes of citrus flavonoids, will attract a great deal of research in the future.

# REFERENCES

Albach, R.F; Redman,G.H. Composition and inheritance of flavanones in citrus fruit. Phytochem. **1969**, *8*, 127–143.

Armentano, L.; Bentsath, A; Beres, T; Rusznyák, S; Szent-Györgyi, A. Über den einfluss von substanzen der flavorgruppe auf die permeabilitat der kapillaren, Vitamin P. Deut. Med. Worchshr. **1936**, *62*, 1326–1328.

Attaway, J.A.; Buslig, B.S. Antithrombogenic and antiatherogenic effects of citrus flavonoids: Contributions of Ralph C. Robbins. In: *Flavonoids in the Living System*. Manthey, J.A. and B.S. Buslig, Eds.; Plenum Press: New York, 1998.

Bar-Peled, M.; Rluhr, R.; Gressel, J. Juvenile-specific localization and accumulation of a rhamnosyltransferase and its bitter flavonoid in foliage, flowers, and young citrus fruits. Plant Physiol. **1993**, *103*, 1377–1384.

Bar-Peled, M.; Lewinsohn, E.; Fluhr, R.; Gressel, J. UDP-rhamnose:flavanone-7-*O*-glucoside-2"-*O*-rhamnosyltransferase. Purification and characterization of an enzyme catalyzing the production of bitter compounds in *Citrus*. J. Biol. Chem. **1991**, *166*, 20953–20959.

Barrett, H.C.; Rhodes, A.M. Numerical taxonomic study of affinity relationships in cultivated *Citrus* and its close relatives. Systematic Botany **1976**, *1*, 105–136.

Benavente-Garcia, O.; J. Castillo, J.; del Rio, J.A. Changes in neodiosmin levels during the development of *Citrus aurantium* leaves and fruits. Postulation of a neodiosmin biosynthetic pathway. J. Agric. Food Chem. **1993**, *41*, 1916–1919.

Bentsath, A.; Ruszynák, I.; Szent-Györgyi, A. Vitamin P. Nature. **1937**, *139*, 326–7.

Berhow, M.A.; Bennett, R.D.; Kanes, K.; Poling, S.M.; Vandercook, C.E. A malonic acid ester derivative of naringin in grapefruit. Phytochem. **1994**, *30*, 4198–4200.

Berhow, M.A; Vandercook, C.E. Biosynthesis of naringin and prunin in detached grapefruit. Phytochem. **1989**, *28*, 1627–1630.

Castillo, J.; Benavente, O.; del Rio, J.A. Naringin and neohesperidin levels during development of leaves, flower buds, and fruits of *Citrus aurantium*. Plant Physiol. **1992**, *99*, 67–73.

Castillo, J.; Benavente, O.; del Rio, J.A. Hesperetin 7-*O*-glucoside and prunin in *Citrus* species (*C. aurantium* and *C. paradisi*). A study of their quantitative distribution in immature fruits and as immediate precursors of neohesperidin and naringin in *C. aurantium*. J. Agric. Food Chem. **1993**, *41*, 1920–1924.

Chahila, B.P.; Sastry, G.P.; Rao, P.R. Chemical investigation of *Citrus reticulata* Blanco. Indian J. Chem. **1967**, *5*, 239–241.

Chang, S.-H. Flavonoids, coumarins, and acridone alkaloids from the root bark of *Citrus limonia*. Phytochem. **1990**, *29*, 351–353.

Chen, J.; Montanari, A.M.; Widmer, W.W. Two new polymethoxylated flavones, a class of compounds with potential anticancer activity, isolated from cold pressed Dancy tangerine peel oil solids. J. Agric. Food Chem. **1997**, *45*, 364–368.

Creuzet, S.; Ravanel, P.; Tissut, M.; Kaouadji, M. Uncoupling properties of three flavonols from plane-tree buds. Phytochem. **1988**, *27*, 3093–3099.

Engler, A. Rutaceae. In: *Die Natulichen Pflanzenfamilien*, Second Edition. Vol. 19a; Engler, A., Prantl, K., Eds; Engelmann: Leipzig, **1931**; pp 316–359.

Erickson, L.C. The general physiology of citrus. In: *The Citrus Industry*, vol 2, Reuther, W., Batchelor, L.D., Webber, H.J. Eds.; University of California Press: Berkeley, **1968**; pp 86–126.

Gerritsen, M.E. Flavonoids inhibit cytokine-induced endothelial cell adhesion protein gene expression. In: *Flavonoids in the Living System*, Manthey, J.A., Buslig, B.S. Eds.; Plenum Press: New York, 1998.

Grieve, C. M.; Scora, R.W. Flavonoid distribution in the Aurantioideae (Rutaceae). Systematic Botany **1980**, *5*, 39–53.

Horowitz, R.M. The citrus flavonoids. In *The Orange*, Sinclair, W.B. Ed; Univ.of California, Div. Agric. Sci: Berkeley, 1961; pp 334–372.

Horowitz, R.M.; Gentili, B. Flavonoids of citrus. VI. The structure of neohesperidose. Tetrahedron **1963**, *19*, 773–782.

Horowitz, R.M; Gentili, B. Flavonoid constituents of *Citrus*. In *Citrus Science and Technology, Vol. 1*., Nagy, S., Shaw, P.E., Veldhuis, M.K. Eds.; Avi Publishing Company, Inc: Westport, Connecticut, 1977; pp 397–426.

Ito, C.; Sato, K.; Oka, T.; Inoue, M.; Ju-Ichi, M.; Omura, M.; Fukukawa, H. Two flavanones from *Citrus* species. Phytochem. **1989**, *28*, 3562–3564.

Jourdan, P.S.; McIntosh, C.A.; Mansell R.L. Naringin levels in *Citrus* tissues. II. Quantitative distribution of naringin in *Citrus paradisi* Macfad. Plant Physiol. **1985**, *77*, 903–908.

Kanes, K.; Tisserat, B.; Berhow, M.; Vandercook, C. Phenolic composition of various tissues of Rutaceae species. Phytochem. **1993**, *32*, 967–974.

Kumamoto, H.; Matsubara, Y.; Iizuka, Y.; Okamoto, K.; Yokoi, K. Structure and hypotensive effect of flavonoid glycosides in Kinkan (*Fortunella japonica*) peelings. Agric. Biol. Chem. **1985a**, *49*, 2613–2618.

Kumamoto, H.; Matsubara, Y.; Iizuka, Y.; Okamoto, K.; Yokoi. K. Structure and hypotensive effect of flavonoid glycosides in Sudachi peelings II. Agric. Biol. Chem. **1985b**, *49*, 2797–2798.

Kumamoto, H.; Matsubara, Y.; Iizuka, Y.; Okamoto, K.; Yokoi. K. Stucture and hypotensive effect of flavonoid glycosides in Yuzu (*Citrus junos* Sieb.) peelings. Nippon Nogeikagaku Kaishi, **1985c**, *59*, 683–687.

Kumamoto, H.; Matsubara, Y.; Iizuka, Y.; Okamoto, K.; Yokoi. K. 1985d. Structure and hypotensive effect of flavonoid glycosides in lemon peelings (part II). Nippon Nogeikagaku Kaishi, **1985d**, *59*, 677–682.

Lebreton, P. J. Pharm. Chim. Paris, 1828, 14, 377.

Lewinsohn, E.; Berman, E.; Mazur, Y.; Gressel, J. 1986. Glycosidation of exogenous flavanones by grapefruit (*Citrus paradisi*) cell cultures. Phytochem. **1986**, 25:2531–2535.

Lewinsohn, E.; Britsch, L.; Mazur, Y.; Gressel, J. Flavanone glucoside biosynthesis in *Citrus*. Plant Physiol. **1989**, *91*, 1323–1328.

Maier, V.P.; Hasegawa, S. L-Phenylalanine ammonia-lyase activity and naringenin glycoside accumulation in developing grapefruit. Phytochem. **1970**, *9*, 139–144.

Matsubara, Y.; Kumamoto, H.; Iizuka, Y.; Murakami, T.; Okamoto, K.; Miyake, H.; Yokoi, K. Structure and hypotensive effect of flavonoid glycosides in *Citrus unshiu* peelings. Agric. Biol. Chem. **1985**, *49*, 909–914.

McIntosh, C.A; Mansell, R.L. Biosynthesis of naringin in *Citrus paradisi*: UDP-glucosyltransferase activity in grapefruit seedlings. Phytochem. **1990**, *29*, 1533–1538.

McIntosh, C.A.; Latchinian, L.; Mansell, R.L. Flavanone specific 7-*O*-glucosyltransferase activity in *Citrus paradisi* seedlings. Purification and characterization. Arch. Biochem. Biophys. **1990**, *282*, 50–57.

Middleton, E. Jr.; Kandaswami, C. The effect of plant flavonoids on immune and inflammatory cell function. In: *Flavonoids in the Living System*. Manthey, J.A., Buslig, B.S. Eds; Plenum Press: New York; 1998.

Nishiura, M.; Kamiya, K.; Esaki, S. Flavonoids in citrus and related genera. Part III. Flavonoid pattern and citrus taxonomy. Agric. Biol. Chem. **1971**, *35*, 1691–1706.

Ravanel, P. Uncoupling activity of a series of flavones and flavonols on isolated plant mitochondria. Phytochem. **1986**, *25*, 1015–1020.

Raymond, R.D.; Maier, V.P. Chalcone cyclase and flavonoid biosynthesis in grapefruit. Phytochem., **1977**, *16*, 1535–1539.

Rusznyák, S.; Szent-Györgi, A. Vitamin P: flavonols as vitamins. Nature **1936**, *138*, 27.

Schneider, H. The anatomy of citrus. In: *The Citrus Industry, vol 2*, Reuther, W., Batchelor, L.D., Webber, H.J. Eds.; University of California Press: Berkeley, 1968; pp 1–85.

Sugiyama, S.; Umehara, K.; Kuroyanagi, M.; Ueno, A.; Taki, T. Studies on the differentiation inducers of myeloid leukemic cells from *Citrus* species. Chem. Pharm. Bull. **1993**, *41*, 714–719.

Swingle, W.T. The botany of citrus and its wild relatives of the orange subfamily (Family Rutaceae, Subfamily Aurantioideae) In: *The Citrus Industry, vol 1*, Webber H.J. Batchelor, L.D., Eds.; University of California Press: Berkeley, 1943; pp 129–474.

Swingle, W.T.; Reece, P.C. The botany of citrus and its wild relatives. In *The Citrus Industry, vol. 1*, Reuther, W., Webber, H.J. Batchelor, L.D. Eds.; University of California Press: Berkely, 1967; pp190–430.

Tanaka, T. Revision Aurantiocearum I. Mem. Tanaka Citrus Exper. Sta. **1932**, *1*, 39–66.

Wagner, A.M.; van Brederode, J. Inhibition of mitochondrial respiration by the flavone aglycone isovitexin causes aberrant petal and leaf morphology in *Silene latifolia*. Plant Cell Rep. **1996**, *15*, 718–722.

Webber, H.J. History and development of the citrus industry. In *The Citrus Industry vol. 1*. Reuther, W., Webber, H.J. Batchelor, L.D. Eds.; University of California Press: Berkely, 1967; pp 1–39.

Wu, T.-S. Flavonoids from root bark of *Citrus sinensis* and *C. nobilis*. Phytochem. **1989**, *28*, 3558–3560.

Wu, T.-S.; Huang, S.-C.; Jong, T.-T.; Lai, J.-S.; Kuoh, C.-S. Coumarins, acridone alkaloids and a flavone from *Citrus grandis*. Phytochem. **1988**, *27*, 585–587.

Wutscher, H.K.; Cohen, M.; Young, R.H. Zinc and water soluble phenol levels in the wood for the diagnosis of citrus blight. Plant Dis. Rep.**1977**, *61*, 572–576.

# 8

# CITRUS FLAVONOIDS: A REVIEW OF PAST BIOLOGICAL ACTIVITY AGAINST DISEASE

## Discovery of New Flavonoids from Dancy Tangerine Cold Pressed Peel Oil Solids and Leaves

Antonio Montanari, Jie Chen, and Wilbur Widmer

State of Florida, Department of Citrus, Citrus Research and Education Center
700 Experiment Station Road
Lake Alfred, Florida, 33850

## 1. INTRODUCTION AND REVIEW OF THE BIOLOGICAL ACTIVITY OF CITRUS FLAVANOIDS

### 1.1. Flavanone Glycosides

Flavonoids are common in fruits and vegetables and many function as natural antioxidants, free radical scavengers, and may also chelate pro-oxidant metals, reducing their capacity to produce free radicals (Affany et al., 1987). It has been estimated that humans consume 1 gram per day of flavonoids (Kuhnau et al., 1976). The majority of citrus flavonoids are flavanones bound as glycosides (Figure 1). Hesperidin, a glycoside of hesperetin, is predominant in orange and tangerine, whereas naringin and narirutin, glycosides of naringenin, are most abundant in grapefruit. Eriocitrin and hesperidin are the predominate flavonoids found in lemon and lime. Some therapeutic effects have been reported for citrus flavanoids (Fisher et al., 1982). In 1936 citrus flavonoids were reported to decrease capillary fragility and improve blood flow and were actually labeled "Vitamin P" (Rusznyak and Szent-Györgyi, 1936). But further research failed to substantiate that abnormal capillary permeability was due to a deficiency in bioflavonoids. In 1950, the Federation of American Societies for Experimental Biology recommended that the term "Vitamin P", which had been applied to flavanoids, be discontinued (Vickery et al., 1950).

Early studies used crude mixtures of citrus flavonoids, and this was the responsible factor for the inconsistencies in biological experimental results. 'Bioflavonoid' refers to crude extracts of several citrus varieties including lemon, orange, mandarin or grapefruit.

*Flavonoids in the Living System*, edited by Manthey and Buslig
Plenum Press, New York, 1998.

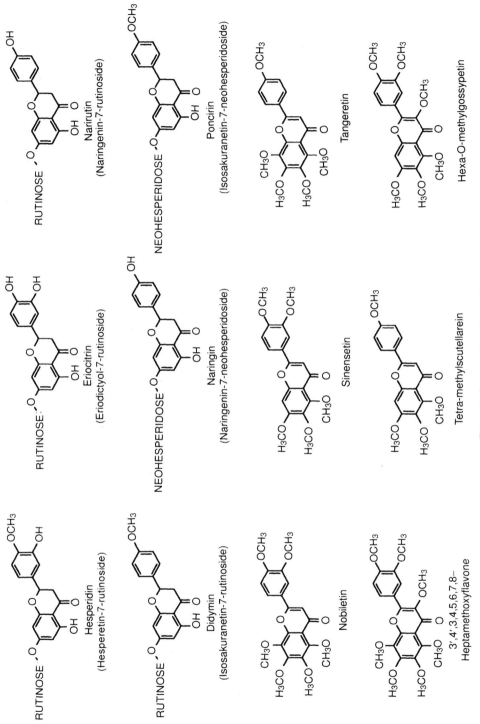

Figure 1. Common citrus flavonoids.

Polymethoxylated flavones (Figure 1) are found as minor components of crude extracts from the peel of orange and mandarin varieties. Grapefruit, lemon and lime peel extracts contain very few polymethoxylated flavones but may contain other flavones such as rutin and diosmin (Figure 1). 'Bioflavonoids' extracted from citrus by definition contains a wide variety of flavonoids in them.

Recent research has focused on the biological activities of purified citrus flavonoids. Hesperetin was found to actively inhibit the replication but not the infectivity of herpes simplex, polio and parainfluenza type viruses (Kaul et al., 1985). Naringin affected red blood cell aggregation *in vitro*. Ingested grapefruit juice was also found to have a beneficial affect in human subjects, lowering elevated hemocrits and regulating blood viscosity *in vivo* (Robbins et al., 1988). These findings suggested naringin was the active component in the grapefruit juice.

It has recently been discovered in several human feeding studies that grapefruit juice, with the suggestion that naringin is the causative agent, has the ability to increase the absorption efficiency of several high blood pressure medications (calcium channel blockers) regulated by cytochrome P450 enzymes. Naringenin is a potent inhibitor of cytochrome P450 in *in vitro* studies (Fuhr et al., 1993), however, humans ingesting naringin alone failed to increase either nisoldipine (Bailey et al., 1993a) or felodipine (Bailey et al., 1993b) (both calcium channel blockers) absorption when compared to grapefruit juice. Although naringin was originally associated with this effect, it was later discovered that a furanocoumarin was responsible for this effect (Edwards et al., 1996).

Naringin and naringenin were found to have a minor protective effect against B16F10 melanoma cells when fed to mice at the rate of 200 nmole/kg body weight (Menon et al., 1995). Fewer lung tumor nodules formed (26%), and survival was reported to be longer than in the control group (27%). Catechin, epicatechin and rutin had a great protective effect whereas ellagic acid, quercetin and morin were inactive. Naringenin and hesperetin were found to be more effective in the inhibition of MDA-MB-435 human breast cell cancer proliferation *in vitro* than genistein, an isoflavone in soybeans (Guthrie et al., 1995). An *in vivo* study of rats fed grapefruit juice, naringin, naringenin and orange juice against induced mammary tumorigenesis by dimethylbenzanthrancene demonstrated that tumor development was delayed in rats given orange juice, yielding fewer and smaller tumors then the control animals (Carroll et al., 1995). The group fed orange juice also had a better weight gain compared to any other test group. The group of rats fed the naringin supplemented diet also showed a delay in tumor development, and also had less weight gain than the control group.

The chemopreventive effects of hesperidin, curcumin, and ß-carotene has been demonstrated against oral carcinogenesis induced by 4-nitroquinoline 1-oxide (4NQ) in rats (Tanaka et al., 1994). The compounds tested were administered at 0.05% in the diet during or after treatment with 4NQ. The ability of the test compounds to modify the initiation and post initiation phases of these cancers were determined. Curcumin protected nearly completely during the initiation phase of carcinogenesis in this model. Animals fed hesperidin experienced one half the number of oral carcinomas and hyperplastic lesions when compared to the control group, and hesperidin was only slightly less effective than ß-carotene. When phytochemical treatments were started one week after exposure to 4NQ, hesperidin exhibited a slight protective effect while ß-carotene and curcumin provided a significant protective effect with the post initiation treatments.

Naringin was effective in protecting rats against gastric mucosal lesions induced by ethanol (Martin et al., 1994). Naringin, 400 mg/kg body weight, administered an hour before ingestion of ethanol resulted in a 70% reduction in ulceration. Protection by naringin was theorized to be accomplished by increasing the gel viscosity of mucus in the stomach.

## 1.2. Flavones

*1.2.1. Polymethoxylated Flavones in Citrus.* A class of flavones known as polymethoxylated flavones are present as minor components in crude bioflavonoid extracts prepared from citrus peel. Polymethoxylated flavones also occur as minor components in citrus fruit and juices (Figure 1) and they are more hydrophobic than hydroxylated flavones. Polymethoxylated flavones are associated exclusively with the oil glands located in the outer layer of peel (flavedo) of citrus fruit. These compounds become incorporated into citrus juices during extraction or with addition of peel oils for flavor enhancement. Polymethoxylated flavones are most abundant in tangerines, and in oranges they comprise 0.1–0.5% of the peel on a dry weight basis. Commercial tangerine peel oils contain 0.5–1% methoxyflavones. Most abundant in commercial tangerine peel oils are the polymethoxylated flavones tangeretin and nobiletin, with smaller amounts of sinensetin, heptamethoxyflavone, tetra-*O*-methyl scutellarein, and hexa-*O*-methyl gossypetin. Commercial orange peel oils contain nobiletin and heptamethoxyflavone as the most abundant methoxyflavones. Total polymethoxylated flavone content in commercial orange oil ranges from 0.2–0.4%.

Polymethoxylated flavones demonstrated greater anti-adhesive effects on red blood cells and platelets than flavanone glycosides (Robbins, 1974). The reduction of red blood cell clumping may be important in the prevention of coronary thrombosis. Polymethoxylated flavones have demonstrated anti-inflammatory properties and reduced histamine release (Middleton and Dzrewiecki, 1982; Middleton et al., 1987). The polymethoxyflavones were much more active as anti-inflammatory agents than naringin, hesperidin, naringenin and hesperetin. Of 17 flavonoids evaluated for antimutagenic activity, the glycosylated flavonoids naringin, hesperidin and rutin were only weakly active, while tangeretin and nobiletin were considerably more active (Wall et al., 1988).

*1.2.2. Anticancer Activity of Methyoxylated Flavones.* A considerable amount of research has been done on the protective effects of polymethoxylated flavones as cancer prevention agents. Tumors are cell populations with uncontrolled and abnormal growth and are classified as benign or malignant. Of less concern are benign tumors because they do not invade surrounding tissue and usually they are easily removed. Malignant tumors aggressively invade surrounding tissues, altering their natural function. When malignant tumor cells spread to the lymph and blood circulatory systems, metastatic cascade begins, spreading cancer cells throughout the body (Wattenberg, 1985).

Control of cancer may be accomplished with "suppressing", "blocking" or "transforming" agents (Wattenberg, 1985, 1992). "Suppressing" agents prevent the formation from procarcinogens other organic compounds capable of promoting new cancer formation. "Blocking" agents prevent the carcinogenic compounds from reaching the critical reactive sites in cells where cancer begins. "Transformation" agents act to metabolize carcinogenic components into less toxic materials or prevent expression of the carcinogen. Another method for controlling cancer involves the blocking of the metastatic cascade. This is done by inhibiting the invasion of cancerous cells into surrounding tissues or inhibiting cancer cell mobility in circulatory systems (Wattenberg, 1985, 1992).

Tangeretin, nobiletin, hesperidin, naringin, (+) catechin, and (+) epicatechin were tested for anti-invasive activity (Bracke et at., 1991). The assay used healthy chicken embrio heart tissue exposed to MO4 malignant mouse tumor cells. Invasion was inhibited greater than 50% by tangeretin at a concentration of 10 $\mu$M. Tangeretin and nobiletin demonstrated almost total anti-invasive activity at 100 $\mu$M, while (+) catechin and epicatechin showed no effect until present at a concentration of 500 μM. Hesperidin and naringin had no activity at any concentration.

Tangeretin was also effective in inhibiting invasion of MCF-7 human breast cancer cells into chick embryo heart tissue when present at both 10 μM and 100 μM (Bracke et al., 1994). In this assay nobiletin, hesperidin, and naringin were inactive. The binding of selected flavonoids to extracellular matrix materials (ECM) of the MCF-7 assay was tested in culture for 4 days. The ECM binding tests showed catachin and epicatachin (or their metabolites) underwent a high degree of binding to the ECM. Tangeretin and nobiletin showed poor affinity for binding with the ECM and the concentrations remained constant in the culture media. Tangeretin and nobiletin must therefore exert their anti-invasive protection differently from catechin.

Epithelial cells are the origin for approximately 80% or more of all human cancers. Cancerous epithelial cells are connected to each other by a calcium dependent cell-cell adhesive glycoprotein molecule called E-cadherin (Takeichi, 1991; Edelman and Crossin, 1991). Epithelial cell invasion can result from a breakdown in the function of E-cadherin between adjacent cells (Chen and Obrinck, 1991; Mareel et al., 1994). It was discovered that tangeretin can restore the function of E-cadherin and restore cell-cell adhesion between cancerous epithelial cells. Restoration of function did not require protein synthesis, therefore, some other mechanism such as an enzymatic glycosylation, or phosphorylation reaction may be responsible. Restoration of epithelial tissue integrity would prevent metastatic cascade by limiting cell motility. Cancerous epithelial cells would be inhibited from entering the blood and lymph systems and spreading to remote areas of the body.

Proliferation of human squamous carcinoma HTB43 cells *in vitro* were inhibited by both nobiletin and tangeretin (Kandaswami et al., 1991). Tangeretin or nobiletin inhibited cell growth by 70% or greater at concentrations of 20 μM when compared to controls. Less effective inhibition was provided at concentrations of 5 μM.

Nobiletin and tangeretin were active as growth inhibitors of MDA-MB-435 estrogen receptor negative (ER-) and MCF-7 estrogen receptor positive (ER+) human breast cancer cells *in vitro* (Guthrie et al., 1996). Tangeretin or nobiletin (0.5 μg/mL) inhibited growth by 50% in ER- cells, with similar results for ER+ cells. The most interesting part of the study was the synergistic interactions of tamoxifin and/or tocotrienols in combination with tangeretin or nobiletin. Testing tangeretin or nobiletin as 1:1 mixtures with various tocotrienols resulted in tangeretin and γ-tocotrienol in combination having an $IC_{50}$ of 0.05 μg/ml in ER- cells and 0.02 μg/ml in ER+ cells. When tangeretin was removed from the mixture, γ-tocotrienol had an $IC_{50}$ in excess of 30 μg/ml in ER- cells and 2 μg/ml in ER+ cells. Combining tangeretin: γ-tocotrienol: tamoxifin (1:1:1) resulted in a lowering of the $IC_{50}$ to 0.01 μg/ml in ER- cells but did not affect the $IC_{50}$ of ER+ cells. The lowest achieved $IC_{50}$ was 0.001 μg/ml in ER+ cells reported by combining nobiletin: δ-tocotrienol: tamoxifin (1:1:1) (nobiletin alone 0.8 mg/ml for ER+ cells).

Polymethoxylated flavones, like tangeretin and nobiletin, have demonstrated the ability to selectively inhibit the growth of leukemia cell lines but not the growth of normal cells *in vitro* (Sugiyama et al., 1993). Polymethoxylated flavones (27 total) isolated from citrus were tested against M1 and HL-60 leukemic cell lines (Sugiyama et al., 1993). These products were isolated from mandarin peel in the following proportions; tangeretin 22%, and nobiletin 64%. Orange peel yielded the following ratio of polymethoxyflavones; heptamethoxyflavone 30%, nobiletin 30%, tangeretin 15%, and sinensetin 15%. Also included in this isolation were polymethoxylated flavones with hydroxyl groups and two flavone glycosides. All the isolated flavones inhibited the growth M1 cell line from 30 - 80% at the test concentration of 50 μM, except the two flavone glycosides. Induced phagocytosis of the M1 and HL-60 cells into differentiated cells, which do not undergo further division, was accomplished by tangeretin, nobiletin, heptamethoxyflavone, sinensetin, and 11 other flavones. Tangeretin and heptamethoxyflavone inhibited growth by greater than 50% and induced cell differentiation at greater than 25% in both the M1 and HL-60 cell lines. Nobiletin and sinensetin inhibited growth by 78% and 38% respectively, but were less effective in inducing cell differentiation at the 50 $\mu$M concentration. The effect of substitution of hydroxyl and methoxyl groups on growth inhibition and induced differentiation appears to be complex and not easily explained. Concentrations as low as 5 $\mu$M level are effective in growth inhibition and phagocytosis with the effect being dose dependent.

In a similar study, the effectiveness of tangeretin in inhibiting growth of HL-60 leukemic cells compared to normal lymphocytes was confirmed (Hirano et al., 1995). Cultured HL-60 leukemic cells with 27 $\mu$M tangeretin *in vitro* were significantly inhibited. Tangeretin was not effective as a growth inhibitor for the MOLT-4 cell line. These polymethoxylated flavones did not effect the growth or division of normal cell lines. Dose dependent DNA fragmentation was observed in HL-60 cells treated with as little as 2.7 $\mu$M tangeretin while healthy cells showed no DNA fragmentation after the same treatment, implying that these polymethoxylated flavones may have very little toxicity to healthy cell tissue.

## 2. DISCOVERY OF NEW FLAVANOIDS FROM DANCY TANGERINE

### 2.1. Materials and Methods

*2.1.1. General Chromatography.* Normal phase thin-layer chromatography (TLC) was performed on high performance silica gel plates (Whatman 4807–400), using two solvent systems: solvent A- chloroform:methanol (95:5), solvent B- benzene:ethyl acetate (90:10). Normal phase plates were visualized by observation under multiband uv-lamp (uv-254/366 nm) and/or by spraying with $H_2SO_4$-ethanol (1:4) followed by charring with a heat gun. Reverse-phase TLC was performed on octadecyl ($C_{18}$) plates (Whatman 4803–600) with three solvent systems: solvent C- methanol:water (80:20), solvent D- ethanol:water (60:40), solvent E- acetonitrile:tetrahydrofuran (90:10), solvent F- methanol:water (90:10). Reversed phase plates were observed under multiband uv-lamp (uv-254/366 nm). Flash chromatography was performed on silica gel (Baker Analyzed 40$\mu$ Lot G42353) using solvent B.

*2.1.2. High-Performance Liquid Chromatography (HPLC).* Preparative HPLC was carried out on a LDC analytical system (Thermo Separation Products Inc. Riviera Beach, Florida) consisting of a ConstaMetric 3200 pump, a model IV refractive index detector, a Rheodyne model 7125 injection valve and a Gilson FC203B fraction collector (Gilson Medical Electronics, Middleton, WI). The column used was a Rainin (Woburn, MA) Dynamax 60A 8 $\mu$m $C_{18}$, 21.4 mm i.d. x 25 cm coupled to a Dynamax 60A guard column 8$\mu$m $C_{18}$, 21.4 mm i.d.x 5 cm. The HPLC separations were done with either solvent (C) or solvent (D) described above.

*2.1.3. Mass, Nuclear Magnetic Resonance, and Ultraviolet Spectroscopy.* The mass spectral analyses were carried out on a Finnigan MAT 95Q magnetic sector mass spectrometer (Finnigan MAT, San Jose, CA), with electron ionization at 70 eV and low resolution scan. $^1$H-NMR and $^{13}$C-NMR spectra were taken on a General Electronic QE-300 spectrometer, chemical shifts in $CDCl_3$ are reported relative to internal TMS. UV spectra were recorded on a PERKIN-ELMER Lambda 5 spectrophotometer (Norwalk, CT).

*2.1.4. Extraction and Isolation.*

2.1.4.1. Dancy Tangerine Cold Pressed Peel Oil Solids. The isolation of these products are outlined in a previous publication using chromatographic systems (Chen et al. 1997).

2.1.4.2. Dancy Tangerine Leaves. Leaves of *Dancy Tangerine* were extracted with methanol-chloroform (1:1) and the concentrated extract was initially separated by use of a vacuum flash silica gel chromatography, eluted with a gradient of iso-octane (TMP) : iso-propanol (IPA) (100% TMP to 100% IPA) with a final column rinse of 100% methanol (total 12 fractions). The 12 fractions were each applied to a flash $C_{18}$ column for chromatography, with a gradient solvent system of ethanol/water (60% EtOH to 95% EtOH). Compounds **III**, **V**, and **XI** (Figure 2) were purified with this column. The unpure fractions were then subjected to $C_{18}$ preparative HPLC for purification, with either solvent system C or system F (listed above).

# 3. RESULTS AND DISCUSSION

## 3.1. Dancy Tangerine Cold Pressed Peel Oil Solids

The isolation and characterization of natural products from Dancy tangerine cold pressed peel oil solids yielded ten polymethoxylated flavones, determined by $^1$H-NMR, $^{13}$C-NMR, MS and UV spectra. Cold pressed tangerine peel oil solids were obtained from the peel of processed Dancy tangerine fruit. After the fractionation with the silica gel column chromatography and preparative high-performance liquid chromatography, ten compounds were purified (**I~X**) (Figure 2). Compounds **II** and **VII** are novel natural products (Chen et al. 1997). The uniqueness of compounds **II** and **VII** comes from the positioning of the hydroxyl group at the 7 position on the A ring. The hydroxyl group was determined to be at the 7 position rather than any other position by the use of NaOMe induced bathochromic shift of the major UV band of both of these compounds.

## 3.2. Dancy Tangerine Leaves

Dancy tangerine leaves methanol/chloroform (1:1) extracts were separated by a combination of silica gel, $C_{18}$ flash column chromatography and $C_{18}$ preparative HPLC to afford ten compounds in crystalline form (Figure 2). Both cold pressed peel oil solids and leaves of Dancy tangerines contained compounds **I, III, IV, V, IX**. Compounds **I, III-VII, XIV, XV** are polymethoxylated flavones; compounds **XVI** and **XVII** are polymethoxylated flavanones. Compounds **III-V, XII, XVII** had a fully oxygenated A-ring (5, 6, 7, 8-tetraoxygenated) system, whereas, compounds **I, XI, XVI** have a trioxygenated A-ring system; compounds **XIV** and **XV** have a chlorine attached to the A-ring. All compounds possessed a 3', 4'-dioxygenated B-ring system, except for compounds **V** and **XV** where the B-ring was 4'-monoxygenated.

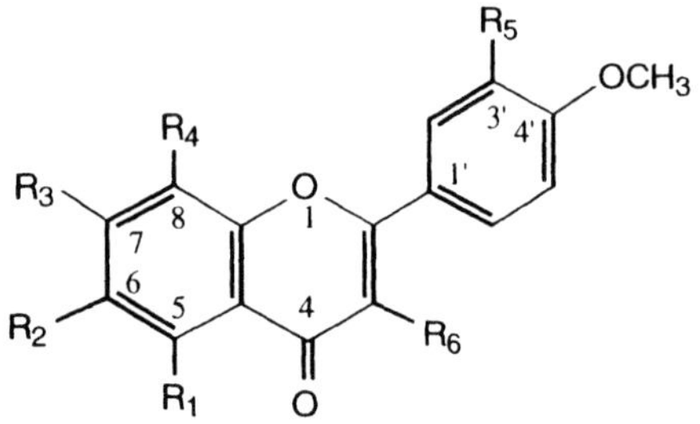

|      | $R_1$ | $R_2$ | $R_3$ | $R_4$ | $R_5$ | $R_6$ |
|------|-------|-------|-------|-------|-------|-------|
| I    | OCH$_3$ | OCH$_3$ | OCH$_3$ | H | OCH$_3$ | H |
| II   | OCH$_3$ | OCH$_3$ | OH | H | OCH$_3$ | OCH$_3$ |
| III  | OH | OCH$_3$ | OCH$_3$ | OCH$_3$ | OCH$_3$ | H |
| IV   | OCH$_3$ | OCH$_3$ | OCH$_3$ | OCH$_3$ | OCH$_3$ | H |
| V    | OCH$_3$ | OCH$_3$ | OCH$_3$ | OCH$_3$ | H | H |
| VI   | OCH$_3$ | H | OCH$_3$ | OCH$_3$ | H | H |
| VII  | OCH$_3$ | OCH$_3$ | OH | OCH$_3$ | OCH$_3$ | OCH$_3$ |
| VIII | OCH$_3$ | OCH$_3$ | OCH$_3$ | H | H | H |
| IX   | OCH$_3$ | OCH$_3$ | OCH$_3$ | OCH$_3$ | OCH$_3$ | OCH$_3$ |
| X    | OCH$_3$ | H | OCH$_3$ | OCH$_3$ | OCH$_3$ | H |
| XI   | OH | OCH$_3$ | OCH$_3$ | H | OCH$_3$ | H |
| XII  | OH | OCH$_3$ | OH | OCH$_3$ | OCH$_3$ | H |
| XIII | OCH$_3$ | OCH$_3$ | OCH$_3$ | OCH$_3$ | H | OCH$_3$ |
| XIV  | OCH$_3$ | OCH$_3$ | Cl | OCH$_3$ | OCH$_3$ | OCH$_3$ |
| XV   | OCH$_3$ | OCH$_3$ | Cl | OCH$_3$ | H | OCH$_3$ |

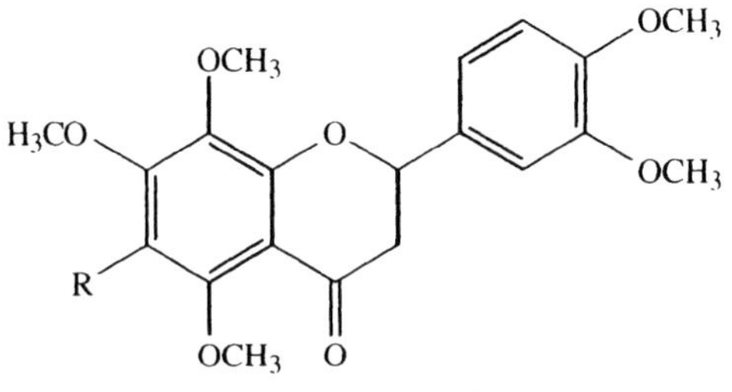

XVI: R=H
XVII: R=OCH$_3$

**Figure 2.** Polymethoxylated flavones and flavanones isolated from Dancy Tangerine cold pressed peel oil solids and leaves.

The presence of a chelated 5-OH, for compounds **III, XI, XII**, was indicated by a sharp singlet at δ 12.58 in the $^1$H-NMR that disappeared on deuteration. The spectral data of compounds **III** and **XI** were in good agreement with those of 5-*O*-desmethylnobiletin (Kinoshita et al., 1996) and 5-hydroxy-6, 7, 3', 4'-tetramethoxyflavone (Sugiyama et al., 1993) respectively. Compound **XII**, had the molecular weight of [M]$^+$ 374 for $C_{19}H_{18}O_8$ based on mass spectral data, corresponding to a flavone containing four methoxyl groups and two hydroxy groups. The fragments (due to retro-Diels-Alder cleavage) present at 212 (15.14) and 162 (4.30) indicated the two methoxyl groups and two hydroxy groups are attached to A-ring and the remaining two methoxyl groups are attached to B-ring (Figure 3). The $^1$H-NMR spectrum of compound **XII** showed that three ABX type aromatic coupling characteristic of 3'-4'-methoxylated B-ring. The presence of significant bathochromic shifts by NaOMe shift reagents in the UV spectrum revealed the presence of a free hydroxyl at the 7-position. From the above evidence the structure of compound **XII** was elucidated as 5',7'-dihydroxy-6,8,3',4'-tetramethoxyflavone.

The $^1$H-NMR spectrum of compound **XVI** showed three one-proton double doublets indicated that **XVI** had a flavanone skeleton (Iinuma et al., 1993; Lin et al., 1995). Five methoxyl signals were observed in the $^1$H-NMR. One A-ring aromatic signals was assigned to the H-6 of the flavanone. The EI-mass spectrum of **XVI** showed a [M]$^+$ at m/z 374 (base peak) and significant peaks at m/z 344, 211, 210 and 164 attributed to the typical retro-Diels-Alder fragmentation of a flavonoid (Figure 4), these fragmentation supported a structure of flavanone with a dimethoxylated B-ring and a trimethoxylated A-ring. Based on the above evidence, compound **XVI** was characterized as 5,7,8,3',4'-pentamethoxyflavanone.

The B- and C-ring for compounds **XVI** and **XVII** were thought to be similar by the comparison of $^1$H-NMR spectral data. Compound **XVII**, $C_{21}H_{24}O_8$ from EI-MS, differed from **XVI** in the absence of one A-ring aromatic proton (δ 6.37 at $^1$H-NMR) and the presence of one more methoxyl group (δ 4.05 at $^1$H-NMR) which suggested that the A-ring of compound **XVII** is fully methoxylated. Thus, **XVII** was determined to be 5,6,7,8,3',4'-hexamethoxyflavanone.

The $^1$H-NMR spectrum of **XIV** showed six methoxyl signals and ABX type coupled aromatic protons assignable to H-6', H-2' and H-5'. Four methoxyl signals in the $^{13}$C-NMR spectrum were characteristic for the methoxyl groups located on A-ring and C-3

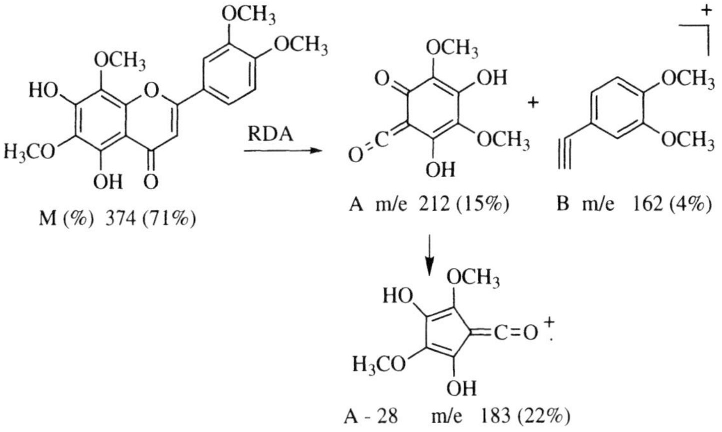

Figure 3. Retro-Diels-Alder MS fragmentation of compound **XII**.

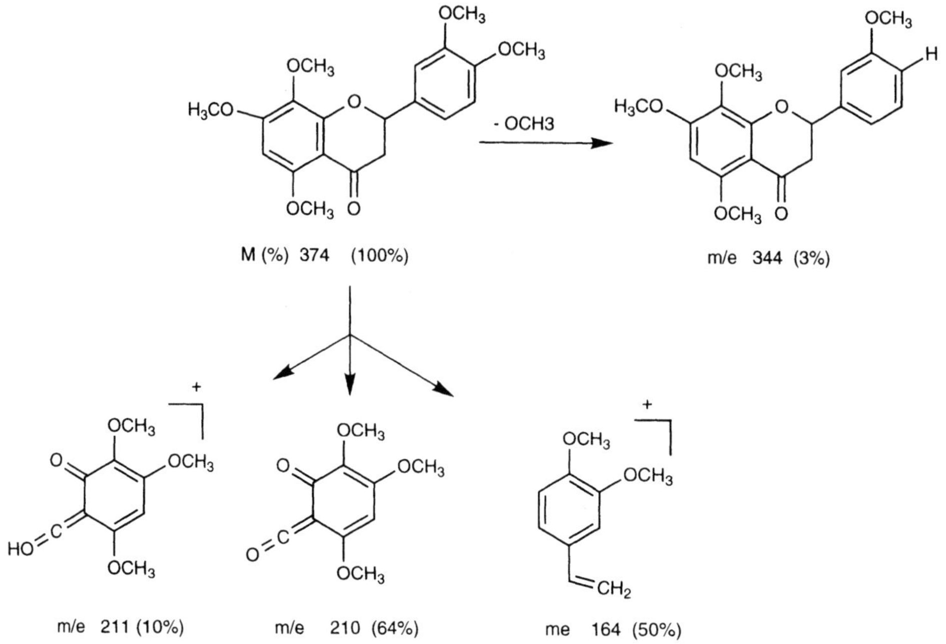

Figure 4. Retro-Diels-Alder MS fragmentation of compound **XVI**.

(Panichpol et al., 1978; Roitman et al., 1985). EI-mass spectrum of **XIV** gave [M]$^+$ at m/z 436 with an isotope M+2 peak observed. The ratio of M+2 peak to the molecular ion peak is 1/3 which suggested the presence of a chlorine atom in compound **XIV**. The HRMS established the molecular formula of **XIV** as $C_{21}H_{21}ClO_8$. The experimentally measured masses of 35 and 37 Cl isotopes were 436.091 (calcd. 436.092) and 438.089 (calcd. 438.090) respectively. The fragments of the retro-Diels-Alder pathway indicated that the chlorine is attached A-ring. Comparing the $^{13}$C-NMR spectrum of a known compound **IX** with **XIV** demonstrated three major changes in chemical shifts were found in C-6, C-7 and C-8 (Table 1). The symmetrical shifts of C-6 and C-8 indicated that the chlorine attached to C-7. Thus the compound **XIV** was identified as 7-chloro-3,5,6,8,3',4'-hexamethoxyflavone.

Compound **XV** was immediately determined to be another 7-chloro polymethoxylated flavone by the NMR and HRMS data. The presence of an $A_2B_2$ pattern in the B-ring ($^1$H-NMR), typical of *para*-substituted benzene ring, fixed the structure of compound **XV** as 7-chloro-3,5,6,8,3'-pentamethoxyflavone. Confirmation of the chlorine attached to C-7 was determined by comparing the $^{13}$C-NMR spectra of **XIII** and **XV** at positions C-6, C-7, and C-8.

## 4. CONCLUSION

Two new polymethoxylated flavones (PMF) were isolated from the cold pressed peel oil solids of Dancy tangerine (**II** and **VII**). In Dancy tangerine leaves, one new PMF (**XII**), two chlorinated PMFs (**XIV** and **XV**) and two new polymethoxylated flavanones (PMFA) (**XVI** and **XVII**) were discovered. Preliminary tests against human cancer cell

Table 1. $^{13}$C-NMR data of compounds **IX, XIII, XIV, AND XV**

| C | Compound **IX**[b] | Compound **XIV** | Compound **XIII**[c] | ($\delta$)[a]<br>Compound **XV** |
|---|---|---|---|---|
| C-2 | 151.1 | 151.7 | 151.3 | 151.7 |
| C-3 | 140.8 | 141.0 | 140.7 | 143.8 |
| C-4 | 173.9 | 172.3 | 174.0 | 173.5 |
| C-5 | 143.9 | 144.7 | 143.9 | 144.6 |
| C-6 | 137.8 | 158.1 | 138.0 | 156.8 |
| C-7 | 151.3 | 117.3 | 153.5 | 117.3 |
| C-8 | 137.8 | 157.2 | 138.0 | 157.7 |
| C-9 | 148.2 | 149.2 | 148.2 | 148.5 |
| C-10 | 115.1 | 113.4 | 115.2 | 114.0 |
| C-1' | 123.5 | 123.9 | 123.4 | 123.8 |
| C-2' | 110.9 | 110.9 | 130.0 | 130.9 |
| C-3' | 148.8 | 147.0 | 114.2 | 114.7 |
| C-4' | 153.0 | 151.8 | 161.5 | 161.9 |
| C-5' | 111.0 | 111.1 | 114.2 | 114.6 |
| C-6' | 121.9 | 123.1 | 130.0 | 130.8 |
| OCH$_3$ | 62.3 | 62.3 | 62.4 | 62.3 |
| | 61.9 | 62.2 | 62.1 | 61.9 |
| | 61.8 | 61.9 | 61.9 | 61.8 |
| | 61.7 | 61.7 | 61.7 | 61.6 |
| | 59.9 | (3,5,6,8) | 58.0 | (3,5,6,8) |
| | (3,5,6,7,8) | 56.1 | (3,5,6,7,8) | 55.4 |
| | 56.0 | 56.0 | 55.0 | (4') |
| | 55.9 | (3',4') | (4') | |
| | (3',4') | | | |

[a] Solvent, CDCl$_3$
[b] Data of $^{13}$C-NMR from Chen, et al., (1997)
[c] Data of $^{13}$C-NMR from Sugiyama, et al., (1993)

lines have determined that compounds **II** and **VII** have some activity, which will be reported in full at a later time. The compounds discovered in Dancy tangerine leaves are being reisolated for testing. The PMFs have been known in the literature for some time, not only from the *Citrus sp.*, but also other species in the Family Rutaceae (Harbourne, 1984). Citrus is the primary source for the PMFs. The PMFs are not found in any other commonly consumed food product. The consumption of all the PMFs in orange juice is on the order of 1–2 milligrams per 240ml (8floz) (Widmer et al., 1996). The level of the major PMFs in Dancy tangerine peel has been determined by HPLC (Widmer et al., 1995). As detailed in the review portion of this chapter, the PMFs have an impact on *in vitro* animal and human cancer cell lines and capillary fragility. It is widely accepted that citrus foods are health promoting foods (National Research Council, 1989). The isolation of these natural products will provide materials for further testing. This will enable us to better determine the contribution made by the polymethoxylated flavones to human health.

## 5. DANCY TANGERINE LEAVES

5,7-dihydroxy-6,8,3',4'-tetramethoxyflavone (**XII**). Yellow needles (MeOH) mp 210–211°C; t$_R$ 21.5 min (prep-HPLC with system A); R$_f$ 0.45 (TLC with system A); UV (MeOH) $\lambda_{max}$ 364, 274sh, 206 nm; $\lambda_{max}$ (MeOH+NaOMe) 397, 330sh, 208 nm; $^1$H-NMR $\delta$:

12.58 (1H, s, OH-5, D$_2$O exchangeable), 7.52 (1H, dd, j = 8.5, 2.4 Hz, H-6'), 7.40 (1H, d, j = 2.4 Hz, H-2'), 7.10 (1H, d, j = 8.5 Hz, H-5'), 6.60 (1H, s, H-3), 6.05 (1H, br, OH-7, D$^2$O exchangeable), 4.10 (3H, s, OMe), 3.99 (3H, s, OMe), 3.97 (3H, s, OMe), 3.94 (3H, s, OMe). $^{13}$C-NMR δ: 160.2 (C-2), 106.8 (C-3), 177.4 (C-4), 150.0 (C-5), 134.0 (C-6), 148.9 (C-7), 137.5 (C-8), 144.5 (C-9), 110.4 (C-10), 123.7 (C-1'), 114.6 (C-2'), 145.5 (C-3'), 149.5 (C-4'), 110.5 (C-5'), 121.5 (C-6'), 62.3 (OMe), 61.7 (OMe), 56.1 (OMe), 56.0 (OMe). EIMS m/z (%): 374 (M$^+$, 70.66), 359 (M-CH$_3$, 84.97), 343 (M-OCH$_3$, 4.00), 212 (15.14), 183 (21.75), 162 (4.30).

5,7,8,3',4'-pentamethoxyflavanone (**XVI**). pale yellow needles (MeOH); mp 177–178 °C; t$_R$ 14.5 min (prep-HPLC with system B) R$_f$ 0.23 (TLC with system B); $^1$H-NMR δ: 7.02 (1H, d, j= 2.5 Hz, H-2'), 7.00 (1H, dd, j= 9.0, 2.5 Hz, H-6'), 6.90 (1H, d, j= 9.0 Hz, H-5'), 6.37 (1H, s, H-6), 5.35 (1H, dd, 13.0, 3.0 Hz, H-2). 3.02 (1H, dd, j= 17.5, 13.0 Hz, H-3$_{ax}$), 2.75 (1H, dd, j= 13.0, 3.0 Hz, H-3$_{eq}$), 3.95 (3H, s, OMe), 3.92 (3H, s, OMe), 3.90 (3H, s, OMe), 3.87 (3H, s, OMe), 3.82 (3H, s, OMe). $^{13}$C-NMR δ: 79.0 (C-2), 45.6 (C-3), 189.2 (C-4), 156.2 (C-5), 89.5 (C-6), 157.7 (C-7), 132.0 (C-8), 156.8 (C-9), 107.5 (C-10), 131.6 (C-1'), 114.7 (C-2'), 147.0 (C-3'), 148.2 (C-4'), 112.1 (C-5'), 118.0 (C-6'), 61.4 (OMe), 56.5 (OMe), 56.2 (OMe), 56.1 (OMe), 56.0 (OMe). EIMS m/z (%): 374 (M$^+$, 100), 343 (M-OCH$_3$, 2.51), 237 (4.50), 211 (10.5), 210 (63.76), 195 (62.17), 167 (53.63), 164 (50.44).

5,6,7,8,3',4'-hexamethoxyflavanone (**XVII**). Pale yellow needles (MeOH); mp 164–165 °C; t$_R$ 15.2 min (prep-HPLC with system B) R$_f$ 0.26 (TLC with system B); $^1$H-NMR δ: 7.02 (1H, d, j= 2.5 Hz, H-2'), 7.00 (1H, dd, j=9.0, 2.5 Hz, H-6'), 6.90 (1H, d, j= 9.0 Hz, H-5'), 5.40 (1H, dd, j= 13.0, 3.0 Hz, H-2), 3.05 (1H, dd, j= 17.5, 13.0 Hz, H-3$_{ax}$), 2.84 (1H, dd, j= 13.0, 3.0 Hz, H-3$_{eq}$), 4.05 (3H, s, OMe), 3.90–3.88 (9H, 3 x OMe), 3.85–3.82 (6H, 2 x OMe). $^{13}$C-NMR δ: 78.0 (C-2), 45.6 (C-3), 190.2 (C-4), 151.2 (C-5), 139.5 (C-6), 154.7 (C-7), 141.6 (C-8), 150.5 (C-9), 112.0 (C-10), 131.4 (C-1'), 114.3 (C-2'), 146.7 (C-3'), 148.1 (C-4'), 112.1 (C-5'), 118.0 (C-6'), 62.0 (OMe), 61.8 (OMe), 61.7 (OMe), 60.8 (OMe), 56.1 (OMe), 56.0 (OMe). EIMS m/z (%): 404 (M$^+$, 38.50), 374 (M-OCH$_3$, 4.50), 241 (9.62), 240 (100), 225 (61.75), 210 (10.85), 197 (43.23), 164 (36.01), 149 (17.41).

7-chloro-3,5,6,8,3',4'-hexamethoxyflavone (**XIV**). Yellow needles (MeOH); mp 145–146 °C; t$_R$ 15.8 min (prep-HPLC with system B); R$_f$ 0.35 (TLC with system B); $^1$H-NMR δ: 7.68 (1H, dd, j= 9.0, 2.2 Hz, H-6'), 7.58 (1H, d, j=2.2 Hz, H-2'), 7.02 (1H, d, j= 9.0 Hz, H-5'), 4.12 (3H, s, OMe), 4.0–3.97 (15H, 5 x OMe). EIMS m/z (%): 436 (M$^+$, 35.68), 438 (M+2, 12.01), 421 (M-15), 100), 244 (8.50), 216 (18.23), 165 (14.02). HRMS m/z (%): 436.091 (M$^+$, 25.06) C$_{21}$H$_{21}$Cl O$_8$ calcd 436.092, 438.089 (M+2, 8.24) calcd 438.090 for 37 Cl.

7-chloro-3,5,6,8,4'-pentamethoxyflavone (**XV**). Yellow needles (MeOH); mp 161–162 °C; t$_R$ 14.7 min (prep-HPLC with system B) R$_f$ 0.28 (TLC with system B); $^1$H-NMR δ: 7.94 (2H, d, j= 9.0 Hz, H-2' and H-6'), 7.02 (2H, d, j= 9.0 Hz, H-3' and H-5'), 4.10 (3H,s, OMe), 3.95 (3H, s, OMe), 3.94 (6H, 2 x OMe), 3.89 (3H, s, OMe); EIMS m/z (%): 406 (M$^+$, 36.42), 408 (M+2, 12.40), 391 (M-15, 100), 348 (9.01), 197 (29.1), 135 (26.25). HRMS m/z (%): 406.082 (M$^+$, 28.04) C$_{20}$H$_{19}$Cl O$_7$ calcd 406.082, 408.077 (M+2, 9.21) calcd 408.079 for 37 Cl.

# REFERENCES

Affany, A.; Salvayre, R. Comparison of the protective effect of various flavonoids against lipid peroxidation of erythrocyte membranes induced by cumene hydroperoxide. *Fund. of Clin Pharm.* **1987**, *1*, 451–7.

Bailey, D. G.; Arnold, J. M. O.; Munoz, C.; Spence, J. D. Grapefruit juice - felodipine interaction: Mechanism, predictability, and effect of naringin. *Clin. Pharm. Ther.* **1993a**, *53*, 637–42.

Bailey, D. G.; Arnold, J. M. O.; Strong, H. A.; Munoz, C.; Spence, J. D. Effect of grapefruit juice and naringin on nisoldipine pharmacokinetics. *Clin. Pharm. Ther.* **1993b**, *54*, 589–94.

Bracke, M. E.; Vyncke, O. G.; Foidart, J. M.; Depestel, G.; Mareel, M. Effect of catechins and citrus flavonoids on invasion *in vitro*. *Clin. Expl. Metastasis.* **1991**, *9*, 13–25.

Bracke, M. E.; Bruyneel, E.; Vermeulen, S. J.; Vennekens, K.; Van Marck, V.; Mareel, M. M. Citrus flavonoid effect on tumor invasion and metastasis. *Food Technology* **1994**, *47*, 121–4.

Carroll, K. K.; So, F.; Gutherie, N.; Chambers, A. F. Effects of citrus flavonoids on proliferation of MDA-MB-435 human breast cancer cells and on mammary tumorigenesis induced by DMBA in female Sprague-Dawley rats. In *Dietary Phytochemicals in Cancer Prevention and Treatment*; American Institute for Cancer Research, Ed.; Plenum Press: New York, **1995**; pp 256.

Chen, W.; Obrinck, B. Cell-cell contacts mediated by E-cadaherin (uvomorulin) restrict invasive behavior of L-cells. *J. Cell Biol.* **1991**, *114*, 319–27.

Chen, J.; Montanari, A. M.; Widmer, W. W. Two new polymethoxylated flavones, a class of compounds with potential anticancer activity, isolated from cold pressed dancy tangerine peel oil solids. *J. Agric. Food Chem.* **1997**, *45*, 364–8.

Edelman, G. M.; Crossin, K. L. Cell adhesion molecules: Implications for a molecular histology. *Annu. Rev. Immunol.* **1991**, *60*, 155–90.

Edwards, D. J.; Bellvue, F. H.; Woster, P. M. Identification of 6',7'-dihydroxybergamottin, a cytochrome P450 inhibitor, in grapefruit juice. *Drug Metab. and Distrib.* **1996**, *24*, 1287–1290.

Fisher, K. D.; Senti, F. R.; Allison, R. G.; Anderson, S. A.; Chinn, H. I.; Talbot, J. M. "Evaluation of the health aspects of hesperidin, naringin, and citrus bioflavonoid extracts as food ingredients," Federation of American Societies for Experimental Biology, 1982.

Fuhr, U.; Lkittich, K.; Staib, H. Inhibitory effect of grapefruit juice and its bitter principal, naringenin, on CYP1A2 dependent metabolism of caffeine in man. *J. Clin. Pharmacol.* **1993**, *35*, 431–6.

Guthrie, N.; Gapor, A.; Chambers, A. F.; Carroll, K. K. Synergistic effects in the inhibition of proliferation of MDA-MB-435 human breast cancer cells by tocotrienols and flavonoids. In *Dietary Phytochemicals in Cancer Prevention and Treatment*; American Institute for Cancer Research, Ed.; Plenum Press: New York, **1995**; pp 255.

Guthrie, N.; Gapor, A.; Chambers, A. F.; Carroll, K. K. Effects of palm oil tocotrienols, flavonoids and tamoxifen on proliferation of MDA-MB-435 and MCF human breast cancer cells. *Research Conference of the American Institute for Cancer research: Dietary Fat and Cancer* **1996**, *Poster #8*.

Harborne, J.B. The flavonoids of the Rutales. in Chemistry and Chemical Taxonomy of the Rutales. Waterman P.G.; Grundon M.F., Eds., Academic Press, New York, NY. 1984; pp 149.

Hirano, T.; Abe, K.; Gotoh, M.; Oka, K. Citrus flavone tangeretin inhibits leukemic, HL-60 cell-growth partially through induction of apoptosis with less cytotoxicity on normal lymphocytes. *Br. J. Cancer* **1995**, *72*, 1380–8.

Iinuma, M.; Ohyama, M.; Tanaka, T.; Mizuno, M.; Hong, S. K. Five flavonoid compounds from *Echinosophora Korensis*. *Phytochemistry* **1993**, *33*, 1241–5.

Kandaswami, C.; Perkins, E.; Solonuik, D. S.; Drzewiecki, G.; Middleton, E. Jr. Antiproliferative effects of citrus flavonoids on a human squamous cell carcinoma *in vitro*. *Cancer Letters* **1991**, *56*, 147–52.

Kaul, T. N.; Middleton, E. Jr.; Ogra, P. L. Antiviral effects of flavonoids on human viruses. *J. Med. Virol.* **1985**, *15*, 71–4.

Kinoshita, T.; Firman, K. Highly oxygenated flavonoids from *Murraya Paniculata*. *Phytochemistry* **1996**, *42*, 1207–1210.

Kuhnau, J. The Flavonoids, A Class of Semi-Essential Food Components: Their Role in Human Nutrition. *World Rev. Nutr. Dig.* **1976**, *24*, 117–19.

Lin, C.N.; Lu, C.M.; Huang, P.L. Flavonoids from *Artocarpus Heterophyllus*. *Phytochemistry* **1995**, *39*, 1447–1451

Mareel, M.; Brcke, M.; Van Roy, F. Invasion promoter versus invasion suppressor molecules: the paradigm of E-cadherin. *Mol. Biol. Reports* **1994**, *19*, 45–67.

Martin, M. J.; Marhuenda, E.; Perez-Guerrero, C.; Franco, J. M. Antiulcer effect of naringin on gastric lesions induced by ethanol in rats. *Pharmacology* **1994**, *49*, 144–50.

Menon, L. G.; Kuttan, R.; Kuttan, G. Inhibition of lung metastasis in mice induced by B16F10 melanoma cells by polyphenolic compounds. *Cancer Let.* **1995**, *95*, 221–5.

Middleton, E. Jr.; Dzrewiecki, G. Effects of flavonoids and transitional metal cations on antigen induced histamine release from human basophils. *Biochem. Pharmacol.* **1982**, *31*, 1449–53.

Middleton, E. Jr.; Fujiki, H.; Savliwala, M.; Dzrewiecki, G. Tumor promotor-induced basophil histamine release: effect of selected flavonoids. *Biochem. Pharmacol.* **1987**, *36*, 2048–52.

National Research Council. 1989. Dietary fiber. In: Diet and Health. National Academy Press, Washington, D. C. pp. 291–310.

Panichpol, K.; Waterman, P. G. Novel flavonoids from stem of *Popowia Cauliflora*. *Phytochemistry* **1978**, *17*, 1363–7.

Robbins, R. C. Action of flavonoids on blood cells: trimodal action of flavonoids elucidates their inconsistent results. *Int. J. Vit. Nutr. Res.* **1974**, *44*, 203–16.

Robbins, R. C.; Maritin, F. G.; Roe, M. D. Ingestion of grapefruit lowers elevated hematocrits in human subjects. *Int. J. Vit. Nutr. Res.* **1988**, *58*, 414–7.

Roitman, J. N.; James, L. F. Chemistry of toxic range plants. Highly oxygenated flavonol methyl ethers from *Gutierezia microcephala*. *Phytochemistry* **1985**, *24*, 835–48.

Rusznyak, S.; Szent-Györgyi, A. Vitamin P: flavonols as vitamins. *Nature* **1936**, *27*, 138.

Sugiyama, S.; Umehara, K.; Kuroyanayi, M.; Ueno, A.; Taki, T. Studies on the differentiation inducers of myloid leukemic cells from citrus species. *Chem. Pharm. Bull.* **1993**, *41*, 714–9.

Takeichi, M. Cadherin cell adhesion receptors as a morphogenetic regulator. *Science* **1991**, *251*, 1451–5.

Tanaka, T.; Makita, H.; Ohnishi, M.; Hirose, Y.; Wang, A.; Mori, H.; Satoh, K.; Hara, A.; Ogawa, H. Chemoprevention of 4-nitroquinoline-1-oxide induced oral carcinogenesis by dietary curcumin and hesperidin: Comparison with the protective effect of ß-carotene. *Cancer Res.* **1994**, *54*, 4653–9.

Vickery, H.; Nelson, E.; Almquist, H.; Elvehjem, C. Joint Committee on Nomenclature - Term "Vitamin P" Recommendation to be Discontinued. *J. Med. Virol.* **1950**, *112*, 628.

Wall, M. E.; Wan, M. C.; Manikumar, G.; Graham, P. A.; Taylor, H.; Hughs, T. J.; Walker, J.; McGivney, R. J. Plant antimutagenic agents: Flavanoids. *J. Nat. Prod.* **1988**, *51*, 1084–91.

Wattenberg, L. W. Chemoprevention of cancer. *Cancer Res.* **1985**, *45*, 1–8.

Wattenberg, L.W. Chemoprevention of cancer by naturally occurring and synthetic compounds. In *Cancer Chemoprevention*, (Wattenberg, L.W., Lipkin, M., Boone, C.W., Kelloff, G.J., Eds.), CRC Press, Boca Raton, FL. **1992**, pp. 19–31.

Widmer, W.W.; Barros, S. Flavonoids in Ambersweet orange and the impact on juice adulteration detection. In Forty-Sixth Annual Citrus Processors' Meeting. Braddock R.J.; Widmer W.W., Sims C.A., Eds.; University of Florida, Gainesville, FL, **1995**, 21–26.

Widmer, W. W.; Montanari, A. The potential for citrus phytochemicals in hypernutritious foods. In *Hypernutritious Foods*; J. W. Finley; D. J. Armstrong; S. Nagy and S. F. Robinson, Eds.; AgScience: Auburndale, Florida, **1996**; pp 75–90.

# DIFFERENTIATION OF SOY SAUCE TYPES BY HPLC PROFILE PATTERN RECOGNITION

## Isolation of Novel Isoflavones

Emiko Kinoshita, Yoshinori Ozawa, and Tetsuo Aishima

Research and Development Division
Kikkoman Corporation
399 Noda, Noda, Chiba 278-0037, Japan

## 1. ABSTRACT

Nonvolatile minor components in various brands of Japanese fermented soy sauce were analyzed by gradient RP-HPLC and monitored at 280 nm. Chemometric pattern recognition techniques, such as cluster analysis, linear discriminant analysis (LDA), LDA using genetic algorithm (GA-LDA) and soft independent modelling of class analogy (SIMCA), succeeded in differentiating the resulting HPLC profiles according to soy sauce brands. Three components playing key roles in the differentiation were isolated by preparative HPLC and purified by gel-filtration chromatography, or simply repeated preparative HPLC. FAB-MS, $^1$H-, $^{13}$C-NMR and IR spectra suggested that these three components having molecular weights of 386, 402 and 418 were isoflavone derivatives. By applying HMBC spectral analysis, these isoflavones were identified as conjugated ethers of tartaric acid with daidzein, genistein and 8-hydroxygenistein. These new isoflavone derivatives are produced by some strains of *Aspergillus* fungi.

## 2. INTRODUCTION

Today, fermented soy sauce is widely used not only in Asian countries but also in North America because of its unique appetizing flavor. Most traditional Japanese dishes, such as *sushi, teriyaki* and *sukiyaki*, are cooked or eaten with soy sauce. Various types of soy sauce are used in Japan but dark-colored soy sauce is the most popular followed by light-colored soy sauce. Although hundreds of chemical components have been identified in soy sauce (Forss and Sugisawa, 1981), the mystery of its unique flavor has not been fully understood.

*Flavonoids in the Living System*, edited by Manthey and Buslig
Plenum Press, New York, 1998.

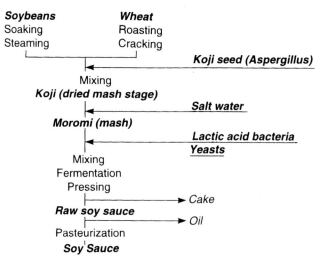

Figure 1. The production process of dark-colored soy sauce.

The production process of dark-colored soy sauce is schematically shown in Figure 1. The main ingredients for the production of soy sauce are soybeans, wheat, salt and water. After the soybeans and wheat are individually heat treated, *Aspergillus* fungi are added to the mixture of soybeans and wheat, called *koji*. The added fungi degrade high molecular weight proteins and starch into low molecular weight components, such as peptides, amino acids and glucose. The hydrolyzed *koji* is then mixed with saline water to make a mash or *moromi*, which is usually fermented for 6–8 months. During the fermentation, lactic acid bacteria and yeasts are added to adjust the pH and to generate the unique flavor. After fermentation, raw soy sauce is expressed from the fermented *moromi*. The following pasteurization of the raw soy sauce adds heated aroma by accelerating the Maillard reaction. The process for production of light-colored soy sauce is similar to that of the dark-colored soy sauce, but heat treatment modifications are introduced to make the color of final product lighter. Thus, the differences in the components of the various soy sauce types are the result of differences in heat treatments of raw materials, microorganisms grown in *koji* and *moromi*, and pasteurization conditions. Furthermore, some minor components originally contained in soybeans and wheat such as isoflavones and unconventional amino acids (Kinoshita *et al.*, 1993) add diversity to soy sauce composition. Examining the many components is the first step in interpreting the quality differences of soy sauces. However, we generally cannot expect to find a single component or a group of components as indicators of differences in closely related product types, as long as the samples belong to the same group of foods. Differences in composition patterns from sample to sample, based on chromatographic or spectroscopic data which contain information on the food's composition, may be exploited to differentiate between samples of varying origins.

Since the late 1960's, various chemometric pattern recognition techniques have been successfully applied to both chromatographic and spectroscopic data sets obtained from various foods (Aishima and Nakai, 1991). Although gas chromatographic data of volatile components in soy sauce have already been studied in detail by pattern recognition analyses (Aishima, 1983; Ishihara *et al.*, 1996), detailed research has not been carried out on nonvolatile components of soy sauce.

Initially, we have attempted to establish the optimum HPLC conditions for soy sauce analysis. This was followed by differentiation of resulting HPLC profiles according to their origins by applying pattern recognition techniques. Isolation of components playing key roles in such differentiation and their structural elucidation was the final goal of this research.

## 3. MATERIALS AND METHODS

### 3.1. Materials

Two types of soy sauce, 51 samples of 8 brands (A-H) of dark-colored fermented soy sauces and 37 samples of 6 brands (I-N) of light-colored fermented soy sauces were purchased monthly in local markets in Japan.

### 3.2. HPLC Analysis

HPLC analysis was carried out with a Shimadzu liquid chromatographic system consisting of a SIL-10A autoinjector, LC-10AD pumps and SPD-10A detector (Shimadzu Corp., Kyoto, Japan). Ten µL of each sample were directly injected on a Wakosil-II 5C18 HG reversed phase column (250 mm x 4.6 mm i.d.; Wako Pure Chemicals Industries, Tokyo, Japan) fitted with a pre-column (30 mm x 4.6 mm i.d.) packed with Wakosil-II 5C18 HG. The column temperature and flow rate were 15°C and 0.8 ml/min, respectively. The peaks were monitored at 280 nm. The chromatography was performed using 0.05% trifluoroacetic acid (TFA) in water (solvent I) isocratically for 20 minutes, followed by a linear gradient of 90% acetonitrile containing 0.05% TFA (solvent II) from 0% to 25% for 270 min, and a further increase of solvent II from 25% to 50% for 50 min. After each analysis, the column was washed with tetrahydrofuran. Data was acquired by Labchart 180 (System Instruments Co., Ltd., Tokyo, Japan).

### 3.3. Pattern Recognition

HPLC data sets of dark-colored soy sauce and light-colored soy sauce were individually analyzed by pattern recognition techniques to extract information on brand differentiation. Fifty main peak areas were quantified and were used as variables. Cluster analysis, an unsupervised patten recognition technique, was used to investigate the mutual relationships among HPLC profiles of soy sauce samples. However, supervised pattern recognition techniques which use brands as a criterion variable are more informative on contribution of each peak to the sample classification. As supervised pattern recognition, linear discriminant analysis (LDA) and soft independent modelling of class analogy (SIMCA) were used to discriminate HPLC profiles according to brands. In addition to the conventional LDA, LDA using genetic algorithm (GA-LDA) (Leardi et al., 1992) was used. In the GA-LDA, the combinations of peaks are optimized to maximize the correct discrimination ratio of samples, similar to the approach of Forrest (1993). Cluster analysis and LDA were performed by SPSS for Windows, ver. 6.1 (SPSS Inc., Chicago, IL). SIMCA was carried out by UNSCRAMBLER ver. 5.3 (CAMO AS, Trondheim, Norway).

## 3.4. Isolation of Components

Peaks indicated by their importance in pattern recognition analysis were isolated and purified by the following procedure. Three liters of soy sauce was extracted three times with 1.5 liters of ethyl acetate at room temperature for 3 minutes. The resulting aqueous layer was adjusted to pH 2.0 with HCl and extracted again three times with 1.5 liters of ethyl acetate for 3 minutes. The upper layer was collected and concentrated to about 11 mL with a rotary evaporator. The concentrate was fractionated by preparative Wakosil-II 5C18 HG HPLC (300 mm x 10 mm i.d fitted with a 50 mm x 10 mm i.d. pre-column). Further purification was carried out by Bio-gel P2 column (380 mm x 20 mm I.D., Bio-Rad Laboratories, Hercules, CA) chromatography, or by simply repeating the preparative reversed phase HPLC. The UV profile of each key component was obtained by a photodiode array detector SPD-M10A (Shimadzu Corp., Kyoto, Japan) and the UV spectra were used as indicators of components in the isolation and purification processes. In addition to the spectra obtained with the photodiode array detector the purity of isolated components were confirmed by TLC on precoated silica gel 60 plates (Merck & Co., Inc., Rahway, NJ) using $H_2SO_4$ visualization.

## 3.5. Instrumental Analysis

Melting points of the purified components were measured by Yanagimoto micromelting point apparatus (Yanagimoto Ltd., Kyoto, Japan). UV spectra of components dissolved in water were measured by a Hitachi Model 557 double wavelength double beam spectrophotometer (Hitachi Ltd., Tokyo, Japan). IR spectra were obtained by KBr discs with a JASCO FT/IR-7300 spectrophotometer (JASCO Corp., Tokyo, Japan). All NMR spectra of each component dissolved in DMSO-$d_6$ were obtained with a JNM-LA400 instrument (JEOL Ltd., Tokyo, Japan). Mass spectra were obtained with a FRIT-FAB JMS-AX2000 LC/MS system (JEOL Ltd., Tokyo, Japan).

## 4. RESULTS

### 4.1. HPLC Analysis

A typical HPLC profile of dark-colored soy sauce is shown in Figure 2. More than 100 peaks were commonly observed in all HPLC profiles, but their peak areas varied considerably from brand to brand. Coefficients of variation (CV) calculated for the 50 major peaks selected for pattern recognition deviated from 1% to 27%, but for the majority of peaks were less than 10%. Areas of peaks in dark-colored soy sauce were generally larger than those of corresponding light-colored soy sauce.

### 4.2. Clustering and LDA of Dark-Colored Soy Sauce

All samples were orderly clustered according to their original brands as shown in Figure 3. This neat clustering indicated that significant differences in composition patterns exist among products in different brands. Supervised pattern recognition succeeded in differentiating eight brands of soy sauce based on their entire HPLC profiles with SIMCA, or with three or four specific peaks in LDA. All samples were separated into 8 groups corresponding to their original brands, A-H, on the basis of peaks 27, 32 and 42 which were selected by LDA as shown in Figure 4.

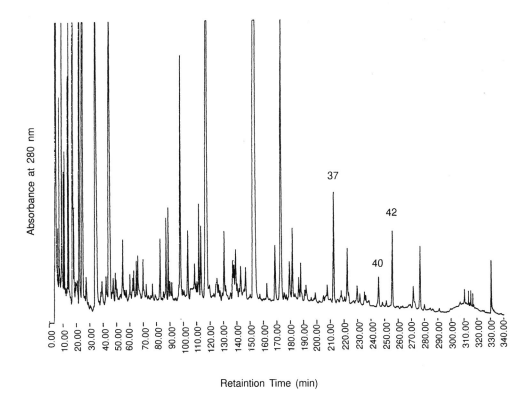

**Figure 2.** Reversed-phase HPLC profile of dark-colored soy sauce.

## 4.3. Clustering and LDA Using Genetic Algorithm of Light-Colored Soy Sauce

HPLC profiles of all light-colored soy sauces were generally accurately classified according to their brands as shown in Figure 5. Only one sample was misclassified. GA-LDA applied to the HPLC data of light-colored soy sauce selected peaks 8, 27, and 40 as the most effective peak combination for brand differentiation. All samples were classified into six groups corresponding to brands (I-N) in the three dimensional scatter plot composed of peaks 8, 27, and 40 as shown in Figure 6.

## 4.4. SIMCA of Light-Colored Soy Sauce

The discrimination power calculated in SIMCA indicates the contribution of each peak to discriminating between two brands. Peaks which contribute highly to the differentiation of light-colored soy sauces were listed in Table 1, along with the means and standard deviations of their discrimination power.

## 4.5. Isolation of Key Components

From the results of LDA, GA-LDA and SIMCA, peaks 37, 40 and 42 were selected as the most important ones. The preparative HPLC chromatogram of the ethyl acetate ex-

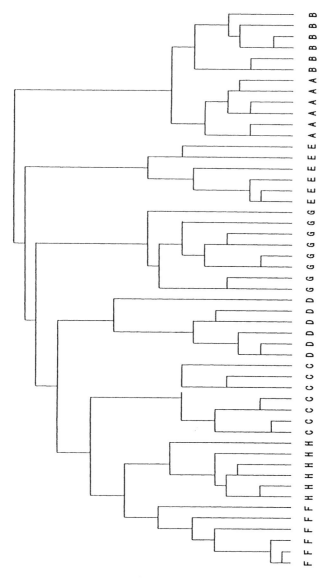

**Figure 3.** Clustering of HPLC profiles of dark-colored soy sauce, A-H: brands.

tract obtained from the aqueous layer at pH 2.0 is shown in Figure 7. Three key components were indicated as compounds **A**, **B**, and **C**, respectively. The yields of **A**, **B**, and **C** from three liters of soy sauce were 33.1, 24.6 and 5.0 mg, respectively.

## 4.6. Structure Elucidation of Three Compounds

Physicochemical properties and the results of $^1$H-NMR and $^{13}$C-NMR spectra of purified **A**, **B** and **C** are summarized in Tables 2 and 3, respectively. Assignments of $^1$H- and $^{13}$C-NMR spectra of compounds **A**, **B** and **C** were carried out on the basis of $^{13}$C-$^1$H COSY and HMBC spectra coupled with $^1$H-$^1$H COSY spectra.

# Differentiation of Soy Sauce Types by HPLC Profile Pattern Recognition 123

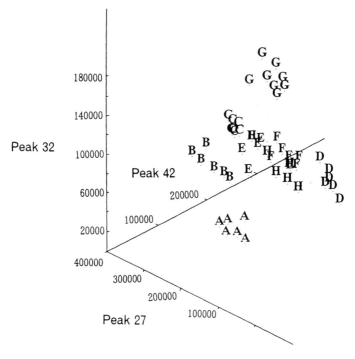

**Figure 4.** Three-dimensional scatter plot of dark-colored soy sauce samples based on peaks selected by LDA.

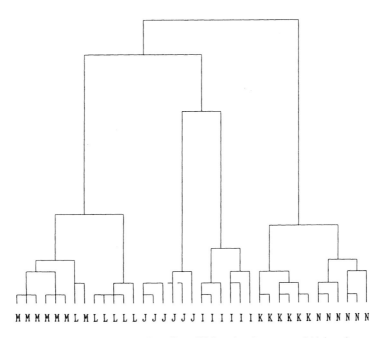

**Figure 5.** Clustering of HPLC profiles of light-colored soy sauce, I-N: brands.

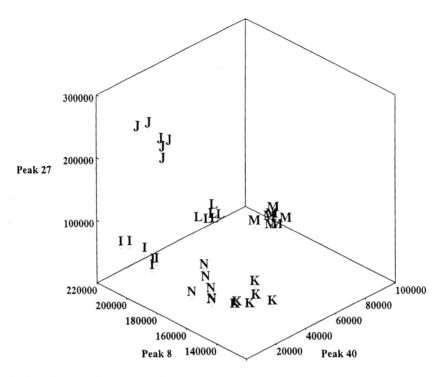

Figure 6. Three-dimensional scatter plot of light-colored soy sauce samples based on peaks selected by GA-LDA.

Compound **A** was obtained as an off-white amorphous powder and its melting point was 212–218°C with decomposition. The $\lambda_{max}$ of **A** in water were at 248 and 302 nm. Absorption bands in the IR spectra were observed at 3420, 1620, 1520, 1450, 1250, 1200, 1110 and 840 cm$^{-1}$. Compound **B** was a pale-yellow amorphous powder and its melting point was 138–144°C with decomposition. The $\lambda_{max}$ of **B** in water was at 260 nm and IR absorption bands were observed at 3450, 1620, 1580, 1520, 1360, 1250 and 840 cm$^{-1}$. Compound **C** was a pale-yellow amorphous powder and its melting point was 209–212°C with decomposition. The $\lambda_{max}$ of **C** in water was at 265 nm. The absorption bands in the IR spectra were observed at 3450, 1650, 1640, 1620, 1540, 1520, 1510, 1460, 1420, 1220, 1100, 1050 and 840 cm$^{-1}$.

The sharp signals at about 8.3 ppm commonly appearing in the $^{1}$H-NMR spectra of compounds **A**, **B** and **C** were typical for the C-2 protons in isoflavones. The chemical shifts of their $^{1}$H-NMR and $^{13}$C-NMR spectra suggested that daidzein, genistein and 8-hydroxy-

Table 1. Discrimination power of peaks in SIMCA

| Peak no. | Mean | Std. dev. |
|---|---|---|
| 9 | 13.0 | 6.61 |
| 28 | 11.0 | 7.08 |
| 31 | 10.7 | 8.88 |
| 37 | 23.2 | 13.00 |
| 40 | 13.1 | 7.32 |
| 42 | 17.9 | 11.10 |

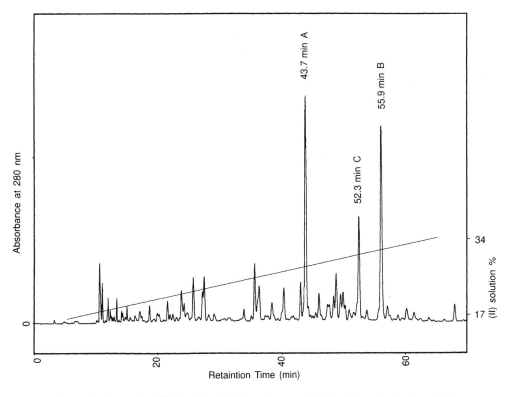

**Figure 7.** Preparative HPLC profile of ethyl acetate extract obtained after adjusting to pH 2.0.

genistein were present as structural components. The [M+H]$^+$ ion peaks at m/z 387, 403 and 419 in the high-resolution FAB-MS of compounds **A**, **B** and **C** indicated that their molecular formulas were $C_{19}H_{14}O_9$, $C_{19}H_{15}O_{10}$ and $C_{19}H_{15}O_{11}$, respectively. Fragments of 253, 269 and 285 in negative FAB-MS [M-H]$^-$ were attributable to the three isoflavone skeletons, daidzein, genistein and 8-hydroxygenistein. Considering the NMR and FAB mass spectral data, the three purified components were determined to be as isoflavone derivatives.

In addition to the signals assigned to isoflavone skeletons, signals of 2 protons and 4 carbons were observed in every $^1$H- and $^{13}$C-NMR spectrum. From the two signals (positions 2" and 3" in Table 3) at 4.6 and 5.2 ppm in $^1$H-NMR spectra and at 71 and 79 ppm in the $^{13}$C-NMR spectra, the existence of two methenyl groups connected to oxygen atoms (-CH-O-) in

**Table 2.** The physico-chemical properties A, B, and C

| | A | B | C |
|---|---|---|---|
| Appearance | White powder | Pale yellow powder | Pale yellow powder |
| M. P. | 212-218 °C (dec.) | 138-144 °C (dec.) | 209-212 °C (dec.) |
| UVmax | 248, 302 nm | 260 nm | 266 nm |
| Negative FAB-MS(m/z) | 385[M-H]$^-$, 253[daidzein-H]$^-$ | 401[M-H]$^-$, 269[genistein-H]$^-$ | 417[M-H]$^-$, 285[8-hydroxygenistein-H]$^-$ |
| HRFAB-MS(m/z) | | | |
| Calcd. | 387.0722[M+H]$^+$ | 403.7061[M+H]$^+$ | 419.0639[M+H]$^+$ |
| Found. | $C_{19}H_{14}O_9$ | $C_{19}H_{14}O_{10}$ | $C_{19}H_{14}O_{11}$ |

**Table 3.** $^1$H(400MHz) and $^{13}$C(100MHz) NMR data of A, B, and C in $d_6$-DMSO

| Position | A δ$^{13}$C(ppm) | A δ$^1$H(ppm) | A J(Hz) | A HMBC | B $^{13}$C(ppm) | B δ$^1$H(ppm) | B J(Hz) | B HMBC | C δ$^{13}$C(ppm) | C d$^1$H(ppm) | C J(Hz) | C HMBC |
|---|---|---|---|---|---|---|---|---|---|---|---|---|
| 2 | 153.1 | 8.33 s | | | 154.4 | 8.35 s | | | 154.3 | 8.50 s | | |
| 3 | 123.7 | | | H-2, 2' | 122.6 | | | H-2, 2' | 122.1 | | | H-2, 2' |
| 4(C=O) | 174.6 | | | H-2, 5 | 180.4 | | | 5-OH | 180.8 | | | H-2 |
| 5 | 127.0 | | | | 161.7 | | | H-6, 5-OH | 152.8 | | | H-6, 5-OH |
| 6 | 115.0 | 8.02 d | 8.8 | H-8 | 98.8 | 6.35 d | 2.2 | H-8 | 98.1 | 6.42 s | | 5-OH |
| 7 | 161.9 | 7.07 dd | 2.2, 8.8 | | 163.5 | | | | 151.6 | | | H-6, 3'' |
| 8 | 102.1 | 7.15 d | H-5, 8, 3'' 2.2 | H-6 | 93.6 | 6.59 d | 2.2 | H-6, 8, 3'' | 126.9 | | | H-6 |
| 9 | 157.0 | | | H-2, 8 | 157.3 | | | H-2, 8 | 145.1 | | | H-2 |
| 10 | 118.1 | | | H-6, 8 | 105.8 | | | H-6, 8 | 105.7 | | | H-6, 5-OH |
| 5-OH | | s | | | | 12.90 | | | | 12.30 br | | |
| 8-OH | | | | | | | | | | 8.72*br | | |
| 1' | 122.3 | | | H-2, 3' | 121.0 | | | H-2, 3' | 121.1 | | | H-2, 3' |
| 2' | 130.0 | 7.40 d | 8.6 | H-3' | 130.1 | 7.39 d | 8.8 | H-3' | 130.1 | 7.39 d | 8.8 | |
| 3' | 115.0 | 8.61 d | 8.7 | H-2' | 115.1 | 6.83 d | 8.8 | H-2' | 115.0 | 6.81 d | 8.8 | H-2' |
| 4' | 157.2 | | | H-2', 3' | 157.5 | | | H-2', 3' | 157.4 | | | H-2', 3' |
| 4'-OH | | 9.45 br | | | | 9.55 br | | | | 9.55*br | | |
| 1''(C=O) | 168.4* | | | H-2'', 3'' | 168.3* | | | H-2'', 3'' | 168.6* | | | H-2'', 3'' |
| 2'' | 72.0 | 4.41 d | 2.2 | H-3'' | 71.0 | 4.56 d | 2.7 | H-3'' | 71.1 | 4.63 d | 2.9 | H-3'' |
| 3'' | 78.8 | 5.02 d | 2.2 | H-2'' | 78.7 | 5.22 d | 2.7 | H-2'' | 79.7 | 5.32 d | 2.9 | H-2'' |
| 4''(C=O) | 171.0* | | | H-2'', 3'' | 171.5* | | | H-2'', 3'' | 171.0* | | | H-2'', 3'' |

* Assignments for these signals within the same column may be interchanged.

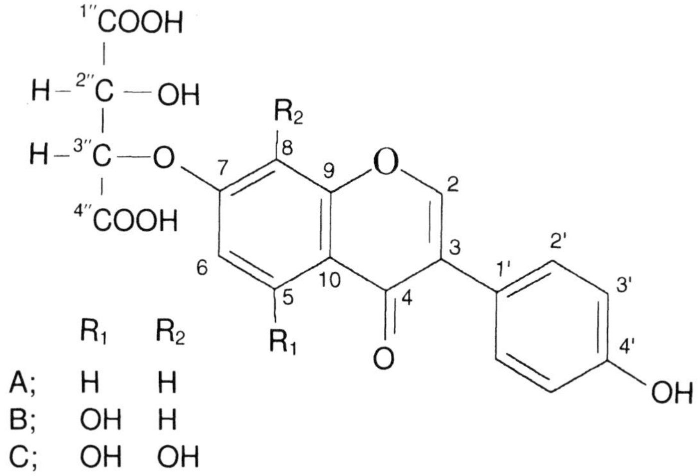

## Figure 8. The chemical structures of compounds A, B and C.

A, B and C were strongly suggested. Further, from the two signals (positions 1" and 4" in Table 3) at 168 and 171 ppm in the $^{13}$C-NMR spectra, we assigned them as two carboxyl groups (-COOH) for the moiety in question in **A**, **B** and **C**. Thus, the partial structures were elucidated as 2,3-dihydroxysuccinic acid or tartaric acid linked to isoflavone skeletons. Further interpretation of the HMBC spectra was made to confirm the bond between the isoflavone skeletons and tartaric acid. Every HMBC spectrum indicated the existence of an ether linkage between the C-7 carbons in isoflavones and C-3" protons in tartaric acid. Finally, the structures of **A**, **B** and **C** were elucidated as shown in Figure 8.

According to quantitative HPLC analysis, quantities of these three isoflavones increased during the *moromi* fermentation as shown in Figure 9.

## 5. DISCUSSION

Pattern recognition analysis applied to HPLC profiles of soy sauces successfully differentiated samples according to their original brands. Although time consuming, HPLC was es-

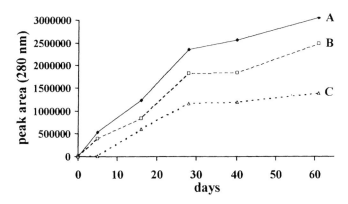

**Figure 9.** Increase of three isoflavone contents during fermentation (in *moromi*) of soy sauce.

tablished as an appropriate method for comparing the composition of nonvolatiles among different brands of soy sauces, as already shown to be in the analysis of many other foods. Further, it was possible to determine a component or a group of components playing key roles in such differentiation as shown by the three new isoflavones identified in this research. To date, the presence of three soy isoflavone aglycones, daidzein, genistein and glycitein, in soybeans and soy foods including soy sauce has been well documented and their quantities in such soybean products have been determined (Wang and Murphy, 1994a, 1994b).

Compounds linked to tartaric acid are not commonly found in foods. However, caffeoyl ester of tartaric acid has been found in peanuts (Snook et al., 1994). An unconventional isoflavone, 6,7,4'-trihydroxyisoflavone (factor 2), was isolated from tempeh, a fermented soy food from Indonesia (György et al., 1964). However, this is the first report on the presence of isoflavone derivatives linked to tartaric acid with an ether linkage. Although isoflavones are predominately present as glucosides in soybean (Coward et al., 1993), their aglycones are liberated from the conjugates during fermentation periods of soy foods (Wang and Murphy, 1996). New isoflavone derivatives, **A**, **B**, and **C**, found in this study seem to be generated during soy sauce fermentation. To date, these isoflavone derivatives have not been found in any other soy foods. Therefore, it could be assumed that compounds **A**, **B**, and **C** were synthesized by the enzymic action of a species of *Aspergillus* fungi. Further, it might be assumed that there were large differences in the synthetic activities of enzymes among strains of *Aspergillus* fungi, since compounds **A**, **B** and **C** were found to be the most important components for differentiating brands.

Recently, many isoflavones transformed by various micro-organisms have been reported (Nakahara et al., 1986; Klaus et al., 1993). However, the transformation mechanisms have not been fully elucidated. The aglycone of compound **C**, 8-hydroxygenistein, was found in a culture of *Aspergillus niger* by Umezawa, et al. (1975). Compound **C** may be synthesized from **B** or from 8-hydroxygenistein, which is possibly synthesized from genistein during the *koji* stage. Additional research indicated varied biological activities of flavonoid compounds, such as antifungal (Naim et al., 1974), antioxidative (Naim et al., 1976) and estrogenic (Murphy, 1982) effects. Other research concerned taste of isoflavones and their bitterness and astringency were also reported (Havesteen, 1983). Future investigations into the biological activities of these three new isoflavones may be conducted in addition to an examination of their biosynthetic mechanisms.

## ACKNOWLEDGMENTS

We thank Mr. M. Mizutani of Japan Tobacco Inc., and Mr. N. Miyabayashi and Ms. S. Kohno of JEOL Ltd. for NMR measurements, Mr. M. Kusama of Japan Tobacco Inc. and Mr. K. Matsuura of JEOL Ltd. for their FAB-MS measurements.

## REFERENCES

Aishima, T. Relationship between gas chromatographic profiles of soy sauce volatiles and organoleptic characteristics based on multivariate analysis. In *Instrumental Analysis of Foods*; Charalambous, G.; Inglett, G., Eds.; Academic Press Inc.: New York, NY, **1983**; Vol. 1, pp. 37–56.

Aishima, T.; Nakai, S. Chemometrics in flavor research. *Food Rev. Internat.* **1991**, *7*, 33–101.

Coward, L.; Barnes, N.C.; Setchell, K. D. R.; Barnes, S. Genistein, daidzein, and their β-glycoside conjugates: Antitumor isoflavones in soy bean foods from American and Asian diets. *J. Agric. Food Chem.* **1993**, *41*, 1961–1967.

Forrest, S. Genetic algorithms: principles of natural selection applied to computation, *Science* **1993**, *261*, 872–878.

Forss, D.A.; Sugisawa, H. Brief reviews of dairy and soy products. In *Flavor Research: Recent advances;* Teranishi, R., Flath, R.A., Sugisawa, H., Eds.; Marcel Dekker: New York, NY, **1981**; pp.356–372.

György, P.; Murata, K.; Ikehata, H. Antioxidants isolated from fermented soy beans (Tempeh). *Nature* **1964**, *203*, 870–872.

Havesteen, B. Flavonoids, a class of natural products of high pharmacological potency. *Biochem. Pharmac.* **1983**, *32*, 1141–1148.

Ishihara, K.; Honma, N.; Matsumoto, I; Imai, S.; Nakazawa, S.; Iwafuchi, H. Comparison of volatile components in soy sauce (*koikuchi shoyu*) produced using *Aspergillus sojae* and *Aspergillus oryzae*. *Nippon Shokuhin Kogaku Kaishi* **1996**, *43*, 1063–1074.

Kinoshita, E.; Yamakoshi, J.; Kikuchi, M. Purification and identification of an angiotensin-I converting enzyme inhibitor from soy sauce. *Biosci. Biotech. Biochem.* **1993**, *57*, 1107–1110.

Klaus, K.; Borger-Papendorf, G; Barz, W. Formation of 6,7,4'-trihydroxyisoflavone (factor 2) from soybean seed isoflavones by bacteria isolated from tempeh. *Phytochemistry* **1993**, *34*, 979–981.

Leardi, R.; Boggia, R.; Terrile, M. Genetic algorithm as a strategy for feature selection. *J. Chemometrics* **1992**, *6*, 267–281.

Murphy, P. A. Phytoestrogen content of processed soybean products. *Food Technol.* **1982**, *36*, 60–64.

Naim, M.; Gestetner, B.; Zilkah, S.; Birk Y.; Bondi, A. Soybean isoflavone characterization, determination, and antifungal activity. *J. Agric. Food Chem.* **1974**, *22*, 806–810.

Naim, M.; Gestetner, B.; Bondi A.; Birk, Y. Antioxidative and antihemolytic activities of soybeans isoflavones. *J. Agric. Food Chem.* **1976**, *24*, 1174–1177.

Nakahara, S.; Tahara, S.; Mizutani, J.; Ingham, J. L. Transformation of dioprenylated isoflavone 2'-hydroxylupalbigenin by *Aspergillus flavus*. *Agric. Biol. Chem.* **1986**, *50*, 863–873.

Snook M. E.; Lynch, R. E.; Culbreath, A. K.; Costello, C. E. 2,3-di-(E)-caffeoyl- (2R,3R)-(+)-tartaric acid in terminals of peanut (*Arachis hypogaea* L.) varieties with different resistances to late leaf spot disease [*Cercosporidium personatum* (Berk. & M.A. Curtis) Deighton] and the insects tobacco thrips [*Frankliniella fusca* (Hinds)] and potato leafhopper [*Empoasca fabae* (Harris)]. *J. Agric. Food Chem.* **1994**, *42*, 1572–1574.

Umezawa, H.; Tobe, H.; Shibamoto, N.; Nakamura, F.; Nakamura, K.; Matsuzaki, M.; Takeuchi, T. Isolation of isoflavones inhibiting dopa decarboxylase from Streptomyces. *J. Antibiotics* **1975**, *28*, 947–952.

Wang, H. -J.; Murphy, P. A. Isoflavone content in commercial soybean foods. *J. Agric. Food Chem.* **1994a**, *42*, 1666–1673.

Wang, H. -J.; Murphy, P. A. Isoflavone composition of American and Japanese soybeans in Iowa: effects of variety, crop year, and location. *J. Agric. Food Chem.* **1994b**, 42, 1674–1677.

Wang, H. -J.; Murphy, P. A. Mass balance study of isoflavones during soybean processing. *J. Agric. Food Chem.* **1996**, *44*, 2377–2383.

# INDUCTION OF OXIDATIVE STRESS BY REDOX ACTIVE FLAVONOIDS

William F. Hodnick,[2] Sami Ahmad,[1] and Ronald S. Pardini[1]

[1]Department of Biochemistry
University of Nevada
Reno, Nevada 89557
[2]Department of Pharmacology
Yale University
New Haven, Connecticut 06520

## IN VITRO PRODUCTION OF REACTIVE OXYGEN SPECIES

Flavonoids have been reported to possess widespread biological activities including antihelminthic (McClure 1975), antimicrobial (McClure 1975; Kühnau 1976; Middleton 1984; Havsteen 1983), antimalarial (Iwu et al 1986; Khalid et al 1986), antineoplastic (Mabry and Ulebelen 1980; Middleton and Kandaswami 1994), cytotoxic (McClure 1975; Kühnau 1976; Havsteen 1983; Middleton and Kandaswami 1994; Kupchan et al 1965; MacGregor 1984; Harborne et al 1975), mutagenic (Middleton 1984; Brown 1980), carcinogenic (Brown 1980), anti-carcinogenic (McClure 1975; DeCross 1982), and anti-oxidant (McClure 1975; Middleton 1984; Havsten 1983; Middleton and Kandaswami 1994; Brown 1980; Salah et al 1995) action. The flavonoids are especially reactive towards redox-active enzymes including inhibition of mitochondrial succinoxidase (Hodnick et al 1986) and NADH-oxidase activities (Hodnick et al 1998), arachidonate metabolism (Middleton 1984), neutrophil NADPH-oxidase (Tauber et al 1984), aldose reductase (Brown 1980), xanthine oxidase (Hayashi et al 1988; Chang et al 1993; Iio et al 1986), DT-diaphorase (Liu et al 1990), mouse brain NADPH-diaphorase (Tamura et al 1994) and glutathione reductase (Elliot et al 1992). The isoflavonoid derivative rotenone is a potent inhibitor of NADH-CoQ reductase and has become a standard inhibitor of mitochondrial respiration (Figure 1).

During our studies on the structural activity relationships of flavonoids for their ability to inhibit beef heart mitochondrial succinoxidase (complexes II, III, and IV) (Hodnick et al 1986; Bohmont et al 1987; Hodnick et al 1994) and NADH-oxidase activity (complexes I, III, and IV) (Hodnick et al 1989; Bohmont et al 1987; Hodnick et al 1994), we

*Flavonoids in the Living System*, edited by Manthey and Buslig
Plenum Press, New York, 1998.

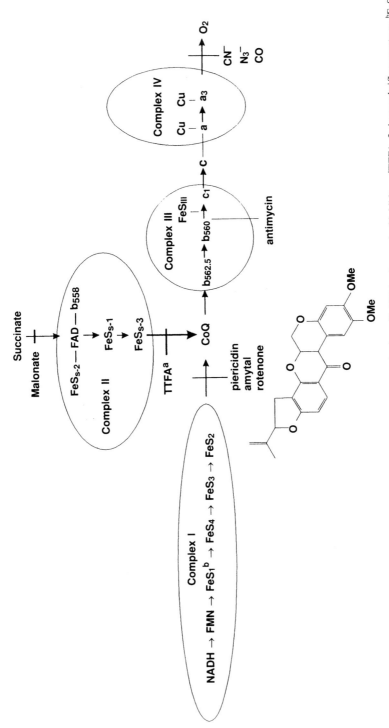

**Figure 1.** Schematic of the mitochondrial electron transport chain including complexes and points of inhibition by standard inhibitors. [a]TTFA: 2-thenoxyltrifluoroacetone; [b]FeS: iron sulfur center.

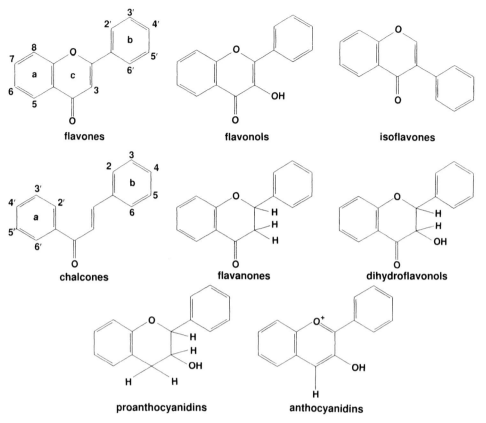

Figure 2. Flavonoid classes and numbering system. Adapted from Hodnick et al. 1986.

included flavonoids from different subclasses possessing identical hydroxyl configurations (Figure 2). In addition we compared various 3,5,7-trihydroxyl flavonoids with different configurations of the B ring hydroxyl for their ability to inhibit mitochondrial respiration. By comparing the flavonoid $I_{50}$ values we observed that NADH-oxidase was more sensitive than succinoxidase and that the structural requisites for inhibition of NADH-oxidase were a $C_{2,3}$ double bond, 4-keto group, and a B ring hydroxylation pattern that supported redox activity. The hydroxylation pattern that supported the greatest inhibition was pyrogallol> catechol> meta-dihydroxy> monohydroxyl> no hydroxyls. We observed that the structural requisites for inhibition of succinoxidase were also those structures that supported redox activity, and the B ring hydroxylation pattern for inhibition of succinoxidase and NADH-oxidase were the same. The sensitivity of succinoxidase towards representatives from each class of flavonoids possessing the same hydroxylation pattern was chalcone> flavone> flavonol> dihydroflavonol> anthrocyanidin, with the proanthrocyanidin catechin being non-inhibitory. The observed structural requisites for inhibition of succinoxidase were similar to those observed for a series of redox active model phenolic compounds including p-hydroquinone, catechol and pyrogallol (Cheng and Pardini 1978; Cheng and Pardini 1979). This indicated that the ability of flavonoids and phenolic compounds to inhibit mitochondrial succinoxidase was related to their ability to undergo redox activity.

Most flavonoids exhibited preferential inhibition of NADH-oxidase suggesting that these compounds inhibit predominantly in complex I. Others preferentially inhibited succinoxidase activity indicating that these compounds inhibit preferentially in complex II. Those flavonoids that inhibit NADH-oxidase and succinoxidase to a similar extent, could either inhibit complexes I and II in a similar manner or inhibit in either complexes III or IV which is the common terminal electron transport portion of NADH-oxidase and succinoxidase.

The specific site and mechanism for flavonoid inhibition of the respiration chain are unknown. However, the structural requisites for inhibition and the ability of exogenous thiols to prevent inhibition is consistent with the flavonoid interacting directly with the respiratory chain to form an $o$-quinone which in turn is capable of interacting with key thiol groups of the respiratory chain to inhibit respiration.

An alternative hypothesis for inhibition of the electron transport chain would be complexation of the flavonoid with FeS centers of complexes I, II or III thereby inhibiting electron transport.

The dinuclear fold at the active site of enzymes utilizing NADH, NADPH, ATP and cAMP forms a hydrophobic pocket which is rather nonspecific in that it will bind aromatic compounds. It has been postulated that flavonoids may interact at this hydrophobic pocket, which may account for flavonoid specificity for inhibiting NADH-oxidase (Fessel et al 1979; Schultz and Shrimer 1979) but wouldn't account for inhibition of succinoxidase.

For those flavonoids that redox cycle, the production of $O_2^{\bullet-}$ and $H_2O_2$ may also oxidize key thiols in the respiratory chain and inhibit mitochondrial electron transport similar to redox active quinones (Pritsos et al 1982; Pritsos and Pardini 1984; Pisani et al 1986). In addition, iron sulfur centers have been shown to be particularly sensitive to ROS, so the mitochondrial FeS centers could also be a target of ROS generated by the flavonoids.

The structural inhibition relationships of flavonoid interaction with the mitochondrial electron transport chain has been discussed in detail (Hodnick and Pardini 1998).

We observed that four of the original fourteen flavonoids tested for inhibition of mitochondrial electron transport produced a substrate-independent increase in oxygen consumption that was insensitive to cyanide ($CN^-$) (Hodnick et al 1986). The data shown in Figure 3 demonstrate that myricetin, quercetagetin, quercetin, and delphinidin all induced a respiratory burst when added to $CN^-$ inhibited mitochondria indicating that terminal electron transport through cytochrome oxidase was not involved in the flavonoid induced respiration (oxygen consumption).

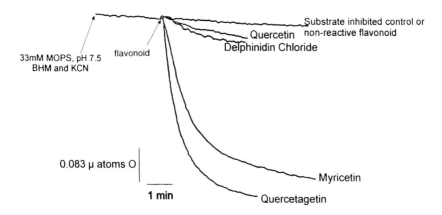

**Figure 3.** Flavonoid induced respiratory bursts in beef heart mitochondria. Adapted from Hodnick et al 1986.

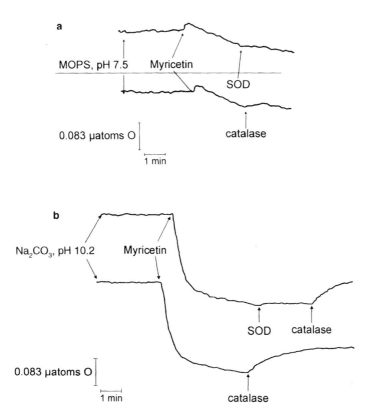

**Figure 4.** a) Non-enzymatic auto-oxidation of myricetin at pH 7.5. b) Non-enzymatic auto-oxidation of myricetin at pH 10.2. Adapted from Hodnick et al 1986.

Next, we attempted to characterize the nature of the cyanide insensitive flavonoid-induced respiratory burst. Initially, we observed that the representative flavonoid, myricetin, underwent a slow rate of auto-oxidation at pH 7.5 as measured by oxygen consumption (Figure 4a). The addition of superoxide dismutase (SOD) and catalase to this reaction mixture resulted in a decrease in oxygen consumption confirming that $O_2^{\bullet-}$ and $H_2O_2$ were produced during the auto-oxidation of myricetin. We also observed that the rate of auto-oxidation was enhanced at pH 10 as were the SOD and catalase effects (Figure 4b) indicating that the increased respiratory burst was associated with increased production of $O_2^{\bullet-}$ and $H_2O_2$ (Figures 4a, 4b). We then observed that, at pH7.5, the rate of auto-oxidation of myricetin and quercetagetin was significantly enhanced by the addition of $CN^-$ (Figure 5). This implies that certain flavonoids undergo a pH sensitive auto-oxidation to produce $O_2^{\bullet-}$ and $H_2O_2$ and interact with $CN^-$ to generate a respiratory burst which also produces $O_2^{\bullet-}$ and $H_2O_2$ (data not shown).

Cyanide has been previously reported to catalyze the oxidation of α-hydroxy aldehydes and related compounds resulting in the production of $O_2^{\bullet-}$ (Robertson et al 1981; Mashino and Fridovich 1987). The proposed mechanism involves tautomerization of the α-hydroxyaldehyde to an enediol, addition of $CN^-$ to form cyanohydrin, and subsequent auto-oxidation to form $O_2^{\bullet-}$ and regeneration of $CN^-$, the catalyst.

Many of the flavonoids which interacted with $CN^-$ possessed the structural features of $C_{2,3}$-double bond, $C_3$ hydroxyl, and a $C_4$ keto group. This configuration resembles a 2-

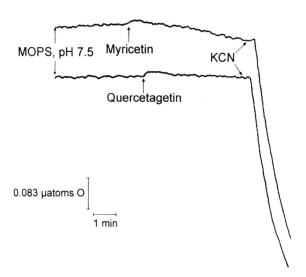

**Figure 5.** Effect of cyanide on the auto-oxidation of myricetin and quercetagetin. Adapted from Hodnick et al 1986.

hydroxyl unsaturated ketone which we propose is tautomerized to form a cyclic enediol which could react with $CN^-$ and auto-oxidize to form ROS.

In an analogous reaction, the reduced form of alloxan, dialuric acid, auto-oxidizes to form $O_2^{\cdot-}$ (Cohen and Heikkila 1984) and was shown to be reactive with $CN^-$ (Robertson et al 1981).

We observed that borate addition almost totally eliminated $CN^-$ stimulated auto-oxidation of rhamnetin and 7,8-dihydroxy flavone (Hodnick et al 1994). Since borate is known to inhibit the $CN^-$ catalyzed oxidation of α-hydroxyaldehydes, by analogy, we suggest that an enediol intermediate is involved in the $CN^-$ induced auto-oxidation of flavonoids.

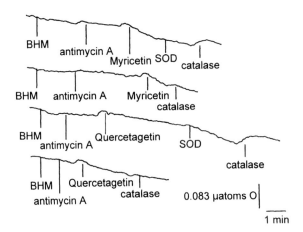

**Figure 6.** Flavonoid-induced respiratory bursts in mitochondria. Adapted from Hodnick et al 1986. BHM: beef heart mitochondria.

In order to assess the redox interactions between flavonoids and the mitochondrial respiratory chain, we inhibited mitochondria with antimycin to eliminate the $CN^-$ induced respiratory burst (occurring independently of the mitochondria (Figures 4 and 5)) and then monitored oxygen consumption polarographically. The data shown in Figure 6 indicate that myricetin and quercetagetin may interact with the mitochondrial respiratory chain to produce a small burst in respiratory activity that is also sensitive to SOD and catalase, implicating the production of $O_2^{\cdot-}$ and $H_2O_2$. The production or regeneration of oxygen was greatest when SOD and catalase were both present suggesting that $H_2O_2$ was a secondary product of the respiratory burst being generated by $O_2^{\cdot-}$ dismutation. Similar results were observed when the mitochondria were inhibited with azide (data not shown). We have shown that in the presence of the electron transport substrate succinate (Figure 7) myricetin can "draw off" electrons from the respiratory chain indicative of true redox cycling and not just autoxidation in the presence of mitochondria.

In summary we have observed that selected flavonoids 1) auto-oxidize, 2) produce a non-enzymatic $CN^-$ stimulated respiratory burst and 3) interact directly with the mitochondrial respiratory chain to produce a respiratory burst that involves the production of $O_2^{\cdot-}$ and $H_2O_2$. More recently, similar findings were observed for rhamnetin, baicalein, robinetin, norwogonin, and 7,8-dihydroxyflavone (data not shown) (Hodnick et al 1994).

Since all respiratory burst generating flavonoids acted similarly with regard to the production of $O_2^{\cdot-}$ and $H_2O_2$, we decided to assess their ability to produce the powerful oxidant $^{\cdot}OH$. The presence of $H_2O_2$ would likely result in the production of $^{\cdot}OH$ through the iron catalyzed Fenton reaction. Thus, in collaboration with Kalyanaraman (Hodnick et al 1989), we employed a spin trap technique in order to evaluate the production of $^{\cdot}OH$ during flavonoid auto-oxidation. We observed that quercetin and the spin trap DMPO + $Fe^{II}$-EDTA in air resulted in a characteristic DMPO-OH signal (Figure 8 panel a). As shown in Figure 8 panel b, the addition of ethanol produced an α-hydroxyethyl-DMPO spectrum. The addition of SOD or $FeCl_3$ enhances the DMPO-OH signal (c), and catalase (d) and the inclusion of the non-reactive $Fe^{III}$-desferrioximine (e) eliminated the signal.

The addition of ethanol and SOD serve as controls to rule out the generation of DMPO-OH from the decomposition of DMPO-superoxide adduct (DMPO-OOH$^-$) and insure that the DMPO-OH adduct derives from $^{\cdot}OH$. If superoxide generated the DMPO-OH

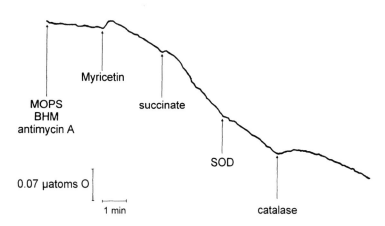

**Figure 7.** Redox cycling of myricetin with mitochondrial succinoxidase enzyme system (Complex II). Adapted from Hodnick *et al* 1986. BHM: beef heart mitochondria, MOPS: 3-(N-morpholino)-propane sulfonic acid.

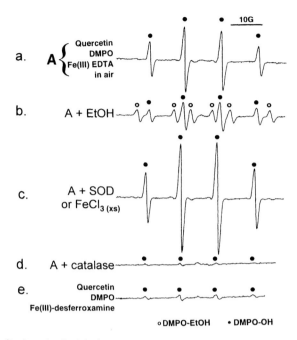

**Figure 8.** Production of hydroxyl radical during auto-oxidation of quercetin. Adapted from Hodnick et al 1989.

signal, then the addition of SOD would be expected to eliminate the signal. Our data show that SOD enhances the signal. In addition, the •OH radical can abstract an electron (actually H•) from ethanol to produce the α-hydroxyethyl radical which is also trapped by DMPO to generate a unique ESR signal distinct from the DMPO-OH adduct. Thus, the DMPO-OH signal comes from •OH. The enhancement of the DMPO-OH signal by SOD or $FeCl_3$ suggests that the flavonoid directly reduces the iron. Reduced iron is required to convert $H_2O_2$ to •OH (Figure 10). Since SOD converts superoxide to $H_2O_2$, the SOD induced increase in DMPO-OH signal indicates that SOD increases the steady state levels of $H_2O_2$ to increase the production of •OH through the Fenton reaction. These findings indicate that during the auto-oxidation of selected flavonoids $H_2O_2$ and iron are involved in the generation of •OH through the Fenton reaction.

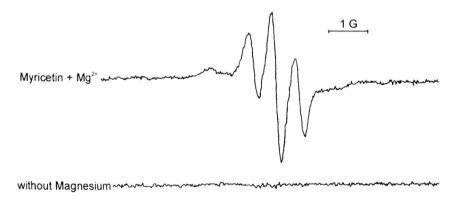

**Figure 9.** ESR spectra of spin-stabilized o-semiquinone species. Adapted from Hodnick et al 1989.

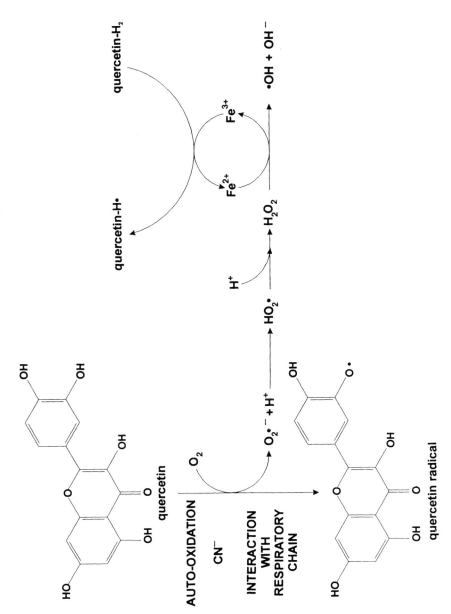

**Figure 10.** Proposed role of redox-active flavonoids in the generation of reactive oxygen species.

**Table 1.** Cyclic voltammetry of flavonoids

| Name | Class | Hydroxylation pattern | E2[1] (V vs SCE) | $\Delta Ep^2$, mV v=100 mV•sec$^{-1}$ | Ip$_a$, µA v=100 mV•sec$^{-1}$ |
|---|---|---|---|---|---|
| Galangin | Flavonol | 3, 5, 7 | 0.34[a] | | 4.6[b] |
| Kaempferol | Flavonol | 3, 5, 7, 4' | 0.17 | 61[c] | 5.9[b] |
| Morin | Flavonol | 3, 5, 7, 2', 4' | 0.14 | 61[c] | 14.7 |
| Quercetin | Flavonol | 3, 5, 7, 3', 4' | 0.06 -0.03[d] | 34 | 18.6 |
| Myricetin | Flavonol | 3, 5, 7, 3', 4', 5' | 0.06[a] | 55 | 26.5 |
| Delphinidin Chloride | Anthocyanidin | 3, 5, 7, 3', 4', 5' | [e] | 58 | 18.0 |
| Fisetin | Flavonol | 3, 7, 3', 4' | 0.14 | 49 | 20.4 |
| Fustin | (-) Dihydroflavonol | 3, 7, 3', 4' | 0.15 | 110 | 18.3 |
| Taxifolin | (+) Dihydroflavonol | 3, 5, 7, 3', 4' | 0.15 | 117 | 14.0 |
| Catechin | (+) Catechin (±) Catechin | 3, 5, 7, 3', 4' 3, 5, 7, 3', 4' | 0.15 0.13 | 66 | 17.0 |
| Cyanidin Chloride | Anthocyanidin | 3, 5, 7, 3', 4' | [f] | | 11.2 |
| Quercetagetin | Flavonol | 3, 5, 6, 7, 3', 4' | 0.06[d] 0.25[a] | 44 | 22.4 |
| Luteolin | Flavone | 5, 7, 3', 4' | 0.18 | 55 | 16.4 |
| Butein | Chalcone | 3, 4, 2', 4' | 0.13 | 28 | 36.0 |

[1] Mid Point Potental (Ep$_a$ + Ep$_c$)/2
[2] Peak potential separation (Ep$_a$ - Ep$_c$)
[a] Ep only (v=100 mV•sec$^{-1}$)
[b] The solubility is less than 5x10$^{-4}$M; a saturated solution was analyzed
[c] At (v=500 mV•sec$^{-1}$)
[d] The most prominent oxidation process
[e] Not electroactive in the investigated range
[f] Ill-defined oxidations
Adapted from Hodnick et al 1988

The addition of Fe$^{III}$-EDTA to flavonoids under anaerobic conditions resulted in a time dependent auxochromic shift in the flavonoid absorption spectrum and the reduction of Fe$^{III}$ for Fe$^{II}$ in the EDTA complex. These findings indicate that redox active flavonoids are capable of directly reducing iron to the ferrous form which would drive the Fenton reaction.

In order to assess the auto-oxidation of myricetin by ESR spectroscopy, Mg$^{++}$ was added to stabilize the myricetin semiquinone, which permitted the detection of the flavonoid semiquinone (Figure 9). The initial four redox active flavonoids act in a similar manner producing •OH and one or more *ortho*semiquinones.

These findings support the sequence proposed in Figure 10 representing quercetin oxidation to its semiquinone via auto-oxidation, CN$^-$-stimulated oxidation and/or oxidation by interaction with the respiratory chain with subsequent interaction of the semiquinone with molecular oxygen to form $O_2^{•-}$, $H_2O_2$, and •OH. In addition, the reduction of iron by the flavonoid quercetin is shown to participate in the Fenton chemistry and production of •OH.

Cyclic voltammetry (Table 1) demonstrated that the flavonoids that auto-oxidized and produced reactive oxygen species (ROS) possessed oxidation potentials ($E_{1/2}$) significantly lower (-30 to +60mV versus SCE (standard calomol electrode)) than those that did not auto-oxidize (+130 to +340mV versus SCE). The most potent flavonoid inhibitors of mitochondrial succinoxidase possessed lower midpoint redox potentials indicating that the electrochemical properties of flavonoids contribute to their ability to inhibit mitochondrial succinoxidase.[29] A similar relationship was observed for a series of structurally related naphthoquinones (Pisani et al 1986).

We have identified nine redox active flavonoids capable of auto-oxidation and production of ROS including myricetin, quercetin, quercetagetin, delphinidin, baicalein, robinetin, norworgonin, rhamnetin, and 7,8-dihydroxyflavone (Figure 11). Since these flavonoids are capable of generating ROS, they have the potential of generating oxidative insult. We elected to test this possibility in our *in vivo* insect model employing a model redox-active flavonoid, quercetin.

## IN VIVO PRODUCTION OF REACTIVE OXYGEN SPECIES

Our phytophagous insect model is comprised of the Cabbage Looper (*Trichoplusia ni*) which naturally feeds mostly on crucifers that contain low levels of quercetin and other pro-oxidants, the Southern Armyworm (*Spodoptera eridania*) which often encounters and consumes plants containing pro-oxidants like quercetin, and the Black Swallowtail Butterfly (*Papilio polyxenes*) which feeds with impunity on many plant species known to contain high levels of pro-oxidants (Pardini et al 1989). The insect model system possessed the anti-oxidant enzymes known to detoxify various intermediates of oxidative stress as shown in Figure 12. These three insects comprise our *in vivo* insect model of low, moderate and high dietary exposure to pro-oxidants. Accordingly, the endogenous activities of the anti-oxidant enzymes SOD and glutathione reductase in our insect model paralleled the natural exposure to dietary pro-oxidants and was inversely related to the toxicity of quercetin (Tables 2a, 2b) (Ahmad and Pardini 1990).

During the quercetin feeding studies, we observed that there was a dose dependent increase in SOD activity (Figure 13a) suggesting that quercetin feeding resulted in an $O_2^{\bullet-}$ challenge and a subsequent *in vivo* response by *S. eridania*. In addition, quercetin feeding resulted in a depression of catalase activity in a dose and time dependent fashion (Figure 13b) and a decrease in glutathione reductase activity in a dose-dependent manner (Figure 13c). These findings indicate that quercetin feeding results in a depression of two critical anti-oxidant enzymes and could enhance the insects' sensitivity toward ROS. The direct inhibition of glutathione reductase by quercetin was corroborated *in vitro* (Elliot et al 1992). Collectively, these observations are consistent with the conclusion that quercetin induces oxidative challenge *in vivo*. Similar findings were observed for *P. polyxenes* (Figure 14) with quercetin feeding resulting in enhanced SOD activity and depression of glutathione reductase (Pritsos 1988). Although quercetin induced an oxidative challenge, it was not acutely toxic to S.eridania or P. polyxenes, implying that their enhanced anti-oxidant enzyme system offered them protection to the pro-oxidant quercetin. In addition, in Figure 15 we report that feeding quercetin at much lower levels to *T. ni* results in a dose and time dependent reduction of GST-peroxidase activity. The inhibitory effect of quercetin towards glutathione reductase and GST-peroxidase could jeopardize their anti-oxidant defense and make them more susceptible to pro-oxidant challenge. In order to evaluate the role of $O_2^{\bullet-}$ in quercetin toxicity, we fed *P. polyxenes* a diet containing a non-toxic dose of quercetin. When the SOD inhibitor diethyldithiocarbamate (DETC) was also fed, we observed a depression in the endogenous SOD activity and an increase in quercetin toxicity (Pritsos et al 1988a; Pritsos et al 1991). These findings link the toxicity of quercetin to the production of ROS, especially in our insect model.

We have previously shown that $CN^-$ interacts non-enzymatically with quercetin to generate ROS (Hodnick et al 1994). So we elected to evaluate the potential synergism of

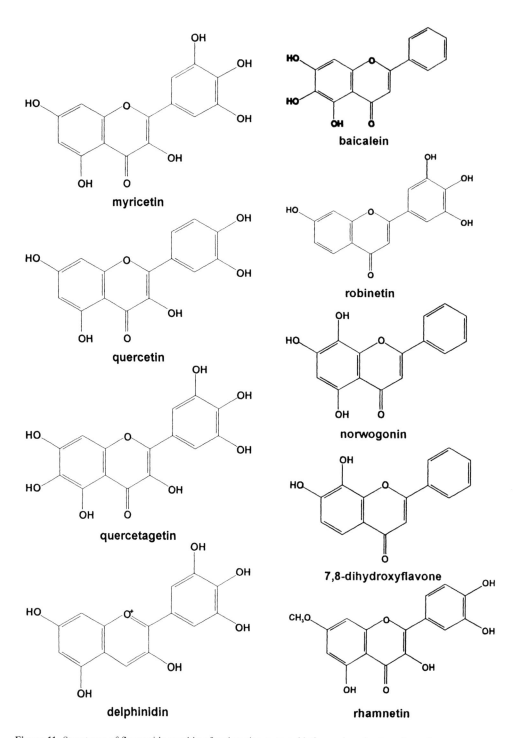

Figure 11. Structures of flavonoids capable of undergoing auto-oxidation and production of reactive oxygen species.

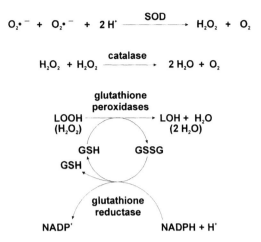

**Figure 12.** Anti-oxidant enzymes of the insect model.

toxicity between CN⁻ and quercetin *in vivo* in *T. ni*. Feeding *T. ni* quercetin enhanced the toxicity of CN⁻ when fed at a non-toxic dose as mortality was the highest when CN⁻ and quercetin were both present in the diet (Figure 16b). Furthermore, feeding *T. ni* both CN⁻ and quercetin resulted in lower larval weights than those fed separately at the same level (Figure 16a). These findings are consistent with a synergistic toxicity between CN⁻ and quercetin. In addition, feeding *T. ni* increasing levels of CN⁻ resulted in a dose dependent enhancement of quercetin induced lipid peroxidation (Figure 17a). Similarly, increasing the levels of dietary CN⁻ resulted in a dose related increase in quercetin induced protein oxidation (Figure 17b). These findings implicate the production of ROS by quercetin and CN⁻ in their synergistic toxicity *in vivo* and support an oxidative-stress mechanism of toxicity.

In summary:

1. Quercetin contains the redox-active group, catechol (Figure 11).
2. Quercetin redox potential is 0.06V vs. SCE, (Table 1).

**Table 2a.** The relative toxicity of quercetin towards the insect model by comparison of the $LC_{50}$ values

| | $LC_{50}$ values | |
| --- | --- | --- |
| *T.ni* | *S. eridani* | *P. polyxenes* |
| Cabbage looper | Southern armyworm | Black swallowtail butterfly |
| 0.0045% | >1% | >1% |

**Table 2b.** Endogenous anti-oxidant enzyme activities of the insect model (3rd instar)

| | *T.ni* (Cabbage looper) | *S. eridani* (Southern armyworm) | *P. polyxenes* (Black swallowtail butterfly) |
| --- | --- | --- | --- |
| SOD[a] | 1.0 ± 0.08 | 5.5 ± 1.20 | 7.4 ± 0.70 |
| Glutathione Reductase | 0.6 ± 0.09 | 0.6 ± 0.15 | 6.8 ± 2.0 |

[a] means ± S.D. as unit·mg$^{-1}$ protein·min$^{-1}$

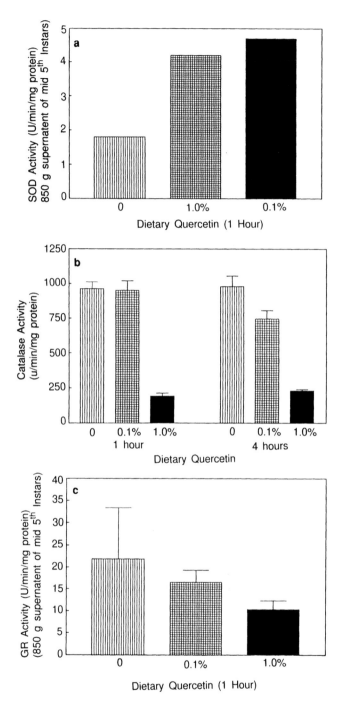

**Figure 13.** a)Superoxide dismutase (SOD) activity of *S. eridania* following dietary exposure to quercetin. b)Catalase (CAT) activity of *S. eridania* following dietary exposure to quercetin. c)Glutathione reductase (GR) activity of *S. eridania* following dietary exposure to quercetin. Adapted from Pritsos *et al* 1990.

# Induction of Oxidative Stress by Redox Active Flavonoids

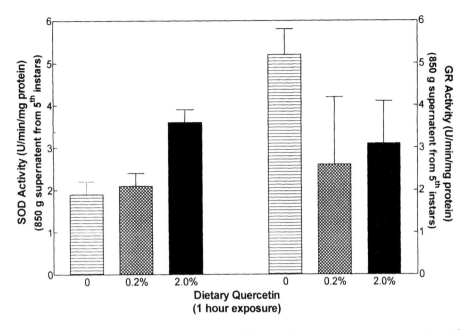

**Figure 14.** Antioxidant enzyme activities of *P. polyxenes* following dietary exposure to quercetin. Adapted from Pritsos *et al* 1988.

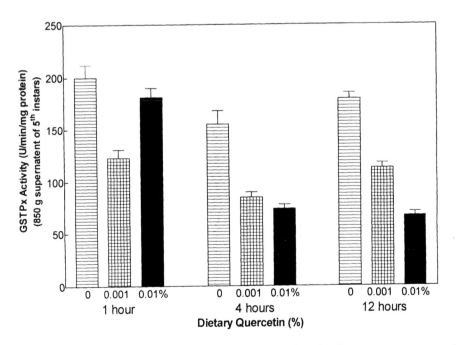

**Figure 15.** Glutathione transferase peroxidase (GSTPx) activity in *T.ni* following dietary exposure to quercetin. Adapted from Ahmad and Pardini 1990.

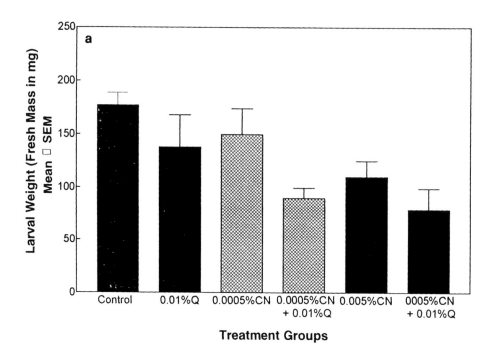

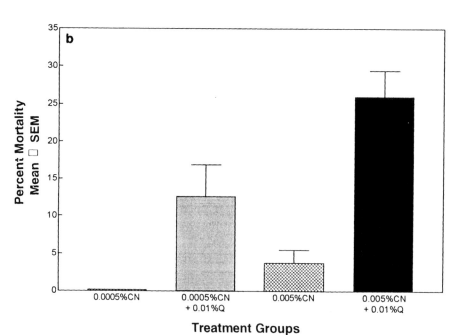

**Figure 16.** a)Larval weight in *T.ni* fed quercetin (Q) ± cyanide (CN⁻). b)Mortality in *T.ni* larvae fed quercetin (Q) ± cyanide (CN⁻).

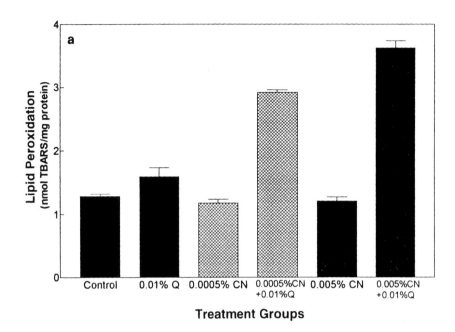

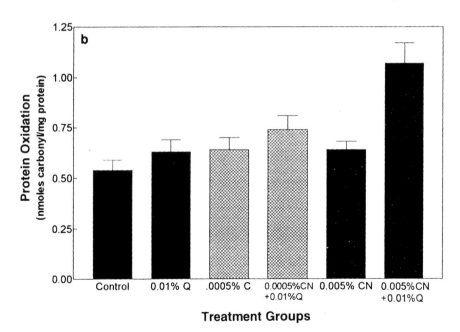

**Figure 17.** CN⁻ potentiated oxidative stress induced by feeding the pro-oxidant allelochemical quercetin to *T.ni*. a)Lipid peroxidation in *T.ni* larvae fed quercetin (Q) ± cyanide (CN⁻). b)Oxidatively modified protein levels in *T.ni* larvae fed quercetin (Q) ± cyanide (CN⁻).

3. Quercetin inhibits mitochondrial respiration in complexes I and II, (Hodnick et al 1986; Hodnick et al 1989).
4. Quercetin auto-oxidizes to form a semiquinone free radical (Figure 9).
5. Quercetin auto-oxidizes to form $O_2^{\bullet-}$, $H_2O_2$, and $^{\bullet}OH$ (Figures 8, 10; Hodnick et al 1986).
6. Quercetin reacts directly with $CN^-$ to generate ROS ($O_2^{\bullet-}$, $H_2O_2$, and $^{\bullet}OH$) (Figure 10; Hodnick et al 1986).
7. Quercetin reacts directly with the mitochondiral respiratory chain to generate a respiratory burst ($O_2^{\bullet-}$, $H_2O_2$, and $^{\bullet}OH$) (Figure 10).
8. Resistance towards the acute toxicity of quercetin in three insect species paralleled the endogenous activities of the anti-oxidant enzymes SOD and glutathione reductase (Table 2).
9. Quercetin inhibits catalase *in vivo* and glutathione reductase *in vivo* and *in vitro* (Figures 13b, 13c, 14) (Elliot et al. 1992).
10. Quercetin feeding increases the activity of SOD in our insect model *in vivo* (Figures 13a, 14)
11. Inhibition of SOD results in increased toxicity of quercetin towards *P. polyxenes* (Pritsos et al 1991)
12. Feeding $CN^-$ increased quercetin-induced lipid and protein oxidation in our insect model (Figure 17).
13. Feeding $CN^-$ increased the toxicity of quercetin in our insect model (Figure 16).

These findings lead to the conclusion that certain redox-active flavonoids, including quercetin, are capable of inducing oxidative stress *in vivo*, and that interactions with other biologically active compounds could exacerbate the oxidative-stress. Cyanide was shown to interact with specific flavonoids, to generate ROS *in vitro* and enhance *in vivo* oxidative-stress and toxicity; thus, care should be encouraged when ingesting redox-active flavonoids especially if exposed to cyanide. Potential sources of cyanide may include exposure of wildlife to cyanide containing mining ponds or perhaps ingestion of cyanogenic glycosides which are known to release cyanide *in vivo*. The cyanogenic glycosides did not interact directly with the redox-active flavonoids (data not shown), thus metabolism to release $CN^-$ would be essential for oxidative toxicity.

Certain anti-oxidants like Vitamin C have been shown to possess anti-oxidant as well as pro-oxidant activities. In this presentation, we have focused on the pro-oxidant effects of flavonoids. Similar to Vitamin C, flavonoids may donate electrons to metal ions which in the presence of $O_2$ can auto-oxidize to generate ROS (Figure 10). Conversely, under certain conditions, flavonoids are capable of donating electrons to electrophilic species like radicals to terminate the radical-chain propogation steps thereby acting in an anti-oxidant capacity. Electron donation could occur efficiently in thermodynamically (electrochemical) and kinetically (steric) favorable conditions. Thus, one would expect the relative concentrations of reactants to influence the direction of the anti-oxidant/pro-oxidant reactions. For example, the ability of flavonoids to act as pro-oxidants or anti-oxidants has been shown to depend both on the concentration and nature of the metal chelates which catalyze the oxidative process (Laughton et al 1989).

Finally, while the *in vitro* anti-oxidant activities of the flavonoids is well documented, the data presented here demonstrate that ingestion of quercetin by our insect model results in oxidative challenge *in vivo*. The oxidative challenge includes induction of SOD, and lipid and protein oxidation in all three insect species tested. acute toxicity was only observed with the oxidative stress susceptible species *T. ni*.

The western diet includes approximately 1 g of mixed flavonoids per day. Since there is currently substantial interest in dietary supplementation with anti-oxidant flavonoids, one should be cautious about possible interactions with other dietary ingredients.

## Collaborators

Dr. Chris A. Pritsos, Dr. B. Kalyanaraman, Craig Bohmont, Dr. Andrew J. Elliot, Dr. Dawn Duval, Dr. Sandra Johnson, Randall S. MacGill, and Dr. James T. MacGregor.

## REFERENCES

Ahmad S and Pardini RS. (1990) Antioxidant defense of the cabbage looper, *Trichoplusia ni*: enzymatic responses to the superoxide-generating flavonoid, quercetin, and photodynamic furanocoumarin, xanthotoxin. *Photochem Photobiol.* 51:305–311.

Bohmont CW, Aaronson LM, Mann K, Pardini RS. (1987). Inhibition of mitochondrial succinoxidase, NADH-oxidase and ATPase by naturally occurring flavonoids. *J Nat Prod.* 50:427–433.

Brown JP. (1980). A review of the genetic effects of naturally occurring flavonoids, anthraquinones, and related compounds. *Muta Res.* 75:243–277.

Chang WS, Lee YJ, Lu FJ, Chiang HC. (1993) Inhibitory effects of flavonoids on xanthine oxidase. *Anticancer Res.* 13:2165–2170.

Cheng SC, Pardini RS. (1978) Structural inhibition: relationships of various phenolic compounds towards mitochondrial respiration. *Pharmacol Res Commun.* 10: 897–910.

Cheng SC, Pardini RS. (1979) Inhibition of mitochondrial respiration by model phenolic compounds. *Biochem Pharmacol.* 28:1661–1667.

Cohen G, Heikkila RE. (1984) Alloxan and 6-hydroxydopamine: cellular toxins. *Meth Enzym* 105:510–516.

DeCrosse JJ. (1982). Potential for chemoprevention. *Cancer.* 50:2250–2253.

Elliot AJ, Scheiber SA, Thomas C, Pardini RS. (1992) Inhibition of glutathione reductase by flavonoids; A structure-activity study. *Biochem Pharmacol.* 44:1603–1608.

Fessell JE Jr, Chang Sing PD, Loew G, King R, Mansour JM, Mansour TE. (1979) Structure/activity studies of flavonoids as inhibitors of cyclic AMP phospho-diesterase and relationship to quantum chemical indices. *Mol Pharmacol* 16:556–568.

Harborne JB, Mabry TJ, Mabry H, eds. (1975). The flavonoids Part 2. New York, Academic Press.

Havsteen B. (1983). Flavonoids, a class of natural products of high pharmacological potency. *Biochem Pharmacol.* 32:1141–1148.

Hayashi T, Sawa K, Kawasaki M, Arisawa M, Shimiza M, Morita N. (1988) Inhibition of cows milk xanthine oxidase by flavonoids. *J. Nat. Prod.* 51:345–348.

Hodnick WF, Kung FS, Roettger WJ, Bohmont CW, and Pardini RS. (1986) Inhibition of mitochondrial respiration and production of toxic oxygen radicals by flavonoids: a structure-activity study. *Biochem Pharmacol.* 35:2345–2357.

Hodnick WF, Milosavljevic EB, Nelson JH, Pardini RS. (1988) Electrochemistry of flavonoids: relationships between redox potentials, inhibition of mitochondrial respiration and production of oxygen radicals by flavonoids. *Biochem. Pharmacol.* 37:2607–2611.

Hodnick WF, Bohmont CW, Capps C, and Pardini RS. (1989). Inhibition of mitochondrial NADH-oxidase (NADH-coenzymeQ oxidoreductase) enzyme system by flavonoids: a structure-activity study. *Biochem Pharmacol.* 36:2893–2874.

Hodnick WF, Kalyanaraman B, Pritsos CP, and Pardini RS. (1989a) The production of hydroxyl and semi-quinone free radicals during the autoxidation of redox active flavonoids. In: Oxygen Radicals in Biology and Medicine, eds: Simic MG, Taylor KA, Ward JF, von Sonntag C. p 149–152. Plenum, New York.

Hodnick WF, Duval DL, Pardini RS. (1994). Inhibition of mitochondrial respiration and cyanide stimulated generated production of reactive oxygen species by selected flavonoids. *Biochem Pharmacol.* 47:573–580.

Hodnick WF, Pardini RS. (1998) Inhibition of mitochondrial function by flavonoids. IN: Flavonoids in health and disease. Rice-Evans CA, Packer L, eds. Plenum: New York. p 179–197.

Iio MY, Ono SK, Fekomoto M. (1986) Effects of flavonoids on xanthine oxidation as well as cytochrome c reduction by milk xanthine oxidase. *J Nutr Sci Vitaminol* (Toyko) 32:635–642.

Iwu MM, Obidoa O, Anazodo M. (1986) Biochemical mechanism of the antimalarial activity of *Azadirachta indica* leaf extract. *Pharmacol Res Commun.* 18:81–91.

Khhnau J. (1976). The Flavonoids. A class of semi-essential food components: their role in human nutrition. *World Rev Nutr Diet.* 24:117–191.

Khalid SA, Farouk A, Geary TG, Jensen JB. (1986) Potential antimalarial candidates from African Plants: an *in vitro* approach using *Plasmodium flaciparum. J Ethnopharmacol.* 15:201–209.

Kupchan SM, Knox JR, Udayamurthy MS. (1965) Tumor Inhibitors VII. Eupatorin, a new cytotoxic flavone from *Eupatorium semiserratum. J Pharm Sci.* 54:929–30.

Laughton MJ, Halliwell B, Evans PJ, Hoult JRS. (1989) Anti-oxidant and pro-oxidant actions of the plant phenolics quercetin, gossypol and myrecitin: effects on lipid peroxidation, hydroxyl radical generation and bleomycin-dependent damage to DNA. *Biochem Pharmacol* 38:2859–2865.

Liu XF, Liu ML, Iyanagi T, Legesse K, Lee TD, Chen SA. (1990) Inhibition of rat liver NAD(P)H: quinone acceptor oxidoreductase (DT-diaphorase) by flavonoids isolated from the chinese herb *Scutaviae radix* (Huang Qin) *Mol Pharmacol.* 37:911–915.

Mabry TJ, Ulebelen A. (1980). Chemistry and utilization of phenylpropanoids including flavonoids, coumarins, and lignans. *J Agric Food Chem.* 28:188–96.

MacGregor JT. (1984) Genetic and carcinogenic effect of plant flavonoids: an overview. In: Nutritional and Toxicological Aspects of Food Safety, ed: Friedman M, p 497–526, Plenum, NY.

Mashino T, Fridovich I. (1987) Mechanism of cyanide catalyzed oxidation of α-ketoaldehydes and α-ketoalcohols. *Arch Biochem Biophys* 252:163–170.

McClure JW. (1975). Physiology and functions of flavonoids. IN: The Flavonoids, Part 2, eds: Harborne B, Mabry TJ, and Mabry H, p 970–1055. Academic Press,New York.

Middleton E, Jr. (1984). The Flavonoids. *Trends Pharmacol Sci.* 5:335–338.

Middleton E, Jr, Kandaswami C. (1994) The impact of plant flavonoids on mammalian biology: implications for immunity, inflammation and cancer. IN: The Flavonoids, Advances in Research since 1986, ed: Harborne JB. p 619–652. Chapman & Hall, London.

Pardini RS, Pritsos CP, Bowen SM, Ahmad S, Blomquist GJ. (1989) Adaptations of plant pro-oxidants in a phytophagous insect model: enzymatic protection from oxidative stress. IN: Oxygen Radicals in Biology and Medicine, eds: Simic MG, Taylor KA, Ward JF, von Sonntag C. p. 725–728. Plenum, New York.

Pisani DE, Elliot AJ, Hinman DR, Aaronson LM, Pardini RS. (1986) Relationship between inhibition of mitochondrial respiration by naphthoquinones, their anti-tumor acitvity and their redox potentials. *Biochem. Pharmacol.* 35:3791–3798.

Pritsos CA, Jensen DE, Pisani D, Pardini RS. (1982) Involvement of superoxide in the interaction of 2,3-dichloro-1,4-naphthoquinone with mitochondrial membranes. *Arch Biochem Biophys* 217:98–109.

Pritsos CA, Pardini RS. (1984) A redox cycling mechanism of action for 2,3-dichloro-1,4-naphthoquinone with mitochondrial membranes and the role of sulfhydryl groups. *Biochem Pharmacol* 33:3771–3777.

Pritsos CA, Ahmad S, Bowen SM, Elliot AJ, Blomquist GJ, Pardini RS. (1988) Antioxidant enzymes of the black swallowtail butterfly, *Papilio polyxenes*, and their response to the prooxidant allelochemical, quercetin. *Arch Insect Biochem Physiol.* 8:101–112.

Pritsos CA, Ahmad S, Bowen SM, Blomquist GJ, Pardini RS. (1988a) Antioxidant enzyme activities in the southern armyworm, *Spodoptera eridania. Comp Biochem Physiol* 90(2):423–427.

Pritsos CA, Ahmad S, Elliot AJ, Pardini RS. (1990) Antioxiodant enzyme level response to pro-oxidant allelochemicals in larvae of the southern armyworm moth, *Spodoptera eridania. Free Rad Res Comm.* 9:127–133.

Pritsos CA, Pastore J and Pardini RS. (1991) Role of superoxide dismutase in the protection and tolerance to the prooxidant allelochemical, quercetin in *Papilio polyxenes, Spodoptera eridania and Trichoplusia ni. Arch Insect Biochem Physiol.* 16:273–282.

Robertson P Jr, Fridovich SE, Misia HP, Fridovich I. (1981) Cyanide catalyzes the oxidation of α-hydroxaldehydes and related compounds: monitored as the reduction of dioxygen, cytochrome c, and nitroblue tetrazolium. *Arch Biochem Biophys* 207:282–284.

Salah N, Miller NJ, Papariga G, Tijburg L, Bolwell GP, Rice-Evans C. (1995). Polyphenolic flavonoids as scavengers of aqueous phase radicals and as chain breaking anti-oxidants. *Arch Biochem Biophys.* 322:339–346.

Schultz GE, Shrimer RH. (1979) Profiles of protein structure. Springer-Verlag: New York.

Tamura M, Kagawa S, Tsuruo Y, Ishimura K, Morita K. (1994) Effects of flavonoid compounds on the activity of NADPH diaphorase prepared from mouse brain. *Jpn J Pharmacol.* 65:371–373.

Tauber AI, Raul FJ, Marletta MA. (1984). Flavonoid inhibition of neutrophil NADPH-oxidase. *Biochem Pharmacol.* 33:1367–1369.

# 11

# FLAVONOIDS IN FOODS AS *IN VITRO* AND *IN VIVO* ANTIOXIDANTS

Joe A. Vinson

Department of Chemistry
University of Scranton
Scranton, Pennsylvania 18510

## 1. INTRODUCTION

In 1936 Szent-Györgi proposed the name of Vitamin P for a mixture of two flavonoids that decreased capillary permeability and fragility in humans and thus began the research on flavonoids and health (Rusznyak and Szent-Györgi, 1936). The flavonoids have been reported to exhibit a wide range of biological effects in animals and man that include antibacterial, antiviral, antiinflammatory, antiallergic, and vasodilatory actions. In addition, flavonoids inhibit lipid peroxidation, protein modification of collagen by peroxynitrite, platelet aggregation, and the activity of enzyme systems including phospholipase $A_2$, glutathione reductase, xanthine oxidase, cyclooxygenase and lipoxygenase (Harborne, 1986).

This chapter presents a review of flavonoids and related compounds and their antioxidant characteristics which are believed to have some relationship to their health properties. In this chapter we will cover the structure and activity relationships, epidemiological evidence for their health benefits in heart disease, *in vitro* antioxidant activity and lastly, *in vivo* absorption and antioxidant activity. Specific examples of some fruits, vegetables and beverages will be discussed in detail.

## 2. CHEMICAL STRUCTURE

### 2.1. Ring Structure of Flavonoids

Flavonoids are polyphenolic substances that are based on the flavan nucleus that consists of 15 carbon atoms arranged in three rings ($C_6$-$C_3$-$C_6$). The rings are labeled A, B and C as shown in Figure 1.

*Flavonoids in the Living System*, edited by Manthey and Buslig
Plenum Press, New York, 1998.

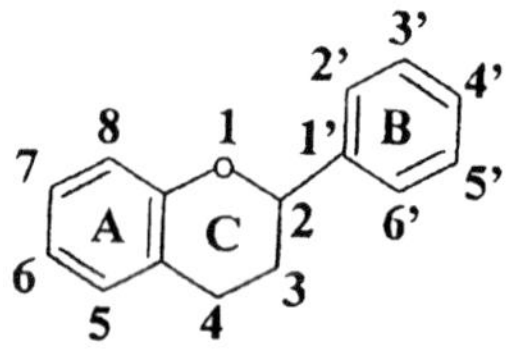

Figure 1. Basic flavonoid ring structure.

## 2.2. Classes of Flavonoids

The major flavonoid classes include flavonols, flavones, flavanones, catechins (flavonols) and chalcones. Differences between the classes consist in the changes in the pyrone ring (absence or presence of double bond, presence of 3-hydroxy and/or 2-oxy groups) and in the number of hydroxyls in rings A and B. In addition we will consider other phenols and polyphenols such as phenolic acids, cinnamic acids and benzoic acids that can also act as antioxidants. Flavonoids may be monomeric, dimeric, or oligomeric and vary greatly in molecular weight. Polymeric derivatives, called tannins, are divided into two groups relating to their structure: condensed and hydrolyzable. Condensed tannins are polymers of flavonoids and hydrolyzable tannins contain gallic acid. Due to the lack of agreement on nomenclature and clarification we will label, in this chapter, the spectrum of compounds as phenols and polyphenols.

## 2.3. Classes of Phenols and Polyphenols and Occurrence

In Table 1 are listed the major classes of plant phenols and polyphenols (Leibovitz, 1992). More than 4000 types of phenol and polyphenol compounds have been identified in vascular plants (Middleton and Kandaswami, 1994). More than 1000 citations are found each year in Chemical Abstracts for these compounds. Plants vary in type and quantity of compounds due to variables in plant growth, soil and weather conditions and the age of the plant. They occur in fruits, vegetables, nuts, seeds, flowers and bark. The hydroxyls of these compounds can be methylated or etherified with various sugars to form $O$-glycosides. Sugars include rhamnose, rutinose, galactose among many, but the most common

Table 1. The major classes of plant phenols and polyphenols

| No. of carbons | Basic skeleton | Class | Example |
|---|---|---|---|
| 7 | C6-C1 | hydroxybenzoic acids | salicylic acid |
|  |  |  | gallic acid |
| 9 | C6-C3 | hydroxycinnamic acids | ferulic acid |
|  |  | coumarins | dicoumarol |
| 10 | C6-C4 | naphthoquinones | juglone |
| 13 | C6-C1-C6 | xanthones | mangiferin |
| 14 | C6-C2-C6 | stilbenes | resveratrol |
| 15 | C6-C3-C6 | flavonoids | quercetin (flavonol) |
|  |  |  | catechin (flavanol) |
|  |  |  | cyanidin (anthocyanin) |
|  |  |  | naringenin (flavanone) |
|  |  |  | luteolin (flavone) |

is glucose. The concentration of these compounds can be as high as 40% of dry weight such as in tea leaves; plums and persimmons contain 5–10% of dry weight, but polyphenols are normally present as levels of 0.1–1.5%. The functions of polyphenols in plants are as antioxidants, antimicrobial, antiviral, photoreceptors, light screening, visual attraction, and feeding repellence.

## 3. EPIDEMIOLOGICAL STUDIES OF HEART DISEASE

### 3.1. Fruits and Vegetables

There is overwhelming epidemiological evidence for a protective effect of vegetable and fruits against cancer and heart disease and most recently, stroke. A highly significant negative association between consumption of total fresh fruits and vegetables and ischemic heart disease was reported in the United Kingdom (Armstrong et al., 1975) and the USA (Rimm et al., 1996). A significant negative relationship between fruit and vegetable consumption and cerebrovascular disease mortality was also found (Acheson and Williams, 1983). More recently strokes in men in the Framingham Study were decreased by eating fruits and vegetables (Gillman et al., 1995). An additional benefit of fruits and vegetables is their negative association with blood pressure (Ascherio et al., 1992), the latter is of course a risk factor for heart disease. The National Cancer Institute and the National Research Council recommend at least five servings of fruits and vegetables per day but only 23% of Americans eat this much (Subar et al., 1992). This 1991 survey found that the average American intake was 3.4 servings per day. The World Health Organization has echoed this with a recommendation of 500 grams/day of fruits and vegetables (Rice-Evans and Miller, 1995). This recommendation is made because of the high level of antioxidants and because fruits and vegetables are good sources of vitamins and minerals and are low in fat and high in fiber.

A recent comparison using an antioxidant score showed that 1 lb. of spinach or strawberries provided the *in vitro* antioxidant activity of 3.0 grams of vitamin E or 2.3 grams of vitamin C (Cao et al., 1996). An *in vivo* comparison of the effect of vitamins A, C, E and beta carotene vs. fruits and vegetables showed that they were equally potent as inhibitors of atherosclerosis in cholesterol-fed rabbits (Singh et al., 1995).

### 3.2. French Paradox

The first indication of flavonoids as protectants for heart disease came with the famous article of 1979 that showed an inverse relationship of coronary artery disease mortality and wine drinking using data from 17 countries (St. Léger et al., 1979). This led to the so-called "French Paradox" in which the French have a very low incidence of heart disease and mortality rates, but have the same average serum cholesterol, saturated fat intake, blood pressure and prevalence of smoking compared with the USA (Renaud et al., 1992). It was hypothesized that the much greater intake of wine by the French was responsible for this effect. As red wine is normally consumed at meals in France and the other countries studied, assuming the flavonoids were responsible was natural since red wines have a high polyphenol content, around 10,000 µM. Similarly wine ethanol had the strongest and most consistent negative correlation with coronary heart disease mortality in 21 developed countries (Criqui and Ringel, 1994). These results were confirmed by a prospective cohort study in Denmark. This showed that individuals who drank 3–5 glasses of wine/day had, independent from age or education, a 50% lower risk of dying from cardiovascular disease during 12 years of follow-up (Grønbaek et al., 1996). Wine drinking was also associated with a lower risk of dying from all causes during follow-up.

## 3.3. Flavonoids

The first epidemiological report that the intake of antioxidant flavonols (as measured in the diet by an HPLC method) reduced the rate of coronary heart disease mortality occurred in 1993 for elderly Dutch men who were followed for five years (Hertog et al., 1993a). Intakes of tea, onions and apples were also inversely related to coronary heart disease mortality. A study of sixteen male cohorts in seven countries including the USA was also published (Hertog et al., 1995). There was a strong inverse association between flavonol plus flavone intake and mortality from coronary heart disease. The composite diets of 1987 were analyzed and used for the correlation study. In multivariate analysis flavonoids explained 8% of the variance in heart disease rates. In fact saturated fat and smoking plus flavonoids explained 90% of the variance in heart disease rates independent of alcohol and antioxidant vitamin intake.

A Finnish study with a 26-year follow-up period showed a significant inverse gradient between dietary intakes of flavonoids and total and coronary mortality (Knekt et al., 1996). Between the lowest and highest quartile of dietary flavonoids, there was 31 and 48% lower risk of total and coronary mortality, respectively. For men there was a 34 and 32% lower risk, respectively. Apple and onion consumption were inversely associated with total and coronary mortality. A recently completed epidemiological study of male health professionals in the USA found a modest but nonsignificant inverse association between intakes of flavonols and flavones and subsequent coronary mortality rates (Rimm et al., 1996). The risk was 37% lower for the highest compared with the lowest quintile intake of flavonoids.

In the Dutch elderly male population dietary flavonoids were inversely associated with stroke incidences after adjustment for potential confounders, including antioxidant vitamins (Keli et al., 1996). This was a study of 552 men and followed up for 15 years. Black tea consumption was inversely associated with stroke incidences. The intake of vitamin C and vitamin E was not associated with stroke risk.

A most interesting article was recently published which gave some more credence to the importance of flavonoids in foods and health. An epidemiology study was done in England and Wales that found that fresh fruit, salads and green vegetables significantly improved lung function in children (Cook et al., 1997). Lung function was measured by forced expiratory volume and standardized for body size and sex. Surprisingly the plasma levels of vitamin C, a well-studied antioxidant in the lungs, were unrelated to lung function. This result indirectly implicates flavonoids as the "active ingredients" in these foods.

## 4. LIPIDS AND OXIDATIVE THEORY OF ATHEROSCLEROSIS

## 4.1. Lipids

Lipids are water-insoluble but have to be transported around the body in the blood. This dichotomy is solved by packaging the lipids into water-soluble particles of lipoprotein. The amphiphatic lipids (phospholipids and some non-esterified cholesterol) and the receptor protein are on the outside of the particles. The hydrophobic lipids (cholesteryl esters, triacylycerols i.e. triglycerides and some non-esterified cholesterol) are hidden from water in the interior of the particles. The four main types of lipoproteins are chylomicrons, very low density lipoprotein (VLDL), low density (LDL) and high density lipoprotein (HDL). Chylomicrons transport mainly triglycerides from the small intestine to the adipose tissue and skeletal muscle, while VLDL transports mainly triglycerides from the liver to the adipose tissue and skeletal muscle. VLDL is converted in the circulation to LDL. Each LDL particle is about 22 nm in diameter on average and contains about 1800 oxidizable molecules in the form of cholesteryl esters and triglycerides in its core. It has about 700 molecules of phospholipid on its surface as a monolayer and approximately 600 molecules of non-esterified cholesterol distributed between its core and its surface. The antioxidant defense contained within the LDL consists of about 20 molecules, mostly the phenolic vitamin E tocopherol with minor amounts of carotenoids and coenzyme $Q_{10}$ (Esterbauer et al., 1992).

## 4.2. Oxidative Theory of Atherogenesis

The oxidative theory of atherogenesis was first promulgated in Steinberg's seminal paper in 1989 (Steinberg et al., 1989). The hypothesis states that it is not LDL or VLDL that is atherogenic but the oxidized form of these lipoproteins. LDL and VLDL enter the arterial wall from the plasma and are oxidized locally within the wall by oxidizing agents derived from cells such as macrophages, smooth muscle cells, endothelial cells and lymphocytes. The oxidized LDL and VLDL are then believed to bind to scavenger receptors on macrophages and internalized rapidly by receptor-mediated endocytosis. The amount of cholesterol from the lipoproteins entering the cells overwhelms the capacity of the macrophages to release it and the cholesterol accumulates inside the cells and converts them to foam cells that are the first visible evidence of atherosclerosis. This hypothesis has been proved by many different pieces of evidence both *in vitro* and *in vivo*.

# 5. STRUCTURE-ACTIVITY ANTIOXIDANT RELATIONSHIPS OF POLYPHENOLS

## 5.1. Antioxidant Definition

In order for a phenol to be classified as antioxidant it must possess two properties: first, when present in low concentration compared with the substrate to be oxidized, it can delay or prevent the autoxidation or free radical-mediated oxidation; secondly, the free radical formed after scavenging must be stable (through intramolecular hydrogen bonding) to further oxidation (Shahidi and Wanasundara, 1992). The substrate, which is oxidizable, includes almost everything found in foods and living tissues including lipids, carbohydrates, proteins and DNA. Free radicals implicated in disease and with which flavonoids are reactive, include superoxide ($O_2^{\cdot-}$), hydroxy ($HO^{\cdot}$), their organic analogs peroxyl ($ROO^{\cdot}$), alkoxyl ($RO^{\cdot}$), and nonradical species such as singlet oxygen ($^1O_2$), peroxynitrite ($ONOO^-$) and hydrogen peroxide ($H_2O_2$).

## 5.2. Polyphenol Antioxidants

Quercetin, the most ubiquitous flavonol and the gallocatechins found in teas have a lower oxidation potential than vitamin E radical and can thus reduce it, effectively recycling vitamin E as an antioxidant (Jovanovic et al., 1996). This is especially important in atherogenesis because LDL oxidation first consumes the antioxidants within the particle of which vitamin E is the most prevalent. Additionally flavonoids have the ability to chelate $Fe^{+3}$ and $Cu^{+2}$ ions that can be changed into strong oxidants when they are free, unbound to proteins such as albumin and ceruloplasmin. However this is probably not polyphenols' mechanism of action for heart disease due to the reasonably high chelating protein levels in the fluid of the subendothelial space where LDL is oxidized and becomes atherogenic. It is likely that water soluble antioxidants in the extracellular subendothelial fluid can prevent the initation of lipid peroxidation since LDL oxidation is most likely initiated from the aqueous environment (Frei, 1995). Thus, both the hydrophobic polyphenols bound to LDL and the amphiphilic polyphenols in the aqueous fluid may be *in vivo* antioxidants.

It is generally agreed from both thermodynamic and kinetic radical studies and *in vitro* studies with many different models that the flavonoid structural requirements for effective radical scavenging and subsequent flavonoid radical stability are a diOH group in the C ring conjugated to a catechol radical in the B ring through the 2,3-double bond.

## 6. QUALITY AND QUANTITY OF POLYPHENOL ANTIOXIDANTS IN FOODS AND BEVERAGES

### 6.1. Quantity of Polyphenol Antioxidants

*6.1.1. Previous Work.* The quantity of polyphenols in foods was first comprehensively analyzed in 1976 using thin layer chromatography and found to be one gram/day in the US (Kühnau, 1976). Using an acid hydrolysis and HPLC separation, Hertog measured the total aglycone content of five flavonols and flavones in fruits, vegetables (Hertog et al., 1992) and beverages (Hertog et al., 1993b). Using this assay method the average US diet contains 20 mg/day of these five compounds (Rimm et al., 1996).

*6.1.2. Author's Research.* The quantity of polyphenols in mixtures was measured in foods after an acid hydrolysis to liberate the free phenols and polyphenols (Hertog et al., 1992) and followed by reaction with the Folin-Cocialteu reagent which assays oxidizable polyphenols vs. a catechin standard. In Table 2 the quantity of polyphenols in foods and beverages is shown. Pinto and kidney beans contained the greatest quantity of polyphenol antioxidants of all the foods and beverages studied. In examining the beverages, the red wines on the average contain more antioxidants than the white wines (Vinson and Hontz, 1995a). Teas contain the greatest concentration of polyphenols of the beverages examined followed by red wines.

### 6.2. Quality of Polyphenol Antioxidants

*6.2.1. Introduction.* The antioxidant quality of pure flavonoids and mixtures such as present in foods and beverages has been measured by various *in vitro* models. Some had biologically relevant oxidizable substrates with antioxidants at a single concentration or in dose-response study for a fixed time. Some measured the kinetics of direct interaction of a single concentration of antioxidants with a radical that may or may not be biologically relevant. The use of artificial radical trapping assays without an oxidizable substrate or with artificial radical generators has been criticized in the literature (Satué-Gracia et al., 1997).

**Table 2.** Antioxidant quantity and quality

| Antioxidant | Total polyphenol concentration* | Antioxidant quality $IC_{50}$ (μM) |
|---|---|---|
| Vitamin C (ascorbic acid) | – | 1.45 |
| Vitamin E (tocopherol) | – | 2.40 |
| Provitamin A (β-carotene) | – | 4.30 |
| Hesperetin (citruis fruit) | – | 3.66 |
| Hesperidin (citrus fruit) | – | >16 |
| Quercetin (fruits and vegetables) | – | 0.22 |
| Rutin (fruits and vegetables) | – | 0.51 |
| Myrcetin (fruits and vegetables) | – | 0.48 |
| Kaempferol (fruits and vegetables) | – | 1.82 |
| Epigallocatechin gallate (tea) | – | 0.075 |
| Cyanidin chloride (fruits) | – | 0.21 |
| Caffeic acid (fruits and vegetables) | – | 0.24 |
| Chlorogenic acid (fruits and vegetables) | – | 0.30 |
| Resveratrol (wines) | – | 0.33 |
| Orange juices (n=5) | 768 | 0.42 |
| Red wines (n=12) | 6192 | 0.72 |
| White wines (n=8) | 602 | 0.71 |
| Black teas (n=4) | 21060 | 0.86 |
| Green teas (n=4) | 13751 | 0.52 |
| Oranges (n=2) | 1400 | 0.34 |
| Grapes (n=2) | 6700 | 0.20 |
| Apples (n=2) | 6400 | 0.31 |
| Beans pinto and kidney (n=4) | 30100 | 0.61 |
| Onions red and yellow (n=4) | 3180 | 0.49 |
| Potatoes (n=2) | 1160 | 0.53 |

*After acid hydrolysis using catechin as a standard and the Folin colorimetric assay.
(μm/L for beverages and μmol/kg for fruits and vegetables fresh wgt)

*6.2.2. Author's Research.* Our particular model has been an *in vitro* oxidation of LDL+VLDL with cupric ion using a fixed concentration of lipoproteins and cupric ion at pH 7.4 for six hours at 37°C in the presence or absence of different concentrations of polyhenols. The dose-response inhibition of oxidation is calculated and a concentration to inhibit the oxidation 50% ($IC_{50}$) is then calculated (Vinson and Hontz, 1995a).

In Table 2 are shown quality measurements ($IC_{50}$) of some common vitamins and polyphenols in foods (Vinson et al. 1995b) and some foods and beverages. The vitamin antioxidants are relatively weak antioxidants compared with the polyphenols listed in the table. In examining the quality of the pure polyphenol antioxidants, the presence of a sugar molecule as in hesperetin vs. hesperidin and rutin vs. quercetin apparently decreases the antioxidant potency, i.e., increases the $IC_{50}$. Interestingly, chlorogenic acid that has a glucose group is a strong polyphenol antioxidant. Also anthocyanins such as cyanidin chloride and phenolic acids such as caffeic acid, which are ubiquitous components of fruits and vegetables, are very powerful antioxidants. The best pure antioxidant among those listed is epigallocatechin gallate. It is the major polyphenol in green tea and is present in black tea.

All of the food products have quite low $IC_{50}$ values indicative of strong antioxidants. Among the food products the teas stand out as the highest source of polyphenolic antioxidants. We have found that teas contain more polyphenols than any other commonly consumed beverage. Onions are a very good source of vegetable antioxidants as they have a large quantity of polyphenols and high quality antioxidants. Grapes and apples are an excellent source of fruit antioxidants as they also have a high quantity and quality of polyphenol antioxidants. The juices, beverages, and fruits and vegetables are shown to have polyphenol antioxidants as potent as pure polyphenols probably because of additivity of individual polyphenol components' activity and/or synergism.

## 7. *IN VIVO* ABSORPTION OF POLYPHENOLS

### 7.1. Introduction

Absorption of pure polyphenols in animals and man has been known for 20 years by using labeled compounds and monitoring the radioactivity in blood and urine. The metabolic pathway of several flavonoids in man and animals has been determined using tracers. Extensive degradation clearly takes place either in the digestive tract and/or passage of the blood through the liver. However, the direct analysis of polyphenols in man following food or beverage consumption has only recently been shown.

### 7.2. Tea

The first proof of polyphenol absorption from foods or beverages was found after tea consumption using nonspecific colorimetric analyses in feces, blood and urine after 14 days of either green or black tea (He and Kies, 1993). Tea polyphenols were found in their conjugated forms (glucuronide or sulfate) in human plasma and urine after subjects drank 1.2 g of a dried green tea extract containing 88 mg of epigallocatechin gallate (EGCG), 82 mg of epigallocatechin (EGC) and 32 mg of epicatechin (EC). Detection was by HPLC with electrochemical detection after enzyme hydrolysis and extraction of plasma or urine. EGCG at 46–268 ng/ml, EGC at 82–206 ng/ml and 48–80 ng/ml of EC were detected in plasma one hour after drinking the tea extract. In the urine only EGC and EC were detected and the cumulative urinary excretion was only 3.4 to 5.5 mg after 24 hours. Polyphenols were also found in urine after black tea consumption (Lee et al., 1993). Another group in Japan found EGCG in human serum at a peak concentration of 60–140 ng/ml after ingestion of five grams of green tea powder. HPLC with electrochemical detection was used for the analysis (Unno et al., 1996). Another Japanese group measured EGCG in human plasma (peak 156 ng/ml) after green tea (97 mg EGCG) using HPLC with a chemiluminescent detector (Nakagawa and Miyazawa, 1997). No EGCG could be detected in the plasma of fasted subjects. Another group detected polyphenols after green tea consumption using HPLC with UV detection (Maiani et al., 1996) The peak concentrations of 1–3 µg/ml of EGCG found were much higher than the previous study that used a more selective detector.

### 7.3. Wine

Polyphenols in LDL were detected after red wine ingestion (Fuhrman et al., 1995). In this study 400 ml of red wine was ingested for 2 weeks. The detection of polyphenols used a colorimetric method. Total polyphenols in LDL were increased 2.3-fold after wine consumption. The basal polyphenols in LDL corresponded to about 10 µg/ml of plasma polyphenols or about 300 µM, an impossible high number as shown below in Section 7.5. Vitamin E, the most prevalent physiological phenol is only present at about 20 µM in plasma, which shows the non-specificity of the colorimetric method.

## 7.4. Vegetables

Several studies have shown absorption of polyphenols from vegetables. In healthy ileostomy volunteers a Dutch group found 52% of quercetin glucosides found in onions as conjugates were absorbed. Only 17% of quercetin given as a pure compound was absorbed (Hollman et al., 1995). This group found that the quercetin in onions was still detectable in the plasma 48 hours after ingestion of 215 grams of fried onions that provided 64.2 mg of quercetin equivalents (Hollman et al., 1996). Most recently a Danish group found that 500 grams of broccoli ingested for 12 days produced detectable keampferol in the urine, 52–57 ng/ml. The daily dose of kaempferol from 250 grams of broccoli was 12.5 mg (Nielsen et al., 1997).

## 7.5. Fruit Juices

Absorption of citrus flavonoids in humans was studied after 500 mg consumption of naringin and hesperidin and after multiple dosing with grapefruit and orange juice (Ameer et al., 1996). The combined juices contained 323 mg of naringenin and 44 mg of hesperitin equivalents in the 1250 ml consumed. Urinary recovery was less than 25% as determined by tandem mass spectroscopy. These flavanones are poorly absorbed from juice and undergo glucuronidation before urinary excretion. The polyphenols phloretin (probably absorbed from apples) and quercetin (from fruits, fruit juices and vegetables) were detected by HPLC at 0.5 to 1.6µM in unsupplemented individuals (Paganga and Rice-Evans, 1997).

# 8. *IN VIVO* ANTIOXIDANT ACTIVITY

## 8.1. Initial Work

The first evidence of an *in vivo* antioxidant activity from food occurred in 1994 when it was found that red wine increased the serum antioxidant activity in humans (Maxwell et al., 1994). Red wine or tea ingestion caused an increase in antioxidant potential in human plasma using the total radical-trapping antioxidant (Serafini et al., 1994). The *in vivo* antioxidant effect of green and black tea was compared (Serafini et al., 1996). None of the these studies were specific in that they did not measure an antioxidant effect of a specific ingested compound or metabolite in plasma or serum.

## 8.2. Red Wine

More specific proof of polyphenols' *in vivo* antioxidant activity was first demonstrated for red wine (Furhrman et al., 1995). Consumption of 400 ml of red wine for 2 weeks resulted in a 20% decline in plasma oxidizability with a free radical generator. In addition, the oxidizability of LDL with cupric ion was significantly decreased as measured by 3 different assays. The lag time of LDL oxidation with cupric ions increased 4-fold after wine consumption. These beneficial effects occurred without any change in plasma antioxidant vitamins. However a more recent study found no effect on LDL oxidizability after a 4-week ingestion of 550 ml of red wine by 13 volunteers (de Rijke et al., 1996). The differences between this and Fuhrman's study can be ascribed to differences in sample preparation of LDL, frequency of wine consumption, differences in type of red wine consumed and, perhaps most important, to the fact that the Dutch used partially dealcoholized wine containing 3.5% alcohol. Alcohol has been suggested to promote absorption of polyphenols in the intestine (Ruf et al., 1995). An *in vitro* study with albumin showed that alcohol inhibits the binding of polyphenols in red wine to proteins. This inhibition would promote absorption (Serafini et al., 1997).

## 8.3. Black Tea

Most recently a Japanese study showed that black tea ingestion, 750 ml/day for four weeks, produced a significant 15% increase in the lag time of LDL oxidation with cupric ions (Ishikawa et al., 1997). As in the wine study the plasma antioxidant vitamins' concentration did not change with tea. The authors suggest that the large quantities of tea consumed in China and Japan protect the LDL from oxidation and keep the mortality rates of coronary artery disease low in spite of the very high rates of smoking. Despite the extremely low plasma polyphenol concentration (less than 1μM), the LDL was significantly protected.

## 8.4. Author's Work

We studied the lipoprotein-bound ability of antioxidants in beverages by means of *ex vivo* spiking experiments using our previously published method (Vinson et al., 1995c). Tocopherol (vitamin E) was used as a comparison. The pure compound or beverage (phenols measured by Folin-Cocialteu using catechin as a standard) was added at 50 μM to a pool of normocholesterolemic plasma from a single individual. The solution was allowed to equilibrate for one hour at $37^0C$. The LDL+VLDL containing any bound antioxidant was isolated by affinity column and oxidized with cupric ions under standard conditions. The oxidation was followed by monitoring conjugated diene formation with time. Lag time, a measure of oxidative resistance, was determined and compared to a control with no added antioxidant. The lag time increased linearly with concentration (data not shown). The results are shown in Table 3.

Among the single antioxidants, tocopherol and epigallocatechin gallate, the major polyphenol in tea, were the most potent, i.e., they increased the lag time the most. Tocopherol is very lipophilic, whereas epigallocatechin gallate is amphiphilic. Quercetin is quite lipophilic and is a good lipoprotein-bound antioxidant. However hesperidin and hesperetin, which are very lipophilic, had no effect. They are probably bound to LDL+VLDL but are very poor antioxidants as seen in Table 2 for the *in vitro* study. There was no effect of ascorbic acid, a water-soluble antioxidant, on lag time.

The non citrus beverages all produced an increase in lag time after spiking with red wine producing the largest increase. Red wine in fact was better than any single antioxidant. Orange juice and grapefruit juice were not lipoprotein-bound antioxidants. Thus it appears that all polyphenols, if they are *in vitro* LDL+VLDL antioxidants, bind to these lipoproteins and make them less susceptible to oxidation.

As a confirmation of the *ex vivo* study, a single subject then consumed 200 ml of each of several beverages at least 2 weeks apart after fasting overnight. Plasma was collected before and 1.5 to 2 hours after drinking when plasma antioxidant activity was maximal for tea and red wine (Serafini et al., 1994). The results are displayed in Table 3. The

**Table 3.** *Ex vivo* and antioxidant activity of pure polyphenols and *in vivo* antioxidant activity of beverages

| Pure antioxidant or beverage | % increase in lag time | |
|---|---|---|
| | After spiking | After consumption |
| Ascorbic acid | 0% | |
| Tocopherol | 55% | |
| Resveratrol (red wine) | 26% | |
| Epicatechin (fruits and vegetables) | 13% | |
| Epigallocatechin gallate (tea) | 44% | |
| Quercetin (fruits and vegetables) | 25% | |
| Rutin or quercetin glucoside (fruits and vegetables) | 24% | |
| Hesperidin (citrus fruits) | 0% | |
| Hesperetin (citrus fruits) | 0% | |
| Cyanidin (fruits) | 24% | |
| Chlorogenic acid (fruits and vegetables) | 31% | |
| Grapefruit juice | 0% | |
| Orange juice | 0% | 0% |
| Red wine (dealcoholized Bordeaux) | 127% | 202% |
| Black tea | 49% | 68% |
| Green tea | 8% | |
| Grape juice | 24% | 67% |

results of the supplementation study correlate well with the *ex vivo* spiking study. The dose of polyphenols was the following: dealcoholized red wine 2.28 mmoles, black tea 2.39 mmoles, grape juice 1.34 mmoles and orange juice 1.40 mmoles (600 ml was ingested). A graph of the oxidation time curve for the beverages is shown in Figure 2. Dealcoholized red wine produced the greatest increase in lag time after consumption and was the only beverage to show a diminished slope during the propagation phase when the absorbance was rapidly changing. Orange juice did not affect the lag phase or slope (graph not shown). Grape juice also was also shown to be an *in vivo* antioxidant and produced just as large a lag time increase as black tea with a dose of polyphenols only 60% that of black tea. This shows that grape juice polyphenols are probably better absorbed than from black tea since black tea is a better lipoprotein-bound antioxidant as shown in the Table. These short term studies with wine and black tea showed the same results as the previously published long term supplementation studies, namely that the polyphenols from

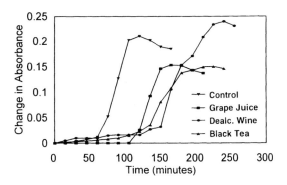

**Figure 2.** Oxidation-time curve of LDL + VLDL isolated from plasma after ingestion of 200 ml of water (control) or various beverages.

these beverages are absorbed and protect the LDL and VLDL from oxidation by enrichment of the lipoproteins.

## 9. SUMMARY

The polyphenol antioxidants in foods and beverages are shown to be powerful *in vitro* antioxidants. The polyphenols in foods and beverages enrich low density lipoproteins and decrease their oxidizability *ex vivo* after spiking and *in vivo* after absorption of the polyphenols following ingestion of beverages. These studies provide mechanisms to explain the epidemiological study which shows that consumption of fruits, vegetables and beverages reduce the risk of cardiovascular and cerebrovascular disease.

## REFERENCES

Acheson, R.M.; Williams, D.R.R. Does consumption of fruit and vegetables protect against stroke? *Lancet* **1983**, *1*, 1191–1193.

Ameer, B.; Weintraub, R.A.; Johnson, J.V.; Yost, R.A.; Rouseff, R.L. Flavone absorption after naringin, hesperidin, and citrus administration. *Clin. Pharmacol Ther.* **1996**, *60*, 34–40.

Armstrong.; B.K.; Mann, J.I.; Adelstein, A.M.; Eskin, F. Commodity consumption and ischemic heart disease mortality, with special reference to dietary practices. *J. Chron. Dis.* **1975**, *28*, 455–469.

Ascherio, A.; Rimm., E.B.; Giovannucci, E.L.; Colditz, G.A.; Rosner, B.; Willett, W.C.; Sacks, F.; Stampfer, M.J.A. A prospective study of nutritional factors and hypertension among US men. *Circulation* **1992**, *86*, 1475–1484.

Cao, G.; Sofic, E.; Prior, R.L. Antioxidant capacity of tea and common vegetables. *J. Agric. Food Chem.* **1996**, *44*, 3426–3421.

Cook, D.G.; Carey, I.M.; Walker, M. Effect of fresh fruit consumption on lung function and wheeze in children. *Thorax* **1997**, *52*, 628–633.

Criqui, M.H.; Ringel, B. Does diet or alcohol explain the French Paradox? *Lancet* **1994**, *344*, 1719–1723.

Esterbauer, H.; Gebicki, J.; Puhl, H.; Jürgens, G. The role of lipid peroxidation and antioxidants in oxidative modification of LDL. *Free Rad. Biol. Med.* **1992**, *13*, 341–390

Frei, B. C. Cardiovascular disease and nutrient antioxidants: role of low-density lipoprotein oxidation. *Crit. Revs. Food Sci. Nutr.* **1995**, *35*, 83–97.

Fuhrman, B.; Lavy, A.; Aviram, M. Consumption of red wine with meals reduces the susceptibility of human plasma and low-density lipoprotein to lipid peroxidation. *Am. J. Clin. Nutr.* **1995**, *61*, 549–554.

Gillman, M.W.; Cupples, L.A.; Gangnon, D.; Posner, B.M.; Ellison, R.C.; Castelli, W.P.; Wolf, P.A. Protective effect of fruits and vegetables on development of stroke in men. **1995**, *J. Am. Med. Assoc.* **1995**, *273*, 1113–1117.

Grønbaek, M.; Deis, A.; Sørensen, T.I.A.; Becker, U.; Schnohr, P.; Jensenen, G. Mortality associated with moderate intakes of wine, beer or spirits. *Br. Med. J.* **1995**, *310*, 1165–1169.

Harborne, J.B. Nature, distribution and function of plant flavonoids. In *Plant Flavonoids in Biology and Medicine;* Cody, V., Middleton, E. Harborne J.B. Eds.; Liss, New York, l986, p.15–24.

He, Y.H.; Kies, C. Green and black tea consumption by humans; impact on polyphenol concentrations in feces, blood and urine. *Plant Foods Hum. Nutr.* **1994**, *46*, 221–229.

Hertog, M.G.L.; Hollman, P.c.H.; Katan, M.B. Content of potentially anticarcinogenic flavonoids of 28 vegetables and 9 fruits commonly consumed in the Netherlands. *J. Agric. Food Chem.* **1992**, *40*, 2379–2383.

Hertog, M.G.L.; Feskens, E.J.M.; Hollman, P.C.H.; Katan, M.B.; Kromhout, D. Dietary antioxidant flavonoids and risk of coronary heart disease; the Zuphen elderly study. *Lancet* **1993a**, *342*, 1007–1011.

Hertog, M.G.L.; Hollman, P.C.H.; van de Putte, B. Content of potentially anticarcinogenic flavonoids of tea infusion, wines and fruit juices. *J. Agric. Food Chem.* **1993b**, *41*, 1242–1246.

Hertog, M.G.L.; Kromhout, D.; Aravanis, C.; Blackburn, H.; Buzina, R.; Fidanza, F.; Giampaoli, S.; Jansen, A.; Menotti, A.; Nedeljkovic, S.; Pekkarinen, M.; Bozida, S.S.; Toshima, H.; Feskens, E.J.M.; Hollman, P.C.H.; Katan, M.B. Flavonoid intake and long-term risk of coronary heart disease and cancer in the seven countries study. *Arch. Int. Med.* **1995**, *155*, 381–386.

Hollman, P.C.H.; Gaag, M.V.D.; Mengelers, M.J.B.; van Trijp, J.M.P.; de Vries, J.H.M.; Katan, M.B. Absorption and disposition kinetics of the dietary antioxidant quercetin in man, *Free Rad. Biol. Med.* **1996**, *21*, 703–707.

Hollman, P.C.H.; de Vries, J.H.M.; van Leeuwen, S.D.; Mengelers, M.J.B.; Katan, M.B. Absorption of dietary quercetin glycosides and quercetin in healthy ileostomy volunteers. *Am. J. Clin. Nutr.* **1995**, *62*, 1276–1282.

Ishikawa, T.; Suzukawa, M.; Toshimitsu, I.; Yoshida, H.; Ayaori, M.; Nishiwaki, M.; Yonemura, A.; Hara, Y.; Nakamura, H. Effect of tea flavonoid supplementation on the susceptibility of low-density lipoprotein to oxidative modification. *Am. J. Clin. Nutr.* **1997**, *66*, 261–266.

Jovanovic, S.V.; Steenken, S.; Hara, Y.; Simic, M.G. Reduction potential of flavonoid and model phenoxyl radicals. Which ring in flavonoids is responsible for antioxidant activity? *J. Chem. Soc. Perkin Trans. 2* **1996**, 2497–2504.

Keli, S.O.; Hertog, M.G.L.; Feskens, E.J.M.; Kromhout, D. Dietary flavonoids, antioxidant vitamins, and incidence of stroke. *Arch. Int. Med.* **1996**, *156*, 637–642.

Knekt, P.; Järvinen, R.; Reunanen, A.; Maatela, J. Flavonoid intake and coronary mortality in Finland; a cohort study. *Br. Med. J.* **1996**, *312*, 478–481.

Kühnau, J. The flavonoids. A class of semi-essential food components: their role in human nutrition. *World Rev. Nutr. Dietet.* **1976**, *24*, 117–191.

Leibovitz, B.E. Polyphenols and bioflavonoids: the medicines of tomorrow. *Nutr. Update* **1992**, *6*, 1–13.

Lee, M.-J.; Wang, Z.-Y.; Li, H.; Chen, L.; Sun, Y.; Gobbo, S.; Balentine, D.A.; Yang, C.S. Analysis of plasma and urinary tea polyphenols in human subjects. *Cancer Epidemiol. Biomarkers Prev.* **1995**, *4*, 393–399.

Maiani, G.; Serafini, M.; Salucci, M.; Azzini, E.; Ferro-Luzzi, A. Application of a new high-performance liquid chromatographic method for measuring selected polyphenols in human plasma. *J. Chromatog. B* **1997**, *692*, 311–317.

Maxwell, S.; Cruickshank, A.; Thorpe, G. Red wine: antioxidant activity in serum. *Lancet* **1994**, *344*, 193.

Middleton, E.; Kandaswami, C. The impact of plant flavonoids on mammalian biology: implications for immunity, inflammation and cancer. In *The Flavonoids: Advances in Research since 1986*, Ed. Harborne, J.B., Chapman and Hall, London, p. 619–652.

Nakagawa, K.; Miyazawa, T. Chemiluminescence high-performance liquid chromatographic determination of tea catechin, (-)-epigallocatechin 3-gallate, at picomole levels in rat and human plasma. *Anal. Biochem.* **1997**, *248*, 41–49.

Nielsen, S.E.; Kall, M.; Justesen, U.; Schou, A.; Dragsted, L.O. Human absorption and excretion of flavonoids after broccoli consumption. *Cancer Lett.,* **1997**, *114*, 173–174.

Paganga, G.; Rice-Evans, C.A. The identification of flavonoids as glycosides in human plasma. *FEBS Lett.* **1997**, *401*, 78–82.

Renaud. S.; de Lorgeril, M. Wine, alcohol, platelets, and the French paradox for coronary heart disease. *Lancet* **1992**, *339*, 1523–1526.

Rice-Evans, C.; Miller, N.J.; Antioxidants-the case for fruits and vegetables in the diet. *Br. Food J.* **1995**, *97*, 35–40.

Rimm, E.B.; Katan, M.B.; Ascherio, A.; Stampfer, M.J.; Willett, W.C. Relation between intake of flavonoids and risk for coronary heart disease in male health professionals. *Ann. Intern. Med.* **1996**, *125*, 384–389.

Ruf, J.-C.; Berger, J.-L.; Renaud, S. Platelet rebound effect of alcohol withdrawn and wine drinking in rats. Relation to tannins and lipid peroxidation. *Arteriosler. Thromb. Vasc. Biol.* **1995**, *15*, 140–144.

Rusznyàk, S.; Szent-Györgi, A. Viamin P; flavonols as vitamins. *Nature* **1936**, *138*, 27.

Satué-Gracia; Heinonen, M.; Frankel, E.N. Anthocyanins as antioxidants on human low-density lipoprotein and lecithin-liposome systems. *J. Agric. Food Chem.* **1997**, *45*, 3362–3367.

Serafini, M.; Ghiselli, A.; Ferro-Luzzi, A. Red wine, tea, and antioxidants. *Lancet* **1994**, *344*, 626.

Serafini, M.; Ghiselli, A.; Ferro-Luzzi, A. *In vivo* antioxidant effect of green and black tea in man. *Eur. J. Clin. Nutr.* **1996**, *50*, 28–32.

Serafini, M.; Maiani, G.; Ferro-Luzzi, A. Effect of alcohol on red wine tannin-protein (BSA) interactions. *J. Agric. Food Chem.* **1997**, *45*, 3148–3154.

Shahid, F.; Wanasundara. Phenolic antioxidants. *Crit. Rev. Food Sci. Nutr.* **1992**, *32*, 67–103.

Singh, R.B.; Niaz, A.M.; Ghosh, S.; Agarawal, P.; Shmad, S.; Begum, R.; Nuchi, Z.; Kummerow, F.A. Randomized, controlled trial of antioxidant vitamins and cardioprotective diet on hyperlipidemia, oxidative stresss, and development of experimental atherosclerosis: the diet and antioxidant trial on atherosclerosis. *Cardiovasc. Drugs Ther.* **1995**, *9*, 763–771.

St Léger, S., Cochrane, A.L.; Moore, F. Factors associated with cardiac mortality in developed countries with particular reference to the consumption of wine. *Lancet* **1979**, *12*, 1017–1020.

Steinberg, D.; Parathasarathy, S.; Carew, T.E.; Khoo, J.C.; Witzum, J.L. Beyond cholesterol; modification of low-density lipoprotein that increases its atherogenicity. *New Engl. J. Med.* **1989**, *320*, 915.

Subar A.; Heimendinger, J.; Krebs-Smith, S.; Patterson, B.; Keller-Pivonka, E. 5 A Day for Better Health: A Baseline Study of American Fruits and Vegetable Consumption. National Cancer Institute, Washington, D.C., 1992.

Unno, T.; Kondo, K.; Itakura, H.; Takeo, T. Analysis of (-)-epigallocatechin gallate in human serum obtained after ingesting green tea. *Biosci. Biotech. Biochem.* **1996**, *60*, 2066–2068.

Vinson, J.A.; Hontz, B.A. Phenol antoxidant index: comparative antioxidant effectiveness of red and white wines. *J. Agric. Food Chem.* **1995a**, *43*, 401–403.

Vinson, J.A.; Dabbagh, Y.A.; Serry, M.M.; Jang, J. Plant polyphenols, especially tea flavonols are powerful antioxidants using an *in vitro* oxidation model for heart disease. *J. Agric. Food Chem.* **1995b**, *43*, 2800–2802.

Vinson, J.A., Jang, J.; Dabbagh, Y.A.; Serry, M.M.; Cai, S. Plant polyphenols exhibit lipoprotein-bound antioxidant activity using an *in vitro* model for heart disease. *J. Agric. Food Chem.* **1995c**, *43*, 2798–2799.

# ANTITHROMBOGENIC AND ANTIATHEROGENIC EFFECTS OF CITRUS FLAVONOIDS

## Contributions of Ralph C. Robbins

John A. Attaway and Béla S. Buslig

Florida Citrus Consultants International, Inc.
Lake Alfred, Florida 33850
and Florida Department of Citrus
Lake Alfred, Florida 33850

This presentation is intended to commemorate the life's work of a singular individual, Dr. Ralph C. Robbins, who was a pioneer among researchers studying the health benefits of flavonoids, particularly flavonoids such as hesperidin, tangeretin, nobiletin and others which occur in citrus fruits. Dr. Robbins entered the field in 1960, a time when research on health effects of flavonoids seemed to raise more questions than answers. To put Dr. Robbins' research in perspective, let us first look very briefly at the history of flavonoid research.

The early research on flavonoid effects on health may be said to begin in 1936 when Hungarian scientist and Nobel Laureate, the late Dr. Albert Szent-Györgyi, and his collaborators reported the presence of a substance in lemon peel believed to be a flavonoid, which appeared to have beneficial therapeutic effects on abnormal capillary fragility. Szent-Györgyi described the properties of this substance as "vitamin like" and named it *vitamin P* (Rusznyák and Szent-Györgyi, 1936). This discovery prompted immediate interest in the scientific community and by 1948 more than 600 research papers had been written about *vitamin P* (Sokoloff and Redd, 1948). An important, but seemingly unrelated event, which occurred during this time period and culminated in 1948 was the development and patenting of a process for the manufacture of frozen concentrated orange juice by scientists employed by the Florida Citrus Commission (MacDowell *et al.*, 1948). As a result of the development of this process, the demand and production of orange juice soared by almost 5-fold. In 1946, Florida had only 250,000 acres of orange trees and processed only 645,000 tons of oranges. However, by 1951, Florida processors were squeezing 1.8 million tons of oranges, and by 1960, the number had soared to 3.2 million tons. This

*Flavonoids in the Living System*, edited by Manthey and Buslig
Plenum Press, New York, 1998.

led to a need for further research to: (1) find uses for the million plus tons of citrus peel being generated at Florida juice plants, (2) find new byproducts, and (3) to demonstrate new health benefits for citrus juices beyond the well-documented claim for vitamin C.

The impact of these developments were immediate in Florida, where the excess of seemingly waste peel put a great burden on plants processing citrus fruit into concentrated forms. Florida researchers, Sokoloff and Redd had conducted extensive research on the recovery of *vitamin P* from citrus peel as early as 1947. They addressed the issue of health benefits (Sokoloff and Redd, 1949) and reported effects on capillary fragility and *vitamin P* protection against radiation in 1950 in articles in *Science* and the *Proceedings of the Society for Experimental Biology and Medicine* (Sokoloff, et al., 1950a,b). In the late 1940's the prospect of a new citrus vitamin was very exciting. However, as we entered the 1950's other investigators could not confirm the claims made by the proponents of *vitamin P*, and as the decade progressed the status of flavonoids as a vitamin fell into disrepute, and the term *vitamin P* was finally discontinued (Vickery et al., 1950), although research by Szent-Györgyi, Sokoloff and others continued on the flavonoid compounds which were found in the *vitamin P* complex.

At this time Ralph Robbins came on the scene. Dr. Robbins had received his Ph.D. degree from the University of Illinois in 1958, after which he moved to the Medical College of South Carolina, Charleston, South Carolina as a Postdoctoral Research Associate. In 1959 he joined the staff of the Department of Food Science, University of Florida, Gainesville, as an Assistant Professor with both teaching and research responsibilities.

Early in his research career in Gainesville, Dr. Robbins became interested in the continuing puzzle over the role of flavonoids, particularly in relation to blood flow, capillary fragility and major problems such as coronary thrombosis and atherosclerosis in which the earlier observations of the effects of flavonoids remained unanswered in the aftermath of the demise of term: *vitamin P*. He organized his research toward making a contribution to a better understanding of the activity of the flavonoids which he felt had important health benefit properties even though no longer labeled as vitamins.

During the initial phase of Robbins' work (Robbins, 1966a,b, Brooks and Robbins, 1968), he developed techniques to study blood cell aggregation and capillary resistance, as well as the use of well-defined synthetic diets to induce experimental thrombosis and atherosclerosis in laboratory animals. Obviously, he needed controllable ways to induce and measure these conditions before he could study the effects of flavonoids without other dietary factors confounding the results.

In his initial paper on these subjects, supported by a grant from NIH, he demonstrated an inverse relationship between blood cell aggregation and capillary resistance (Robbins, 1966a). His following work (Robbins, 1966b) showed highly significant ($P<0.01$) increases in incidence and severity of blood cell aggregation in rats fed a thrombogenic regimen and a very highly significant ($P<0.001$) increase in rats fed an atherogenic regimen (Table 1).

With the techniques established, Robbins then continued his research by applying the possible inhibitory effects of flavonoids on the processes induced by these diets. First he studied the possible role of mixtures of vitamin C and flavonoids in guinea pigs and found that addition of vitamin C to the diet of vitamin C deficient guinea pigs resulted in decreased blood cell aggregation and a trend toward increased capillary resistance, and the addition of the flavonoids rutin, hesperidin or naringin (Table 2) significantly enhanced the action of the vitamin C (Robbins, 1966c).

Simultaneously, Robbins carried out a study of the effect of flavonoids on the longevity of rats fed the atherogenic or thrombogenic diets studied earlier (Robbins, 1967).

**Table 1.** Comparison of degree of blood cell aggregation in rats fed commercial chow, a thrombogenic diet and an atherogenic diet

| Diet | Number of determinations | Distribution of aggregation[a] | | | | |
|---|---|---|---|---|---|---|
| | | 0 | + | ++ | +++ | ++++ |
| Laboratory chow | 100 | 35 | 64 | 0 | 1 | 0 |
| Thrombogenic diet | 100 | 15 | 57 | 21 | 7 | 0[*] |
| Atherogenic diet | 100 | 2 | 37 | 46 | 14 | 1[**] |

[a] 0: No aggregation observed; +: Fine particles observed but blood flow normal; ++: Masses of blood cells discernible, blood flow slightly reduced; +++: Aggregated masses of cells sufficient to reduce blood flow to 2/3 or ¾ normal; ++++: Large aggregated cell masses present with blood flow reduced by ½ or more. Frequent plugging of vessels.
[*] Highly significant increase (P<0.01) over commercial chow.
[**] Very highly significant increase (P<0.001) over commercial chow.
Adapted from Robbins (1966a).

The flavonoids employed were hesperidin, rutin, naringin or tangeretin. The results showed that all 16 of 16 groups of rats fed the thrombogenic diet with flavonoids showed an average increase in survival time over the controls, while only 2 of 4 groups fed the atherogenic diet with flavonoids showed an increase in survival time over the controls. However, hesperidin feeding in the diet showed consistently increased survival times with both thrombogenic and atherogenic regimens.

Robbins' next approach in thrombosis research was to develop an *in vitro* method to study blood cell aggregation with the hope that this would simplify and make it possible to advance the research and shorten the time required to achieve results (Brooks and Robbins, 1968). He accomplished this by using a smear technique with a densitometer to establish quantitative and reproducible points of reference, thus reducing visual observer bias. The smear-densitometer technique was proposed to develop series of standards for better comparison of results from various laboratories.

The next step in his research program was a cooperative study with Dr. L. M. Morrison of the Institute for Arteriosclerosis Research at the Loma Linda University School of Medicine in Los Angeles, California. This study was directed toward studying the effect of

**Table 2.** Effect of three vitamin C levels with and without flavonoids on blood cell aggregation in vitamin C deficient guinea pigs

| Treatment mg/animal/day | | Frequency distribution of grades of blood cell aggregation[2] | | | | |
|---|---|---|---|---|---|---|
| Vitamin C | Flavonoids[1] | 0 | 1+ | 2+ | 3+ | 4+ |
| 0.25 | 0 | 0 | 0 | 20 | 49 | 31 |
| 0.25 | 20 | 0 | 3 | 41 | 50 | 6[b] |
| 3.0 | 0 | 42 | 33 | 16 | 2 | 7[a] |
| 3.0 | 20 | 55 | 19 | 8 | 8 | 10[b] |
| 8.0 | 0 | 19 | 17 | 17 | 27 | 20[a] |
| 8.0 | 20 | 36 | 10 | 23 | 26 | 5[b] |

[1] Data combined on rutin, hesperidin, and naringin.
[2] Frequency distribution for a 24 day period, shown in percent.
[a] Statistically significant reduction (P<0.01) in blood cell aggregation due to vitamin C.
[b] Statistically significant reduction (P<0.01) in blood cell aggregation due to flavonoid.
Adapted from Robbins (1966c).

**Table 3.** Effect of chondroitin sulfate A (CSA), rutin or hesperidin on serum cholesterol and erythrocyte sedimentation rate (ESR) of female rats fed 1,500,000 USP units of vitamin $D_2$ per kg diet

| Treatment | Cholesterol | | ESR | |
|---|---|---|---|---|
| | mg/100 ml | S.D. | mm/hr | S.D. |
| Control | 120 | 14.1 | 0.1 | 0.34 |
| Vitamin $D_2$ | 131 | 5.5 | 1.7 | 1.60 |
| $D_2$ + CSA | 137 | 21.6 | 0.2** | 0.49 |
| $D_2$ + Rutin | 136 | 20.1 | 0.3** | 0.68 |
| $D_2$ + Hesperidin | 130 | 27.9 | 0.3** | 0.68 |

**Significant at the 0.01 level.
Adapted from Robbins and Morrison (1968).

chondroitin sulfate A (CSA) and flavonoids on blood cell aggregation of rats (Robbins and Morrison, 1968). Here, a different approach to induction of atherosclerosis in animals, excessive vitamin D intake, was employed. A comparison of the vitamin D excess induced atherogenesis, with earlier work by several others, (1) vitamin C deficiency, (2) cholesterol, cholic acid and thiouracil administration, or (3) high fat diets which all induce atherogenesis, indicated an increase in blood cell aggregation. The addition of CSA or flavonoids, rutin or hesperidin, at a 1% level in the diet produced highly significant reduction in blood cell aggregation, with no effect observed on the level of serum cholesterol. This work also indicated that erythrocyte sedimentation rate (ESR) may be used as an indicator for cellular aggregation (Table 3). However, it was not established whether aggregation was a factor, or merely an indicator, in the pathogenesis of atherosclerosis.

Robbins then moved ahead to investigate individual compounds and their effect on blood cell aggregation *in vitro*, using ESR as the indicator of effectiveness. With addition of one of 5 flavonoids, low molecular weight dextran (LMD), quinine and caffeine at rates of 50 µmol/L to whole blood he observed reduced aggregation as measured by ESR. The compounds affected ESR in the following order (from high to low): (4',5,6,7,8)-pentamethoxyflavone (tangeretin) > quinine > hesperidin > low molecular weight dextran (LMD) > quercetin > rutin > control = caffeine > naringin in ability to decrease blood cell sedimentation rate (Robbins, 1971). However, only tangeretin (and quinine) showed sufficient activity to merit additional investigation.

As a result of this finding, Robbins elected to turn his attention to the methoxylated flavones. This work proceeded to involve a cooperative effort between the Departments of Food Science, Pharmacy and Veterinary Science at the University of Florida to study the inhibitory effect of methoxylated flavonoids and LMD on erythrocyte aggregation *in vitro* using horse blood, because of its high ESR. They made the very interesting finding that the compounds with the greater numbers of methoxyl groups were the most active. All compounds tested showed activity, decreasing in the following order: sinensetin > nobiletin > heptamethoxyflavone > tangeretin > tetra-*O*-methylscutellarein > LMD > tri-*O*-methyl-apigenin > hesperidin. On a molar basis, compounds with 5 to 7 methoxyl groups were several fold more active. The highly methoxylated flavonoids showed a higher level of activity than LMD, a clinically used erythrocyte disaggregating agent (Robbins *et al.*, 1971).

During the next series of experiments in 1973, while on leave at a Livingston, Tennessee hospital, Robbins had the opportunity to use human blood, but as the individual flavonoids did not have FDA approval for administration to humans, the study was performed *in vitro*. The following three flavonoids with 5, 6 and 7 methoxyl groups, were used as follows: sinensetin (3',4',5,6,7-pentamethoxyflavone), nobiletin (3',4',5,6,7,8--hexamethoxyflavone) and 3',4',3,4,6,7,8-heptamethoxyflavone. At a concentration of 48 µM all 3 flavonoids were found to reduce the erythrocyte sedimentation rate (ESR), and it was suggested that this would coincide with inhibition of blood cell clumping. Again, the results strongly suggested that the reduction in blood cell aggregation is a beneficial effect of flavonoids on microcirculation which alleviates a variety of adverse effects, such as thrombosis and embolism, caused by cell aggregation (Robbins, 1973a). Two additional papers, dealing with the same flavonoids, sinensetin (5 methoxyl groups), nobiletin (6 methoxyl groups) and heptamethoxyflavone, were published in 1973, utilizing blood from hospitalized patients. He reported that the 3 methoxylated flavones reduced the aggregation of cells in blood from patients with a variety of diseases, with no statistically significant difference among the 3 compounds (Robbins, 1973b,c). However, the results suggested that there may be some qualitative differences in anti-aggregative ability between these compound due to specific adhesive forces involved in clumping in various diseases. A specificity concept in flavonoid action is suggested to explain variability of results in clinical experiments with these compounds (Robbins, 1973c). A further study with rats, with results also published in 1973, showed that the hexamethoxylated flavone, nobiletin, exhibited significant antithrombogenic activity at 3.2 mg/kg of body weight in ADP induced thrombogenesis and showed a highly significantly greater protective effect than did heparin at 132 units/kg body weight (Robbins, 1973d).

Continuing his research on flavonoids and their effects on blood cells Robbins reported the effect of 9 flavonoids on the aggregation and sedimentation of erythrocytes in blood from hospitalized patients (Robbins, 1974). He found what he described as a "trimodal action" in that the flavonoids inhibited aggregation and sedimentation in some patients, accelerated it in others, and had no effect in a third group. This type of "trimodal action" was greatest with hesperidin, naringin, quercetin and rutin, long noted for their inconsistent action. The highly methoxylated flavones, sinensetin and nobiletin, exhibited the highest degree of inhibitory activity. The originally proposed specificity of anti-adhesive action of flavonoid types were combined with a concentration dependent component to explain the observed inconsistent action of some of the flavonoids examined (Table 4).

The high level of activity against cell aggregation invited further investigation into the role of methoxylated flavones, and Robbins found that the antiadhesive action of compounds such as sinensetin and nobiletin was significantly associated with calcium, which was consistent with the role of calcium in cell aggregation and disaggregation (Robbins, 1975a). Further investigation of the seemingly anomalous action of flavonoids, such as causing aggregation at times or prevention of aggregation at other times lead to examination of the types and concentration dependence of flavonoid action. The relationship with packed cell volume was also investigated. From this work, the suggestion arose, that the aggregative behavior may be due to a type of regulatory role, permitting blood components, including aggregated cells, to be removed by the liver or spleen to reestablish desirable rheological characteristics of circulating blood (Robbins, 1975b). Blood constituents, known to be associated with increased aggregation and sedimentation of erythrocytes were correlated with ESR rates in both flavone treated and untreated blood. The calcium levels in treated blood showed highly significant negative correlation coefficient level, but not a significant correlation coefficient in untreated blood. Other significant correlations in treated and untreated blood with the exception of blood urea nitrogen, which fell to no significance, showed little change in correlation values. Later, he obtained further evi-

Table 4. Magnitude of inhibition or acceleration of ESR by nine flavonoids at six concentration in human blood

| | Flavonoid concentration in blood, µM | | | | | | | | | | | |
|---|---|---|---|---|---|---|---|---|---|---|---|---|
| | 24 | | 48 | | 96 | | 192 | | 384 | | 768 | | Average | |
| | I | A | I | A | I | A | I | A | I | A | I | A | I | A |
| | Percent inhibition (I) or acceleration (A) of ESR | | | | | | | | | | | | | |
| Sinensetin | 32 | 6 | 48 | 0 | 83 | 0 | 82 | 0 | 80 | 0 | 72 | 0 | 66 | 1 |
| Nobiletin | 34 | 7 | 53 | 0 | 81 | 0 | 74 | 0 | 70 | 0 | 60 | 0 | 62 | 1 |
| Heptamethoxyflavone | 34 | 10 | 35 | 19 | 76 | 24 | 75 | 6 | 73 | 0 | 62 | | 59 | 10 |
| Tangeretin | 14 | 14 | 31 | 17 | 42 | 0 | 53 | 0 | 48 | 0 | 43 | 3 | 36 | 6 |
| Tetra-*O*-methylscutellarein | 26 | 28 | 32 | 19 | 36 | 15 | 52 | 0 | 40 | 6 | 41 | 33 | 38 | 17 |
| Quercetin | 10 | 17 | 21 | 8 | 16 | 17 | 15 | 13 | 26 | 28 | 35 | 13 | 21 | 14 |
| Rutin | 7 | 15 | 20 | 21 | 10 | 23 | 14 | 11 | 42 | 26 | 29 | 13 | 20 | 18 |
| Hesperidin | 22 | 19 | 24 | 30 | 5 | 18 | 5 | 11 | 13 | 15 | 18 | 10 | 15 | 23 |
| Naringin | 37 | 14 | 11 | 16 | 0 | 19 | 14 | 24 | 8 | 19 | 21 | 29 | 15 | 20 |
| Average | 24 | 14 | 31 | 14 | 39 | 13 | 43 | 7 | 44 | 10 | 42 | 11 | 37 | 12 |

Adapted from Robbins (1974)

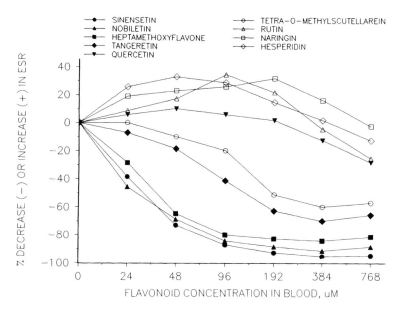

**Figure 1.** Effect of nine flavonoids on ESR in human blood (from Robbbins, 1976).

dence for the trimodal action of some, mainly the hydroxylated flavonoids, which may be causing some increase in ESR acceleration at lower concentrations and higher cell concentrations, but no significant effect at higher concentrations of the flavonoid (Figure 1). He also confirmed that the methoxylated flavones exhibited a highly significant antiadhesive action on erythrocytes and that certain flavones, particularly the methoxylated compounds which may occur in the diet, increase resistance of erythrocytes to aggregation by shifts in blood viscosity and flow. He inferred that the flavonoid compounds which occur in plants and are found in certain foods may permit dietary control of the blood high viscosity syndrome, possibly due to a combination of the cell sequestering action of the hydroxylated forms and the reduction of aggregation by the polymethoxyflavones (Robbins, 1976). The trimodal action of the citrus flavonoids indicated that these compounds may be useful in stabilization of blood flow properties, as the varied ESR accelerating and inhibiting properties, as well as the physiological action on some blood components and their interactions within the liver and the spleen may be utilized to regulate proper viscosity and flow of blood (Robbins, 1977).

In the following years, Robbins produced several comprehensive summaries of the physiological and medical aspects of flavonoids (Robbins, 1980a,b), expanding the knowledge base available. Noting the similarity of effect between aspirin's antiadhesive action, he evaluated the ability of penta-, hexa- and hepta-methoxylated flavones from citrus on the protection against adhesive activity on rat platelets induced by calcium ions and ADP, and found the order of activity from greatest to least was penta-, hexa-, and hepta-methoxylated (Robbins, 1982). This was the same order previously observed in horse blood and against red cell aggregation in human diseases. A decreased antiadhesive activity was associated methoxyl groups at the 3 and 8 positions. Based on the *in vitro* observed activity of the naringin, which at low levels caused red cell aggregation, followed by phagocytic removal of the clumped cells from circulation, grapefruit was fed to human subjects with various levels of hematocrit. He showed that consumption of the fruit over

42 days significantly lowered elevated hematocrit levels. No statistically significant differences were observed between consuming 1/2 or 1 grapefruit/day, but the decrease of hematocrit was statistically significant at the $P<0.01$ level (Robbins, et al., 1988). In a later study, he compared the effect of 2 methoxylated flavones from citrus, sinensetin (pentamethoxy) and heptamethoxy flavone *in vitro* with nobiletin (hexa-methoxy) for protective antiadhesive action on rat platelets. While he did not observe definite differences this time, nevertheless he concluded that the penta and heptamethoxylated flavones exhibited antiadhesive activity similar to that of nobiletin. The results did not show a parallel between the number of methoxyl groups, which suggested some degree of specificity between flavone structure and antiadhesive activity on platelets (Robbins, 1988).

In summary, Dr. Robbins began his research with the idea that the inconsistent activity of *vitamin P* (crude hesperidin) was the probable result of the presence of other flavonoids. In time, he developed methodology to quantitate the effects of the various flavonoids he examined and he identified the highly methoxylated flavones (5 to 7 methoxy groups) as the compounds most responsible for the greatest activity in reduction of cell aggregation and important in prevention or treatment of atherogenesis or thrombogenesis. His research greatly advanced progress in the area of the medical and physiological actions of plant flavonoids. His death prematurely deprived flavonoid research of a productive and imaginative proponent.

# REFERENCES

Brooks, K. M.; Robbins, R. C. An *in vitro* Method for Determination of Degree of Intravascular Aggregation of Blood Cells. *Lab. Invest.*, **1968**, *19*, 580–583.

MacDowell, L. G., E. L. Moore and C. D. Atkins. *Method of Preparing Full-Flavored Fruit Juice Concentrates*, U.S. Patent No. 2,453,109 **1948**.

Robbins, R. C. Relation of Blood Cell Aggregation to Capillary Resistance. *Angiology* **1966**a, *17*, 416–421.

Robbins, R. C. Intravascular Aggregation of the Cellular Elements of Blood in Rats Fed Thrombogenic or Atherogenic Regimens, *J. Atheroscler. Res.*, **1966**b, *6*, 467–473.

Robbins, R. C. Effect of Vitamin C and Flavonoids on Blood Cell Aggregation and Capillary Resistance. *Internat. J. Vit. Nutr. Res.*, **1966**c, *36*, 10–15.

Robbins, R. C. Effect of Flavonoids on Survival Time of Rats Fed Thrombogenic or Atherogenic Regimens. *J. Atheroscler. Res.*, **1967**, *7*, 3–10.

Robbins, R. C. Effects of Phenyl Benzo-γ-Pyrone Derivatives (Flavonoids) on Blood Cell Aggregation: Basis for a Concept of Mode of Action. *Clin. Chem.*, **1971**, *17*, 433.

Robbins, R. C. Effect of Methoxylated Flavones on Erythrocyte Aggregation and Sedimentation in Blood of Normal Subjects: Evidence of a Dietary Role for Flavonoids. *Internat. J. Vit. Nutr. Res.*, **1973**a, *43*, 494–503.

Robbins, R. C. In Vitro Effects of Penta-, Hexa- and Hepta-Methoxylated Flavones on Aggregation of Cells in Blood from Hospitalized Patients. *J. Clin. Pharmacology*, **1973**b, *13*, 271–275.

Robbins, R. C. Specificities Between Blood Cell Adhesion in Human Diseases and Antiadhesive Action in vitro of Methoxylated Flavones. *J. Clin. Pharmacology*, **1973**c, *13*, 401–407.

Robbins, R. C. Antithrombogenic Properties of a Hexamethoxylated Flavonoid. *Atherosclerosis*, **1973**d, *18*, 73–82.

Robbins, R. C. Action of Flavonoids on Blood Cells: Trimodal Action of Flavonoids Elucidates Their Inconsistent Physiologic Effects. *Internat. J. Vit. Nutr. Res.*, **1974**, *44*, 203–216.

Robbins, R. C. Action in Human Blood of Methoxylated Flavones which Confer Disease Resistance on Both Plants and Animals: Concept of a Dietary Conditioned Mechanism of Defense. *Internat. J. Vit. Nutr. Res.*, **1975**a, *45*, 52–59.

Robbins, R. C. Mechanism of Reversed Flavonoid Action on Blood Cells: Significance in Thromboembolism. *Internat. J. Vit. Nutr. Res.*, **1975**b, *45*, 163–174.

Robbins, R. C. Regulatory Action of Phenylbenzo-γ-Pyrone Derivatives on Blood Constituents Affecting Rheology in Patients with Coronary Heart Disease (CHD). *Internat. J. Vit. Nutr. Res.*, **1976**, *46*, 338–347.

Robbins, R. C. Stabilization of Flow Properties of Blood with Phenylbenzo-γ-Pyrone Derivatives (Flavonoids). *Internat. J. Vit. Nutr. Res.*, **1977**, *47*, 373–382.

Robbins, R. C. Medical and Nutritional Aspects of Citrus Bioflavonoids. In *Citrus Nutrition and Quality*, Nagy, S.; Attaway, J. A. eds., **1980a**, pp. 43–59, American Chemical Society, Symposium Series, ACS, Washington, DC.

Robbins, R. C. On Bioflavonoids: New Findings about a Remarkable Plant Defense against Disease and Its Dietary Transfer to Man. *Executive Health* **1980b**, *16*(12), 1–9.

Robbins, R. C. Inhibition of Platelet Aggregation by Methoxylated Flavones Isolated from Citrus. Report to the Florida Department of Citrus, Lakeland, Florida **1982**.

Robbins, R. C. Flavones in Citrus Exhibit Antiadhesive Action on Platelets. *Internat. J. Vit. Nutr. Res.*, **1988**, *58*, 418–421.

Robbins, R. C.; Hammer, R. H.; Simpson, C. F. Methoxylated Phenyl Benzo-γ-Pyrone Derivatives (Flavonoids) that Highly Inhibit Erythrocyte Aggregation. *Clin. Chem.*, **1971**, *17*, 1109–1113.

Robbins, R. C.; Martin, F. G.; Roe, J. M. Ingestion of Grapefruit Lowers Elevated Hematocrits in Human Subjects. *Internat. J. Vit. Nutr. Res.*, **1988**, *58*, 414–417.

Robbins, R. C.; Morrison, L. M. Effect of Chondroitin Sulfate A and Flavonoids on Blood Cell Aggregation of Rats. *Fed. Proc.*, **1968** 27, 646 (Abstract).

Rusznyák, S.; Szent-Györgyi, A. Vitamin P: Flavonols as vitamins. Nature **1936**, *27*, 138.

Sokoloff, B.; Redd, J. B. The Story of Vitamin P. *Citrus Industry* **1948**, *29*(7).

Sokoloff, B.; Redd, J. B. The Health Angle in the Consumption of Canned Orange Juice. *Citrus Industry* **1949**, *30*(5) 5–13.

Sokoloff, B.; Redd, J. B.; Dutcher, R. Vitamin P Protection Against Radiation, *Science* **1950a**, *112*, 112–113.

Sokoloff, B.; Redd, J. B.; Dutcher, R. Capillary Fragility and Vitamin P Protection Against Radiation. *Proc. Soc. Exp. Biol. Med.*, **1950b**, *75*, 6–9.

Vickery, H. B.; Nelson, E. M.; Almquist, H. J.; Elvehjem, C. A. Term Vitamin P Recommended to be Discontinued. *Science* **1950**, *112*, 628.

# EFFECT OF PLANT FLAVONOIDS ON IMMUNE AND INFLAMMATORY CELL FUNCTION

Elliott Middleton, Jr.

Chebeague Island Institute of Natural Product Research
Box 596 Capps Road
Chebeague Island, Maine 04017

## 1. ABSTRACT

The flavonoids are a large group of naturally occurring phenylchromones found in fruits, vegetables, grains, bark, roots, stems, flowers, tea, and wine. Up to several hundred milligrams are consumed daily in the average Western diet. Only limited information is available on the absorption, distribution, metabolism, and excretion of these compounds in man. Some compounds are absorbed, however, and measurable plasma concentrations are achieved which could have pharmacological relevance.

A variety of in *vitro* and *in vivo* experiments have shown that selected flavonoids possess antiallergic, antiinflammatory, antiviral and antioxidant activities. Moreover, acting by several different mechanisms, particular flavonoids can exert significant anticancer activity including anticarcinogenic properties and even a prodifferentiative activity, amongst other modes of action. Certain flavonoids possess potent inhibitory activity against a wide array of enzymes, but of particular note is their inhibitory effects on several enzyme systems intimately connected to cell activation processes such as protein kinase C, protein tyrosine kinases, phospholipase $A_2$, and others. Evidence suggests that only activated cells are susceptible to the modulating effects of flavonoids, i.e. cells which are responding to a stimulus. The stimulated activities of numerous cell types, including mast cells, basophils, neutrophils, eosinophils, T & B lymphocytes, macrophages, platelets, smooth muscle, hepatocytes, and others, can be influenced by particular flavonoids. On balance, a considerable body of evidence suggests that plant flavonoids may be health-promoting, disease-preventing dietary compounds.

## 2. INTRODUCTION

The flavonoids comprise a large group of naturally occurring, low molecular weight compounds that are present in all vascular plants. They are present in fruits, vegetables, nuts, seeds, stems, flowers, bark, tea, and wine, and the average Western diet contains up to approximately 1000 mg per day (Kuhnau, 1976). The remarkable properties exhibited by these compounds make it possible that they should be considered natural dietary biologic response modifiers, i.e. disease-preventing, health-promoting substances. The flavonoids have been shown in both *in vitro* and *in vivo* experimental systems to possess antiallergic, antiinflammatory, antiviral, anticancer, and anticarcinogenic activities. In addition, some of them are potent antioxidants, a subject of much current interest. The reader will find extensive bibliography regarding the biological response modifying properties of flavonoids in two recent reviews (Middleton and Kandaswami, 1993, 1994).

With respect to antiallergic activity it is now well known that certain flavonoids, depending on structure, can inhibit the stimulated release of proinflammatory mast cell, basophil, and eosinophil granular constituents that participate in the pathogenesis of diseases such as asthma, allergic rhinoconjunctivitis, urticaria and others. Quercetin, a planar molecule, is a particularly active inhibitor as described in greater detail below, while taxifolin (dihydroquercetin), a molecule with a nonplanar conformation, lacked activity (Middleton and Drzewiecki, 1982). Antiinflammatory activity can be assessed by the effects of flavonoids on neutrophil function. Certain flavonoids including quercetin have the capacity to inhibit neutrophil activation accompanied by activation of phospholipase $A_2$ (Lee *et al.*, 1982) and NADPH oxidase (Tauber *et al.*, 1984). Activation of the NADPH oxidase results in generation of superoxide anion, a tissue-damaging oxygen radical, by neutrophils. In addition to blocking superoxide anion production, quercetin inhibits the release of 3H-arachidonic acid from prelabelled neutrophils following a nonimmunologic stimulus (zymosan-activated serum). Release of arachidonic acid is catalyzed by phospholipase $A_2$ and is the rate limiting step leading to the production of proinflammatory prostaglandins and leukotrienes.

Although antiviral activity of flavonoids has been demonstrated with quite a few viruses, no clinically or commercially successful flavonoid has been discovered. Selected flavonoids can inhibit viruses such as polio virus type 1, parainfluenza virus type 3, respiratory syncytial virus and herpes virus type 1, amongst others. Four flavonoids were examined by us for their ability to affect viral infection of tissue culture monolayers, or once the monolayers were infected, the ability of the flavonoids to inhibit intracellular replication of the virus (Kaul *et al.*, 1985). Quercetin, a flavonol, exhibited both antiinfective and antireplicative activity against each of the four viruses. Hesperetin (a flavanone; reduced between the C2-C3 positions) showed only antireplicative activity, apparently indicating that the flavonoid or a metabolite could gain entrance to the interior of the cell. Catechin (a flavan; lacks C4 carbonyl) had no antireplicative properties but showed antiinfective activity against respiratory syncytial virus and herpes simplex virus type 1. Naringin (a flavanone glycoside) exhibited neither antiinfective nor antireplicative activity against any of the four viruses. Clearly there are unique structure-activity relationships to be considered with respect to antiviral activity of flavonoids. Of particular note with respect to antiviral activity is the fact that certain flavonoids have activity against three enzymes that are critically involved in the life cycle of the human immunodeficiency virus, namely, anti-reverse transcriptase activity (Spedding *et al.*, 1989), antiintegrase activity (Fesen *et al.*, 1993), and antiprotease activity (Brinkworth *et al.*, 1992).

The antioxidant activity of certain flavonoids has been recognized for some time. Some of the flavonoids are as potent as, or more potent than, familiar antioxidants such as vitamin E and β-carotene. Of current interest is the presence of potent antioxidant flavonoids in red wine. The explanation of the "French paradox" may lie in this observation. The French paradox is that the French population typically eats quite a lot of fatty foods, but does not have very high incidence of coronary artery disease (Maalej et al., 1997; Maxwell, 1997). A possible explanation is that the French consume a fair amount of red wine with their meals and that the antioxidant activity of the red wine may account for the antiatherogenic activity.

Chemically, the flavonoids are phenylbenzo-γ-pyrones with two major categories, i.e. flavones with the phenyl group at the 2 position of the chromone moiety and isoflavones with the phenyl group at the 3 position. There are a number of other chemical variations of the flavonoids, such as the state of oxidation of the bond between the C2-C3 position and the degree of hydroxylation, methoxylation, or glycosylation (or other substituent moieties) in the A, B, and C rings and the presence or absence of a carbonyl at position 4. There are now over 4000 flavonoid compounds in nature that have been structurally characterized. Clearly these compounds must be of importance in plant physiology and biochemistry or they would not have survived evolution. The activity of various flavonoids in different mammalian cell test systems is profoundly affected by structure, including changes in the degree of hydroxylation, methoxylation, the presence or absence of a double bond between C2-C3 and the presence or absence of the C4 carbonyl as mentioned above. Certain flavonoids can undergo dimerization or oligomerization to provide very interesting and novel compounds such as amentoflavone (a biapigenin) and proanthocyanidins. It is of considerable interest that the antiallergic drug cromolyn, which is used in the management of asthma and allergic rhinoconjunctivitis, is structurally very closely related to naturally occurring flavonoids. Cromolyn is a bischromone, that is, 2 chromone moieties (the benzopyrone ring system) attached to each other via a short carbon bridge.

Information accumulated over several decades indicates that a number of mammalian enzyme systems can be affected by selected flavonoids. Some of these enzyme systems are: protein kinase C, protein tyrosine kinases, myosin light chain kinase, phospholipase C, phospholipase $A_2$, lipoxygenase, cyclooxygenase, cyclic nucleotide phosphodiesterase, and ATPases, amongst others. A number of these enzymes such as protein kinase C, protein tyrosine kinase, phospholipase $A_2$ and phospholipase C are intimately involved in cell activation and signal transduction processes that occur in all physiologically stimulated cells.

A number of mammalian cell systems have been studied to establish the effects of flavonoids on their function. These include (a partial listing) B and T lymphocytes, monocytes/macrophages, mast cells, basophils, eosinophils, neutrophils, and platelets. The effects of flavonoids on some of these cell systems will be briefly described.

In the mid 1970's, British investigators showed that rat peritoneal mast cell histamine release stimulated with antigen from sensitized animals, the mitogen concanavalin A or the calcium ionophore A23187, was significantly inhibited by certain flavonoids. We decided to examine the effects of one of the flavonoids, quercetin, on the release of histamine from a somewhat related granule-containing cell, the human peripheral blood basophil (Middleton et al., 1981). In this work, we found that (1) quercetin caused a concentration-dependent (IC50 approximately 10 μM) inhibition of histamine release, (2) quercetin only affected antigen-activated cells (see below), and (3) its onset of action was instantaneous. The inhibitory activity of quercetin was partially reversed by increased buffer calcium concentration (a non-significant effect). We also observed that quercetin was antagonistic to the histamine release-augmenting effect of heavy water ($D_2O$), suggesting an effect of the compound on microtubule assembly and function. The action of quercetin was not potentiated by theophylline, indicating that cyclic AMP-related mechanisms were probably not involved (Middleton et al., 1981). Also, some interesting structure-activity relationships were discerned as described below. A point of particular interest in these experiments was the observation that quercetin only inhibited histamine release when the cells had been activated by antigen. When peripheral blood leukocyte suspensions containing basophils were incubated for 30 minutes with the flavonoid in the absence of antigen and then washed, they responded perfectly normally to the antigenic stimulus with appropriate histamine release (Middleton et al., 1981), indicating that the basophils did not appear to have a tight-binding receptor for quercetin in the unstimulated state. On the other hand, if the antigen was added to a suspension of basophil-containing leukocytes, thereby initiating the histamine release reaction, the addition of quercetin to the actively histamine releasing cells at 2, 5 or 10 minutes caused an abrupt cessation of further histamine release (Middleton et al., 1981). This suggests that a quercetin-sensitive substance is generated in the antigen-activated basophil, which interacting with quercetin abolished further histamine release. This is a rather novel pharmacological mechanism of action and is of considerable interest with respect to drug design. In subsequent experiments with basophil histamine release we found that compounds with a saturated bond between C2-C3, such as taxifolin (dihydroquercetin), were inactive as inhibitors of histamine release. Likewise, the rhamnosylglucoside of quercetin, rutin, also lacked activity as an inhibitor, perhaps by virtue of steric hindrance produced by the glycoside group at position C3. Certain other flavonoids also lacked activity if their structure did not include certain basic features, namely, hydroxylation in the B-ring, double bond at C2-C3 and a carbonyl at C4.

Subsequent experiments were performed to find out whether selected flavonoids had an effect on histamine release from basophils stimulated by other secretogogues, including anti-IgE, concanavalin A, the chemoattractant peptide f-methionyl-leucyl-phenylalanine, the calcium ionophore A23187 and tetradecanoyl phorbol acetate (TPA), which activates protein kinase C directly. Some flavonoids totally lacked activity against these various secretogogues, but others proved to be good inhibitors of histamine release stimulated by these various secretogogues. The concentration-effect curves of flavonoids, such as quercetin, against each one of these secretogogues did differ, however. It is possible that these differences in concentration-effect relationships reflect differences in the signal transduction mechanism employed by the basophil for each of the different secretogogues and shows differential sensitivity to the effects of the flavonoids (Middleton and Drzewiecki, 1982).

Basophil histamine release caused by TPA suggested that activation of protein kinase C is a necessary step in the secretory process. To study this further we prepared a partially purified protein kinase C preparation from rat brain and studied the effects of flavonoids on protein kinase C activation by TPA (Ferriola et al., 1989). Fisetin, quercetin, and luteolin were all very active (at 50 µM), a number of other compounds were less active and some totally lacked activity. Interestingly, the most active inhibitors of protein kinase C were also the most active inhibitors of TPA-induced basophil histamine release. Further kinetic studies with the rat brain preparation indicated that fisetin and quercetin caused inhibition of protein kinase C by virtue of blocking the ATP binding site in the catalytic portion of the enzyme. This could represent one fundamental mechanism of action of the flavonoids (Ferriola et al., 1989).

Other leukocytes, the eosinophils, participate in inflammatory reactions of various sorts, including bronchial asthma, allergic rhinoconjunctivitis, inflammatory bowel disease, and certain dermatoses, to mention a few. The granules of eosinophils contain several very toxic proteins that contribute to the pathogenesis of these clinical disorders. The eosinophil, like the mast cell and basophil, is a secretary cell. Therefore, we were interested to find out if quercetin and taxifolin had any effect on the secretion of eosinophil cationic protein and Charcot-Leyden crystal protein elicited by the calcium ionophore A23187 from the granules of partially purified eosinophils. The results clearly indicated that quercetin (10–50 µM) caused a 70–90% inhibition of eosinophil cationic protein secretion, while taxifolin had a negligible effect (Sloan et al., 1991). Essentially similar results were found with the same two compounds on the secretion of Charcot-Leyden crystal protein. Thus the secretory function of eosinophils, like those of mast cells and basophils, can be inhibited by quercetin.

The development of an inflammatory process must begin by trapping leukocytes in the vascular system of the tissue that is becoming inflamed. That is, activation of the endothelial cells within the capillaries and venules of the tissue, e.g. the lung in asthma, must take place in order for leukocytes to adhere to the endothelium and to migrate through the capillaries and venules into the inflammatory focus within the tissue. The proteins involved in the business of making leukocytes stick to endothelial cells are called adhesion molecules. There are many adhesion molecules now recognized. Some are present on the surface of endothelial cells. Others, called counter receptors, are present on the membranes of leukocytes (Bochner, 1997).

We were interested to know whether quercetin would inhibit the expression of one particular endothelial cell adhesion molecule, known as intracellular adhesion molecule-1 (ICAM-1) following stimulation of the endothelial cells with endotoxin. The cells studied were human umbilical vein endothelial cells (HUVEC) in tissue culture (Anné et al., 1994). Interestingly, quercetin caused a concentration-dependent inhibition of the expression of ICAM-1 in endotoxin-stimulated endothelial cells. Gerritsen et al. (1995) have performed more extensive experiments with a number of other flavonoids. Apigenin proved to be a potent inhibitor of cytokine-induced ICAM-1, VCAM-1 and E-selectin expression on the cells (the cytokines used were IL-1, TNF-α, and interferon-γ). Apigenin also blocked IL-1α-induced prostaglandin generation, as well as IL-6 and IL-8 synthesis stimulated by TNF-α. The possible importance of these observations with respect to diet and inflammation is obvious. Also, their significance with respect to drug design cannot be overemphasized.

Other cell types are also involved in immune function. The effects of flavonoids on these cells will be briefly described. The macrophage/monocyte is a cell critical to the initiation of the immune response. These cells take up antigens and foreign particles, e.g. microbes and partially digest them and present them on the cell surface so that it can interact with the T cell receptor to initiate an immune response. Some investigators have indicated that the process of antigen presentation can be inhibited in these cells by quercetin (Mookerjee et al., 1992). T and B lymphocytes are cells essentially involved in immune function. Certain flavonoids can inhibit the lymphocyte proliferation that is stimulated by phytomitogens such as phytohemagglutinin and concanavalin A. Also, certain flavonoids, depending on structure, have been shown to inhibit the generation of cytotoxic lymphocytes in murine mixed spleen cell cultures (Schwartz and Middleton, 1984). These observations clearly indicate that flavonoids can have an effect on cell-cell interactions and relate directly to the issue of adhesion molecules mentioned above. The B cell is the precursor of the plasma cell which is the principal antibody-producing cell of the immune system. Activation of the B cell antigen receptor causes B cell activation with accompanying phosphorylation of tyrosine residues in several proteins within the B cell. This process can be inhibited by the isoflavone genistein, a prominent soy flavonoid (Lane et al., 1991). Other experiments showed that B cell precursors stimulated with recombinant human IL-7 also resulted in phosphorylation of tyrosine residues in several proteins within the precursor B cell and this was accompanied by phosphatidyl inositol turnover with increased production of inositol triphosphate. Genistein inhibited these effects. Phosphatidyl inositol turnover is a vital process in transmembrane signal transduction and a key enzyme involved in this process is phosphatidyl inositol kinase (PI-4-kinase). This kinase is inhibitable by the isoflavone orobol, as well as by quercetin and fisetin (Nishioka et al., 1989). The epidermal growth factor-induced phosphatidyl turnover in epidermoid carcinoma cells (A431 cells) is inhibited by the isoflavone psi-tectorigenin (Imoto et al., 1988). The effect of flavonoids on antibody formation has also been studied. A unique flavonoid called plantagoside (a flavanone glucoside and an alpha mannosidase inhibitor) caused a concentration-dependent inhibition of the mouse spleen cell antibody response to sheep red blood cells used as antigen (Yamada et al., 1988). Finally, cell mediated immunity or delayed-type hypersensitivity reactions can also be affected by certain flavonoids. As noted, the related phenomenon of mitogenesis stimulated by PHA or conA is inhibitable by certain flavonoids. Recent work indicates that selected flavonoids can inhibit delayed-type hypersensitivity reactions in mice (Gerritsen et al., 1995).

Flavonoids also affect platelet function. The platelet is now appreciated to be a participant in inflammatory processes *via* the release of numerous vasoactive and inflammatory substances. It is of interest, therefore, that platelet aggregation and the release reaction can be inhibited by certain flavonoids as can calcium mobilization and the adhesion of platelets to collagen (Gryglewski et al., 1987). The latter observation once again points to the effect of flavonoids on adhesion molecule-dependent interactions.

It is now widely recognized that diets rich in fruits and vegetables appear to be associated with a reduced frequency of cancer of various organ systems. Clearly there must be substances in fruits and vegetables which have anticancer effects. The flavonoids are known to possess a number of anticancer activities; actually there are eight different mechanisms by which flavonoids affect cancer.

Certain flavonoids exert anticarcinogenic activity. This can take place by induction of enzymes affecting carcinogen metabolism. Also certain flavonoids actually inhibit adduct formation between carcinogens and DNA, and finally, certain flavonoids inhibit *in vivo* experimental carcinogenesis. A good example of the latter is an experiment where female rats that developed mammary carcinoma in response to a particular carcinogen were found to have a 50% reduction in the number of tumors as compared to control when they consumed a diet containing 5% quercetin (Verma et al., 1988).

In addition, certain flavonoids possess antitumor-promoter activity, in which case the flavonoids inhibit the various activities of tumor promoters that are involved in the process of carcinogenesis. Moreover, antitumor activity of flavonoids has been described with a number of different hormone-dependent tumors and certain flavonoids turn out to be very active antiproliferative agents, inhibiting cancer cell proliferation *in vitro* (Middleton, 1996). Additionally, a rather extraordinary process that is stimulated by certain flavonoids is a prodifferentiation effect, that is, certain flavonoids can actually stimulate a malignant cell to develop into a mature phenotype (Middleton, 1996). Other properties of flavonoids beneficial to health include an inhibitory effect of certain flavonoids on the expression of the multi-drug resistance gene and modulation of topoisomerase activity, which can be associated with reduced cancer growth. Finally, through effects on adhesion molecule expression and function certain flavonoids have an antimetastatic activity and reduce the development of metastases (Middleton, 1996).

In summary then, some biochemical properties of the flavonoids include alteration of enzyme activity, antioxidant activity, vitamin C sparing activity, chelation of metal cations, inhibition of lipid peroxidation, radical scavenging activity, and effects on protein phosphorylation. Some basic life processes that are affected by flavonoids include immune mechanisms, inflammation, cellular differentiation, heat shock protein synthesis, cancer, atherosclerosis, metabolism, and perhaps even aging. It seems reasonable to consider seriously the possibility that dietary flavonoids may be very important disease-preventing and health-promoting substances. Further research is definitely warranted.

## REFERENCES

Anné, S.; Agarwal, R.; Nair, M. P.; Schwartz, A.; Middleton, E. Inhibition of endotoxin-induced expression of intercellular adhesion molecule-1 and of leukocyte adhesion to endothelial cells by the plant flavonol quercetin. *J. Allergy and Clin. Immunol.*, **1994**, *93*, 276 (abstract).

Bochner, B. Cellular Adhesion in Inflammation. In *Allergy: Principles and Practice;* E. Middleton; C. E. Reed; E. F. Ellis; N. F. Adkinson; J. W. Yunginger; W. W. Busse, Eds., Mosby Yearbook: St. Louis, MO. **1998**.

Brinkworth, R.; Stoermer, M. J.; Fairlie, D. P. Flavones are inhibitors of HIV-1 proteinase. *Biochem. Biophys. Res. Comm.*, **1992**, *188*, 631.

Ferriola, P. C.; Cody, V.; Middleton, E. Protein kinase C inhibition by plant flavonoids, kinetic mechanisms and structure-activity relationships. *Biochem. Pharmacol.*, **1989**, *38*, 1617.

Fesen, M. R.; Kohn, K. W.; Leteurtre, F.; Pommier, Y. Inhibitors of human immunodeficiency virus integrase. *Proc. Nat. Acad. Sci. U.S.A.*, **1993**, *90*, 2399.

Gerritsen, M.E.; Carley, W. H.; Ranger, G. E. Flavonoids inhibit cytokine-induced endothelial cell adhesion protein gene expression. *Am. J. Pathol.*, **1995**, *147*, 278–292.

Gryglewski, R. J.; Korbut, R.; Robak, J.; Swies, J. On the mechanism of antithrombotic action of flavonoids. *Biochem. Pharmacol.*, **1987**, *36*, 317–322.

Imoto, M.; Yamashita, Y.; Sawa, T.; Kurasawa, S.; Naganawa, H.; Takeuchi, T.; Bao-quan, Z.; Umezawa, K. Inhibition of cellular phosphatidylinositol turnover by psi-tectorigenin. *FEBS Letters* **1988**, *230*, 43–46.

Kaul, T. N.; Middleton, E.; Ogra, P. L. Antiviral effect of flavonoids on human viruses. *J. Med. Virol.*, **1985**, *15*, 71.

Kuhnau, J. The flavonoids: a class of semi essential food components: their role in human nutrition. *World Rev. Nutr. Diet.*, **1976**, *24*, 117–191.

Lane, P. J. L.; Ledbetter, J. A.; McConnell, F. N.; Draves, K.; Deans, J.; Schieven, G. L.; Clark, E. A. The role of tyrosine phosphorylation in signal transduction through surface Ig in human B cells: inhibition of tyrosine phosphorylation prevents intracellular calcium release. *J. Immunol.*, **1991**, *146*, 715–722.

Lee, T.-P.; Matteliano, M. L.; Middleton, E. Effect of quercetin on human polymorphonuclear leukocyte lysosomal enzyme release and phospholipid metabolism. *Life Sci.*, **1982**, *31*, 2765.

Maalej, N.; Demrow, H. S.; Slane, P. R.; Folts, J. D. Antithrombotic effects of flavonoids in red wine. In *Wine: Nutritional and Therapeutic Benefits*; T. R. Watkins, ed.; American Chemical Society, Symposium Series; Washington, **1997**; pp. 247–260.

Maxwell, S. R. J. Wine Antioxidants and their Impact on Antioxidant Activity *in vivo*. In *Wine: Nutritional and Therapeutic Benefits*; T. R. Watkins, ed.; American Chemical Society, Symposium Series; Washington, **1997**; pp. 150–165.

Middleton, E. The Flavonoids as Potential Therapeutic Agents. In *ImmunoPharmaceuticals*; E. S. Kimball, Ed. CRC Press, Boca Raton, FL. **1996**, pp.227–257.

Middleton, E.; Drzewiecki, G. Flavonoid inhibition of human basophil histamine release stimulated by various agents. *Biochem. Pharmacol.*, **1982**, *31*, 1449.

Middleton, E.; Drzewiecki, G.; Krishnarao, D. P. Quercetin: an inhibitor of antigen-induced human basophil histamine release. *J. Immunol.*, **1981**, *127*, 546.

Middleton, E., Jr.; Kandaswami, C. The impact of plant flavonoids on mammalian biology: implications for immunity, inflammation, and cancer. In *Advances in Flavonoid Research*; J. B. Harborne, ed.; Chapman and Hall, London, **1993**; pp. 619–652.

Middleton, E., Jr.; Kandaswami, C. Free radical scavenging and antioxidant activity of plant flavonoids. *Adv. Exp. Med. Biol.*, **1994**, *366*, 251–366.

Mookerjee, B. K.; Lee, T.-P.; Lippes, H. A.; Middleton, E. Some effects of flavonoids on lymphocyte proliferative processes. *J. Immunopharmacol.*, **1986**, *8*, 371.

Nishioka, H.; Imoto, M.; Sawa, T.; Hamada, M.; Naganawa, H.; Takeuchi, T.; Umezawa, K. Screening of phosphatidylinositol kinase inhibitors from streptomyces. *Antibiot.*, **1989**, *42*, 823–825.

Schwartz, A.; Middleton, E. Comparison of the effects of quercetin with those of other flavonoids on the generation and effector function of cytotoxic T lymphocytes. *Immunopharmacology* **1984**, *7*, 115.

Sloan, R.; Boran-Rogotzky, R.; Ackerman, S. J.; Drzewiecki, G.; Middleton, E. The effect of plant flavonoids on eosinophil degranulation. *J. Allergy Clin. Immunol.*, **1991**, *87*, 282 (abst.).

Spedding, G.; Ratty, A.; Middleton, E. Inhibition of reverse transcriptases by flavanoids. *Antiviral Res.*, **1989**, *12*, 99.

Tauber, A. I.; Fay, J. R.; Marletta, M. A. Flavonoid inhibition of human neutrophil NADPH-oxidase. *Biochem. Pharmacol.*, **1984**, *33*, 1367–1369.

Verma, A. K.; Johnson, J. A.; Gould, M. N.; Tanner, M. A. Inhibition of 7,12-dimethylbenz[a]anthracene and N-nitroso methylurea-induced rat mammary cancer by dietary flavonol quercetin. *Cancer Res.*, **1988**, *48*, 5754.

Yamada, H.; Nagai, T.; Takemoto, N.; Endoh, H.; Kiyohara, H.; Kawamura, H.; Otsuka, Y. Plantagoside, a novel alpha-mannosidase inhibitor isolated from the seeds of *Plantago asiatica*, suppresses immune response. *Biochem. Biophys Res. Comm.*, **1989**, *165*, 1292.

# 14

# FLAVONOIDS: INHIBITORS OF CYTOKINE INDUCED GENE EXPRESSION

Mary E. Gerritsen

Bayer Corporation
Pharmaceutical Division
400 Morgan Lane
West Haven, Connecticut 06516

Flavonoids demonstrate a remarkable spectrum of biochemical activities which critically focuses on the immune and inflammatory response, including direct inhibitory effects on tyrosine and serine-threonine protein kinases, phospholipases, cyclooxygenases and lipoxygenases (rev. in Middleton Jr. and Kandaswami (1992)). However, our laboratory recently demonstrated that flavonoids can also exert anti-inflammatory effects by inhibiting cytokine induced gene expression (Gerritsen et al., 1995). In this earlier study we reported that apigenin and certain structurally related hydroxyflavones inhibited tumor necrosis factor (TNF), interleukin-1 (IL-1), lipopolysaccharide, and interferon-γ (IFNγ) induced expression of the adhesion molecules intercellular adhesion molecule-1 (ICAM-1), vascular cell adhesion molecule-1 (VCAM-1), and E-selectin. The effects of apigenin were reversible, occurred in the first 30 minutes of coincubation with the cytokine, and appeared to be at the level of transcription. Additionally, apigenin inhibited the cytokine induced upregulation of several other inflammatory genes including IL-6, IL-8, and cyclooxygenase-2.

Cytokines, a group of polypeptide intercellular signalling molecules secreted by leukocytes and other cell types, orchestrate immune and inflammatory responses. The so-called inflammatory cytokines, tumor necrosis factor (TNF), interleukin-1(IL-1) and the interferon-γ (IFN-γ) upregulate the expression of a number of genes central to the initiation and propagation of immune and inflammatory response (reviewed in Gerritsen and Bloor (1993)). Three major intracellular signalling pathways have been identified which play important roles in the transcriptional activation induced by these cytokines: NF-κB, stress kinase and JAK-STAT.

Endothelial cells, which line the vasculature, play an integral role in the initiation of the inflammatory response. Exposure of the endothelium to inflammatory mediators results in alterations in cell structure and function which pertain to the local regulation of coagulation and inflammation. Among the consequences of endothelial activation by cy-

*Flavonoids in the Living System*, edited by Manthey and Buslig
Plenum Press, New York, 1998.

tokines is the acqusition of surface adhesive properties for neutrophils, monocytes and lymphocytes, the result of the induction and surface expression of cell surface adhesion molecules, E-selectin, ICAM-1 and VCAM-1. Promoter analysis of the 5' untranslated region for each these genes demonstrated the critical role of site(s) which bind to NF-κB. NF-κB is a family of dimeric transcription factor complexes of which the p50/p65 heterodimer is the predominant species in most cell types (reviewed in (Collins et al., 1995)). Cytokine responsiveness of the endothelial cell adhesion molecules, E-selectin, VCAM-1 and ICAM-1 requires the presence of conserved NF-κB sites in the promoters of all three genes, and p50/p65 has been shown to bind the NF-κB sites in ICAM, E-selectin and VCAM-1 (Collins et al., 1995). In addition, NF-κB activation is required for a number of other cytokine activated genes, including IL-6, IL-8 and MCP-1 (reviewed in Collins et al., 1995)).In resting cells NF-κB is maintained in the cytoplasm through its interaction with an inhibitory molecule, IκBα. IκBα prevents nuclear localization of NF-κB by binding to and masking a nuclear localization signal in resting cells. Activation of cells by a variety of stimuli, including cytokines, ionizing radiation, and oxidant stress results in the phosphorylation of IκB. This event targets IκB for rapid ubiquination and proteolysis by the proteasome. Upon IκB degradation, NF-κB translocates to the nucleus where it activates the transcription of various genes (reviewed in Baeuerle and Henkel (1994), Beg and Baldwin (1993) and Verma et al., (1995)).

Activation of the mitogen-activated protein kinase (MAPK) family of serine/threonine proteins kinases also plays important roles in cytokine induced cell activation. TNF activation of the "stress kinase cascade" leads to activation of the MAPK family members known as JNK and p38 kinases. JNK phosphorylates, and thus activates, certain transcription factors such as c-jun and ATF-2. The physiological substrate(s) for p38 are not well defined, but inhibitors of this kinase have been shown to inhibit cytokine induced cyclooxygenase expression and IL-6 production. These parallel pathways may converge with NF-κB to result in maximal expression of certain genes.

Interferon signaling involves a different family of transcription factors whose activity is also regulated by phosphorylation and compartmentalization, the STATs (signal transducers and activators of transcription). A DNA binding activity composed of STAT1α (p91) dimers is induced upon treatment of cells with IFNγ as a consequence of tyrosine phosphorylation of STAT1 by members of the JAK (Janus kinase) family of receptor associated protein tyrosine kinases. The general scheme of this pathway involves aggregation of ligand-occupied receptors, resulting in the formation of multimeric complexes with JAKS. The JAKS catalyze the phosphorylation of the receptors, themselves and the STAT proteins. Tyrosine phosphorylated and dimerized STATS translocate to the nucleus and bind to DNA regulatory elements. Members of the JAK-STAT signaling cascade have been shown to be activated in response to a variety of cytokines and growth factors in addition to the interferons (Bovolenta et al., 1996; Novak et al., 1995; Tsukada et al., 1996; Tweardy et al., 1995).

As discussed above, flavonoid pretreatment or cotreatment of endothelial cells blocks the cytokine-induced expression of both the protein and mRNA for the adhesion molecules ICAM-1, E-selectin, and VCAM-1, as well as the production of the immune/inflammatory proteins IL-6, IL-8, and the inflammatory lipid, $PGE_2$ (Gerritsen et al., 1995). The effect of the flavonoids was reversible, dose- and time-dependent, and structure specific (Gerritsen et al., 1995). As detailed in Table 1, the most potent compounds were trihydroxyflavones, with some activity also demonstrated by structurally related flavonols, but not flavanones or isoflavones. We have attempted to identify the molecular mechanism of this effect.

Table 1. Actions of flavone derivatives on ICAM expression

| Compound | 3 | 5 | 6 | 7 | 8 | 3' | 4' | 5' | IC$_{50}$μM |
|---|---|---|---|---|---|---|---|---|---|
| | | | | Basic structure A | | | | | |
| Apigenin | H | OH | H | OH | H | H | OH | H | 10 |
| Hinokiflavone | H | OH | H | OH | H | H | X | H | 10 |
| Chrysin | H | OH | H | OH | H | H | H | H | 25 |
| Chysoeriol | H | OH | H | OH | H | OCH$_3$ | OH | H | 25 |
| Luteolin | H | OH | OH | H | H | OH | OH | H | 25 |
| Baicalein | H | OH | OH | OH | H | H | H | H | 50 |
| Flavone | H | H | H | H | H | H | H | H | Inactive |
| Apigenin-3-Me | H | OCH$_3$ | H | OCH$_3$ | H | H | OCH$_3$ | H | Inactive |
| | H | OCH$_3$ | H | OCH$_3$ | H | H | C(CH$_3$)$_3$ | H | Inactive |
| Chrysin-2-Me | H | OCH$_3$ | H | OCH$_3$ | H | H | H | H | Inactive |
| | H | OH | H | OCH$_3$ | H | H | H | H | Inactive |
| | H | OH | H | OCH$_3$ | H | H | OCH$_3$ | H | Inactive |
| Tangeretin | H | OCH$_3$ | OCH$_3$ | OCH$_3$ | OCH$_3$ | H | OCH$_3$ | H | Inactive |
| | G1* | OH | H | G2* | H | H | OH | H | Inactive |
| | G3* | OH | H | OH | H | H | OH | OH | Inactive |
| | H | OH | H | G4* | H | H | OH | H | Inactive |
| | H | OH | H | G5* | H | OH | OCH$_3$ | H | Inactive |
| | H | OH | H | OH | G6* | OH | OH | H | Inactive |
| | OH | OH | OH | OH | H | OH | OH | H | Inactive |
| | OH | OH | H | OH | H | OCH$_3$ | OH | H | Inactive |
| Amentoflavone | H | OH | H | OH | H | H | OH | X | Inactive |

| Compound | 3 | 5 | 6 | 7 | 8 | 3' | 4' | 5' | IC$_{50}$ |
|---|---|---|---|---|---|---|---|---|---|
| | | | | Basic structure B | | | | | |
| Naringenin | H | OH | H | OH | H | H | H | H | Inactive |
| Hesperetin | H | OH | H | OH | H | OH | OCH$_3$ | H | Inactive |
| | | | | Basic structure C | | | | | |
| Cyanidin | OH | OH | H | OH | H | OH | OH | H | Inactive |
| | | | | Basic structure D | | | | | |
| Genistein | H | OH | H | OH | H | H | OH | H | Inactive |

| Compound | 3 | 4 | 5 | 6 | 7 | 8 | IC$_{50}$ | | |
|---|---|---|---|---|---|---|---|---|---|
| | | | Basic structure E | | | | | | |
| Coumarin | H | H | H | H | H | H | Inactive | | |
| Daphnetin | H | H | H | H | OH | OH | Inactive | | |
| Esculetin | H | H | H | OH | OH | H | Inactive | | |
| Scopoletin | H | H | H | OCH$_3$ | OH | H | Inactive | | |
| | NH$_2$ | H | H | H | H | H | Inactive | | |

Inactive: Compounds showing no significant inhibitory activity on ICAM-1 expression at a concentration of 50μM. G1:O-D-galactose-1-rhamose, G2: O-L-rhamose, G3: 6-deoxy-α-mannopyranosyl, G4: 7-ß-neohesperidoside; G5: rutinoside; G6: -C-glucoside, x-5,7 4'-Trihydroxyflavone

Human umbilical vein endothelial cells were incubated with 10 ng/ml TNF in the absence (control) or presence of drugs (flavone derivatives tested at doses ranging from 1 to 50 μM) in tissue culture medium. After an incubation period of 16 hr, media were removed, the cells fixed in 3.7% formalin, and surface ICAM-1 expression determined as previously described (Gerritsen et al., 1995). (See Figure 1 for basic structures).

Potential cytotoxicity was ruled out by comparing the effects of the flavonoids on gene expression and cellular metabolism (assessed by the "MTT" assay (Denizot and Lang 1986)). Apigenin and the structurally related flavonoid acacetin inhibited endothelial ICAM, VCAM and E-selectin expression at concentrations devoid of inhibitory activity

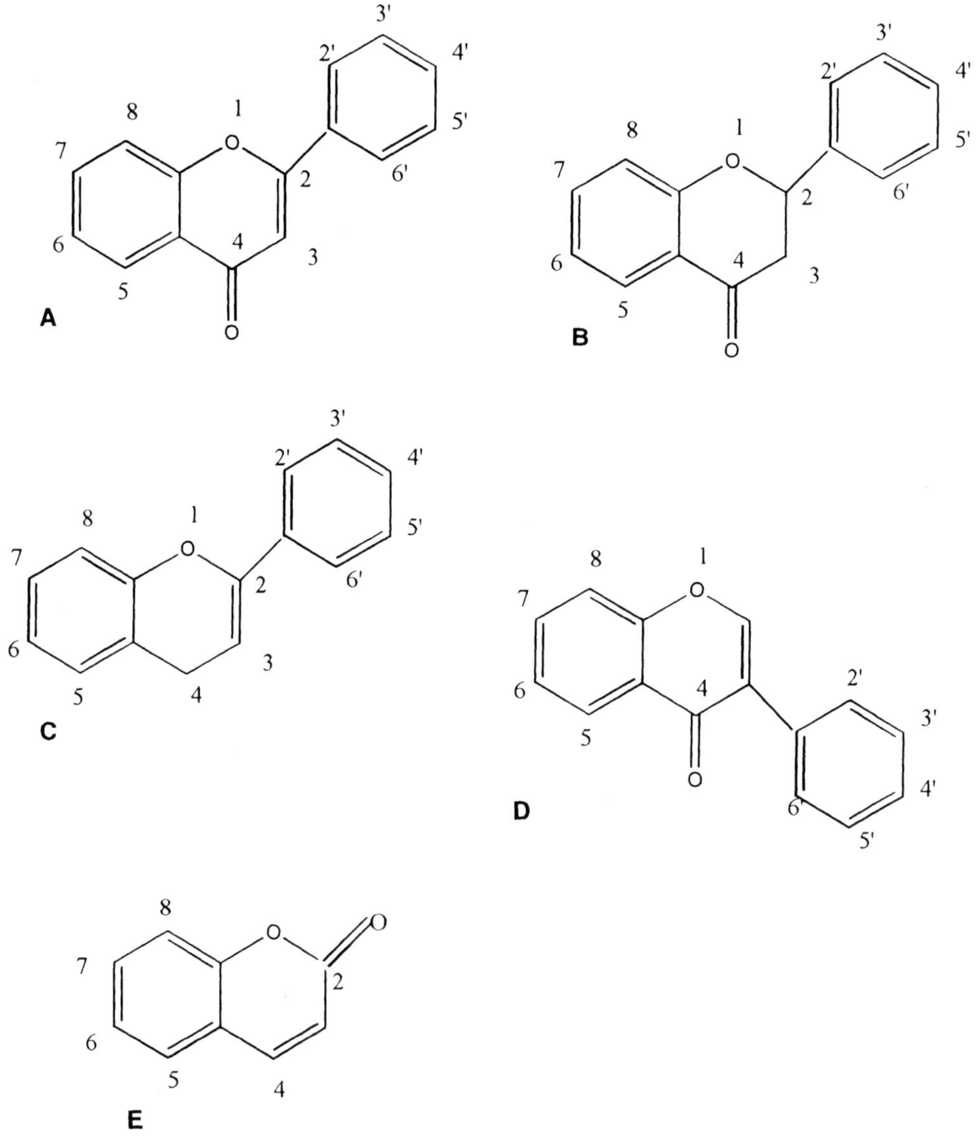

**Figure 1.** Basic structures of flavonoids.

on MTT (Gerritsen et al., 1995). In light of the central role for NF-κB in cytokine induced gene expression, the effect of flavonoids on NF-κB was evaluated. We found that apigenin did not inhibit NF-κB activation as assessed by several different assays. The electrophoretic mobility shift assay assesses the ability of proteins present in the nuclei of basal and cytokine activated cells to bind to specific radiolabelled oligonucleotide probes known to selectively bind NF-κB. TNF treatment of endothelial cells results in the translocation of activated NF-κB (p50/p65) to the cell nucleus; this activity can be monitored in the electrophoretic mobility shift assay. Apigenin did not inhibit the translocation of p65 to the nucleus (Figure 2), nor the binding of the p50-p65 heterodimer to the consensus NF-κB oligonucleotide binding site (Gerritsen et al., 1995). In other experiments we examined the cytoplasmic and nuclear levels of IκB in the presence and absence of flavonoid in basal and TNF treated endothelial cells. Apigenin did not inhibit the TNF induced degradation of IκB (Figure 2). Nonetheless, flavonoid pretreatment of A549 cells stably transfected with a NF-κB reporter construct blocked TNF induced reporter activity (Gerritsen et al., 1995). This observation suggested that apigenin might interfere with the transcriptional activity of NF-κB or could act at a yet to be defined post-transcriptional site.

We also determined the effects of flavonoids on TNF-induced JNK and p38 MAPK family members using an immunoprecipitation kinase assay as previously described (Read et al., 1996). Apigenin, at 10 and 25 μM, had no effect on TNF activation of either of these stress-kinase pathway activators. We also assessed the effect of apigenin on IFN-γ induced STAT-1 (p91) binding to the consensus site. Apigenin did not inhibit this activity (unpublished observations).

In our earlier publication, we demonstrated that the *in vitro* activities of apigenin translated to functional *in vivo* anti-inflammatory effects. Apigenin demonstrated dose-dependent inhibition of paw swelling in the carrageenan-induced paw edema and also inhibited mouse ear inflammation in a delayed type hypersensitivity response (Gerritsen et al.,

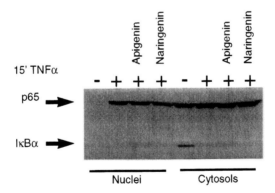

**Figure 2.** Flavonoids do not inhibit TNF-induced translocation of the p65 subunit of NFκB or the TNF-induced degradation of IκBα. Human endothelial cells were incubated without (-) or with 50 U/ml (+) of human recombinant TNFα. After a 15 min incubation at 37°C, cells were harvested and nuclear and cytosolic fractions prepared as previously described by Read et al. (1994). Aliquots of each fraction were subjected to SDS-PAGE followed by Western blotting. Blots were probed with antisera to IκBα and p65 and visualized using as previously described (Read et al., 1994). Note that 15 min following TNF treatment that the p65 subunit is now detectable in the nucleus, in contrast to the absence of this subunit in resting (control) cells. At 15 min following TNF treatment the immunoreactive IκB in the cytosol is no longer detectable due to its rapid degradation. Extracts from cells treated with apigenin (25μM) and TNF were not different from those treated with TNF alone. Data kindly provided by Margaret Read and Tucker Collins (Brigham and Women's Hospital, Boston MA).

1995). These effects could have been due to inhibition of cytokine induced gene expression (adhesion molecules, cyclooxygenase), or direct inhibition of lipid mediator production through inhibitory effects on phospholipase, cyclooxygenase or lipoxygenase activity (Corvazier and Maclouf 1985; Lanni and Becker 1985) , free radical scavenging (Robak and Gryglewski 1988; Yuting et al., 1889), or interference with neutrophil function (Blackburn et al., 1987; Busse et al., 1984; Schneider et al., 1979). To address this more directly, we recently assessed the effects of apigenin and the inactive flavanone analog, naringenin on TNF induced ICAM-1 expression in the rat (Panes et al., 1996). ICAM-1 expression was measured under baseline conditions or 5 hours following treatment with TNF using $^{125}$I labeled anti-rat ICAM-1 monoclonal antibody and $^{131}$I isotype matched control antibody (to correct for non-specific accumulation of the binding antibody). Treatment with apigenin (100 mg/kg) but not placebo or naringenin (100 mg/kg) blocked TNF induced increases in ICAM-1 in different organs, including lung, heart, liver and intestine. Pretreatment with free radical scavengers did not inhibit the ICAM-1 upregulatory response to TNF suggesting that the effect of apigenin in this model could not be explained by its putative anti-oxidant effects (Panes et al., 1996). Based on these observations, we concluded that apigenin inhibited TNF-induced ICAM-1 upregulation *in vivo* by a mechanism unrelated to free radical scavenging or leukocyte function. It would be interesting in future studies to determine the effect of apigenin administration on cytokine induced immune/inflammatory mediators such as IL-6 and IL-8.

What then, is the mechanism of the flavonoid suppression of cytokine-induced gene expression? At this juncture, we can only speculate. However, since apigenin and related flavonoids inhibit a number of genes driven by the transcription factor NF-κB, it is tempting to invoke a mechanism that involves this transcription factor at a level after its activation and translocation to the nucleus. Such mechanisms could include inhibition of the recruitment and/or interaction with other transcription factors or co-activators. The recent report that the transcriptional activity of NF-κB is regulated through phosphorylation of NF-κB p65 by protein kinase A suggests a potential target for flavonoid action (Zhong et al., 1997). As discussed in our earlier paper (Gerritsen et al., 1995), the structure-activity relationship for active flavonoids suggested the critical importance of planarity, a feature consistent with potential kinase inhibitory activity. Alternatively, flavonoids could interfere with gene expression by inhibiting other signalling pathways. For example, Kuo and colleagues (Kuo and Yang 1995) recently reported that apigenin inhibited mitogen activated protein kinase (MAPK) in NIH 3T3 cells. Another possibility worthy of consideration is the role of DNA topoisomerases. Deisher and colleagues (Deisher et al., 1993) recently reported that inhibitors of topoisomerase II can prevent cytokine-induced expression of vascular cell adhesion molecule-1. Flavonoids have been reported to act as DNA topoisomerase antagonists (Azuma et al., 1995; Boege et al., 1996; Constantinou et al., 1995) and DNA topoisomerase I has been suggested to be involved in both repression and activation of transcription, an activity apparently independent of its "topoisomerase" activity (Merino et al. 1993). Further studies of the mechanism of flavonoid inhibition of cytokine-induced gene expression should reveal important insights into the well-known anti-inflammatory actions of this class of natural products.

# REFERENCES

Azuma, Y.; Ohishi, Y.; Y Sato, Y.; H Kizaki. Effects of protein tyrosine kinase inhibitors with different modes of action on topoisomerase activity and death of IL-2-dependent CTLL-2 cells. *J. Biochem. Tokyo* **1995**, *118*, 312–318.

Baeuerle, P.A.; Henkel, T. Function and activation of NF-κB in the immune system. *Annual. Rev. Immunol.* **1994**, *12,* 141–179.

Beg, A.A.; Baldwin, A.S. The IκB proteins: multifunctional regulators of Rel/NF-κB transcription factors. *Genes and Dev.* **1993**, *7,* 2064–2070.

Blackburn, W.D.; Heck, L.W.; Wallace, R.W. The bioflavonoid quercetin inhibits neutrophil degranulation, superoxide production, and the phosphorylation of specific neutrophil proteins. *Biochem. Biophys. Res. Comm.* **1997**, *144,* 1229–1236.

Boege, F.; Straub, T; Kehr, A; Boesenberg, C; Christiansen, K; Andersen, A.; Jakob, F.; Kohrle J. Selected novel flavones inhibit the DNA binding or the DNA religation step of eukaryotic toposisomerase I. *J. Biol. Chem.* **1996**, *271,* 2262–2270.

Bovolenta, C; Gasperini, S.; Cassatella, M.A. Granulocyte colony-stimulating factor induces the binding of STAT1 and STAT3 to the IFNgamma response region within the promoter of the Fa(gamma)RI/CD64 gene in human neutrophils. *FEBS. Lett.* **1996**, *386,* 239–242.

Busse, W.W.; Kopp, D.E.; Middleton Jr. E. Flavonoid modulation of human neutrophil function. *J. Allergy Clin. Immunol.* **1984**, *73,* 801–809.

Collins, T.; Read, M.A.; Neish, A.; Whitley, M.; Thanos, D.; Maniatis, T. Transcriptional regulation of endothelial cell adhesion molecules: NF-κB and cytokine-inducible enhancers. *FASEB J.* **1995**, *9,* 899–909.

Constantinou, A.; Mehta, R.; Runyan, C.; Rao, K; Vaughan, A; Moon, R.. Flavonoids as DNA topoisomerase antagonists and poisons: structure-activity relationships. *J. Natural Products* **1995**, *58,* 217–225.

Corvazier, E.; Maclouf, J. Interference of some flavonoids and non-steroidal anti-inflammatory drugs with oxidative metabolism of arachidonic acid by human platelets and neutrophils. *Biochim. Biophys. Acta* **1985**, *835,* 315–321.

Deisher, T.A.; Kaushansky, K.; Harlan, J.M. Inhibitors of topoisomerase II prevent cytokine-induced expression of vascular cell adhesion molecule-1, while augmenting the expression of endothelial leukocyte adhesion molecule-1 on human umbilical vein endothelial cells. *Cell Adhesion and Comm.* **1993**, *1,* 133–42.

Denizot, F.; Lang, R. Rapid colorimetric assay for cell growth and survival. *J. Immunol. Methods* **1986**, *89,* 271–277.

Gerritsen, M.E.; Bloor, C. Endothelial cell gene expression in response to injury. *FASEB J.* **1993**, *7,* 523–532.

Gerritsen, M.E.; Carley, W.W.; Ranges, G.E.; Shen, C-P; Phan, S.A.; Ligon,G.F.; Perry, C.A. Flavonoids inhibit cytokine-induced endothelial cell adhesion protein gene expression. *Am. J. Pathol.* **1995**, *147,* 278–292.

Kuo, M.L.; Yang, N.C. Reversion of v-H-ras-transformed NIH 3T3 cells by apigenin through inhibiting mitogen activated protein kinase and its downstream oncogenes. *Biochem. Biophys. Res. Comm.* **1995**, *212,* 767–775.

Lanni, C.; Becker, E.L. Inhibition of phospholipase $A_2$ by *p*-bromophenyacyl bromide, nordihydroguaretic acid, 5,8,11,14-eicosatetraynoic acid and quercetin. *Int. Arch. Allergy Appl. Immunol* **1985**, *76,* 214–217.

Merino, A.; Madden, K.R.; Lane, W.S.; Champoux, J.J.; Reinberg, D. DNA topoisomerase I is involved in both repression and activation of transcription. *Nature* **1993**, *365:* 227–232.

Middleton Jr., E.; Kandaswami, C. Effects of flavonoids on immune and inflammatory cell functions. *Biochem. Pharmacol.* **1992**, *43,* 1167–1179.

Novak, U.; Harpur, A.G.; Paradiso, L.; Kanagasundaram, V.; Jaworowski, A; Wilks, A.F.; Hamilton J.A. Colony stimulating factor 1-induced STAT1 and STAT3 activation is accompanied by phosphorylation of Tyk2 in macrophages and Tyk2 and JAK1 in fibroblasts. *Blood* **1995**, *86,* 2948–2956.

Panes, J.; Gerritsen,M.E.; Anderson, D.C.; Miyasaka, M.; Granger, D.N. Apigenin inhibits TNF-induced ICAM-1 upregulation in vivo. *Microcirculation* **1996**, *3,* 279–286.

Read, M.A.; Whitley, M.Z.; Williams, A.J.; Collins, T. NF-κB and IκBα: an inducible regulatory system in endothelial cells. *J. Exp. Med.* **1994**, *179,* 503–512.

Read, M.A.; Whitley, M.Z.; Gupta, S.; Pierce, J.W.; Best, J.; Davis, R.J.; Collins, T. TNFα induced E-selectin expression is activated by the NF-κB and JNK/p38 MAP kinase pathways. *J. Biol. Chem.* **1996**, in press.

Robak, J.; Gryglewski, R.J. Flavonoids are scavengers of superoxide anions. *Biochem. Pharmacol.* **1988**, *37,* 837–841.

Schneider, C.; Berton, G.; Spisani, S.; Traniello, S.; Romeo, D. Quercitin, a regulator of polymorphonuclear leukocyte (PMNL) functions. *Adv. Exp. Med. Biol.* **1979**, *121,* 371–379.

Tsukada, J.; Waterman, W.R.; Koyama, Y.; Webb, A.C.; Auron, P.E. A novel STAT-like factor mediates lipopolysaccharide, interleukin 1 (IL-1), and IL-6 signaling and recognizes a gamma interferon activation-like element in the IL1β gene. *Mol. Cell. Biochem.* **1996**, *16,* 2183–2194.

Tweardy, D.J.; Wright, T.M.; Ziegler, S.F.; Baumann, H.; Chakraborty, A; White, S.M.; Dyer, K.F.; Rubin, K.A. Granulocyte colony-stimulating factor rapidly activates a distinct STAT-like protein in normal myeloid cells. *Blood* **1995**, *86,* 4409–4416.

Verma, I.M.; Stevenson, J.K.; Schwarz, E.M.; Van Antwerp, D.; Miyamoto, S.. Rel/NF-κB/IκB family: intimate tales of association and dissociation. *Genes Dev.* **1995**, *9,* 2723–2735.

Yuting, C.; Rongliang, Z.; Zhongjian, J.; Yong, Y. Flavonoids as superoxide scavengers and antioxidants. *Free Rad. Biol. and Med* **1989**, *9*, 19–21.

Zhong, H.; Su Yang, H.; Erdjument-Bromage, H.; Tempst, P.; Ghosh, S. The transcriptional activity of NF-κB is regulated by the IκB-associated PKAc subunit through a cyclic AMP-independent mechanism. Cell **1997**, *89*, 413–424.

# RECENT ADVANCES IN THE DISCOVERY AND DEVELOPMENT OF FLAVONOIDS AND THEIR ANALOGUES AS ANTITUMOR AND ANTI-HIV AGENTS*

Hui-Kang Wang, Yi Xia, Zheng-Yu Yang, Susan L. Morris Natschke, and Kuo-Hsiung Lee[†]

Natural Products Laboratory
Division of Medicinal Chemistry and Natural Products
School of Pharmacy
University of North Carolina
Chapel Hill, North Carolina 27599

## ABSTRACT

Antitumor and anti-HIV flavonoids and their analogues will be reviewed with emphasis on those discovered in our laboratory. The active antitumor compounds include the antileukemic tricin (**1**) and kaempferol-3-*O*-β-D-glucopyranoside (**2**) from *Wikstroemia indica*, the cytotoxic hinokiflavone (**3**) from *Rhus succedanea*, the cytotoxic isoflavone (**8**) from *Amorpha fruticosa*, two dihydroxypentamethoxyflavones (**9, 10**) from *Polanisia dodencandra*. The development of synthetic 2-phenyl-4-quinolones as potent cytotoxic antimitotic flavonoid analogues and 2-phenylthiochromen-4-ones as potent antitumor flavonoid analogues will be presented. Selected results from other laboratories and antitumor-related biological studies also will be discussed. Flavonoids have also been investigated as potential anti-HIV agents. In our laboratory, acacetin-7-*O*-β-D-galactopyranoside (**131**) from *Chrysanthemum morifolium* and chrysin (**102**), as well as apigenin-7-*O*-β-D-glucopyranoside (**130**), from *Kummerowia striata*, have been found to exhibit anti-HIV activity. In other studies, some flavonoids and related compounds have been investigated

---

* Antitumor agents 179. For 178, see Chen, K.; Kuo, S. C.; Mauger, A.; Lin, C. M.; Hamel, E.; and Lee, K. H. Synthesis and biological evaluation of substituted 2-aryl-1,8-naphthyridin-4-ones as antitumor agents that inhibit tubulin polymerization. *J. Med. Chem.* 1997, *40*, 3049–3056.
† To whom correspondence should be addressed.

as inhibitors of HIV-1 reverse transcriptase, protease, and integrase. The isolation and structural modification of such plant-derived active principles provide a continuing source of potential antitumor and anti-HIV agents.

## 1. INTRODUCTION

The flavonoids, one of the most numerous and widespread groups of natural products, have been known from as early as the alkaloids. The distribution of flavonoids in the plant kingdom has already been reviewed by Harborne (1975, 1986). To date, over 4000 flavonoids (Brandi, 1992) have been identified from both the higher and lower plants, and new structures are being reported at an ever-increasing rate. Unlike alkaloids or terpenoids, whose skeletons are rich and varied, flavonoids have only limited skeletal varieties. However, these relatively unified substances do not only contribute to plant color, taste (Nobel, 1994; Horowitz, 1986), and self-defense, but are also important to mankind because they are widely present in human food and have demonstrated extensive biological activities such as anti-inflammatory and anti-allergic (Gabor, 1986), mutagenic and carcinogenic (MacGregor, 1986; Das et al., 1994), antihepatotoxic (Wagner, 1986), free radical scavenging and antioxidant (Kandaswami and Middleton, 1994), antiviral (Selway, 1986) and antitumor activities.

In the past two decades, the discovery and development of flavonoids and their analogues as antitumor or anti-HIV agents have been widely researched, and significant progress has been achieved. This review will deal only with recently discovered antitumor or anti-HIV compounds and their analogs with a emphasis on the work carried out in our laboratory.

## 2. ANTITUMOR FLAVONOIDS AND THEIR ANALOGS

### 2.1. Flavonoids as Cytotoxic Antitumor Principles

Bioassay-directed isolation and characterization of potent cytotoxic antitumor agents and analogues from plants have been conducted in many laboratories including our own over the past two decades. This methodology is very efficient and has led to the identification of active lead flavonoids for further development as antitumor drugs.

The whole plant of *Wiksteroemia indica* (Thymelaeaceae), known as "Nan-Ling-Jao-Hua" or "Po-Lun" in Chinese folklore, is used as a herbal remedy for the treatment of human syphilis, arthritis, whooping cough, and cancer (Sugi and Nagashio, 1977). Bioassay-directed isolation of the antitumor extract of this plant conducted in our laboratory has led to the characterization of tricin (**1**) and kaempferol-3-*O*-β-D-glucopyranoside (**2**) as the major antileukemic constituents (Lee et al., 1981). Tricin afforded a T/C=133 and 174% when tested at 6 mg/kg and 12.5 mg/kg, respectively, in an *in vivo* P-388 screen. In the same screen, compound **2** exhibited a T/C=122 and 130% when tested at 12.5 mg/kg. The good antileukemic activity demonstrated by **1** at 12.5 mg/kg is noteworthy, as cytotoxic flavonoids seldom show significant *in vivo* activity against P-388 lymphocytic leukemic growth in mice (Edwards et al., 1979). It is also interesting to note that both **1** and **2** possess the same 5,7,4' tri-hydroxylated pattern.

Bioassay-directed fractionation performed in our laboratory led to the isolation and characterization of hinokiflavone (**3**) as the cytotoxic principle (ED$_{50}$ (KB) = 2.0 µg/ml)

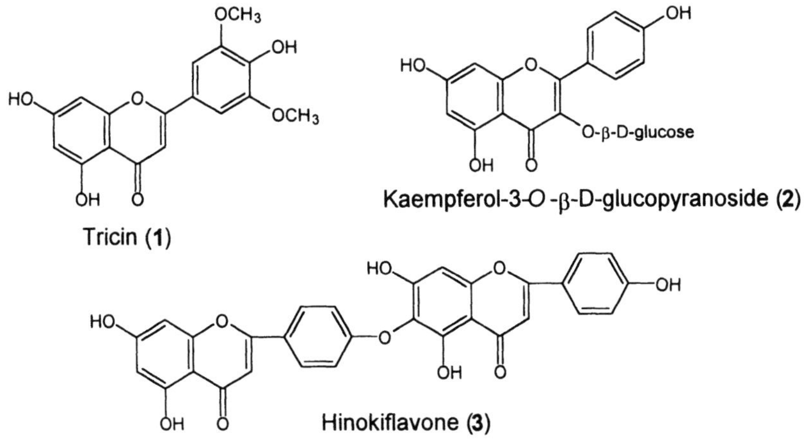

Figure 1. The structures of cytotoxic flavones 1–3.

from the drupes of *Rhus succedaneal* (Anacardiaceae). A comparison of the cytotoxicity of **3** and other related biflavonoids indicates that an ether linkage between two units of apigenin (**86**, see Table 13 for structures of all compounds not appearing as Figures) as seen in **3** is structurally required for significant cytotoxicity (Lin et al., 1989).

Several cytotoxic compounds have been isolated from *Amorpha fruticosa* in our laboratory (Li et al., 1993). One compound, 12aβ-hydroxyamorphigenin (**4**), was first shown to exhibit extremely potent cytotoxicity ($ED_{50} < 0.001$ µg/ml) against six neoplastic cell lines. Amorphigenin (**5**) was also very active in all assays, but its activity was lower than its hydroxylated analogue **4**. Tephrosin (**6**), a structurally similar compound, also showed potent cytotoxicity in all assays. Amorphispironone (**7**), a novel spirone-type rotenoid possessing an unusual spiro A-B ring system, showed potent ($ED_{50} < 1.0$ µg/ml) and

Table 1. Cytotoxic *isoflavone* and rotenoids from *Amorpha fruticosa*

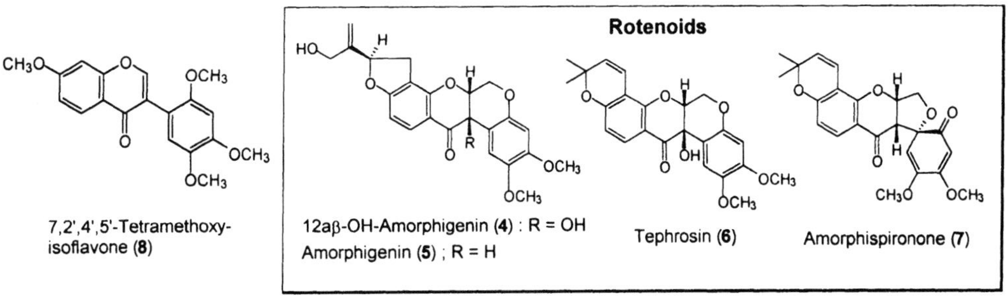

|   | A-549 | HCT-8 | RPMI-7951 | TE671 | KB | P388 |
|---|---|---|---|---|---|---|
| 4 | < 0.001 | < 0.001 | < 0.001 | < 0.001 | < 0.001 | < 0.001 |
| 5 | 0.05 | 0.03 | 0.05 | < 0.01 | 0.04 | 0.04 |
| 6 | < 0.001 | 0.09 | 0.07 | 0.05 | 0.36 | 0.06 |
| 7 | --- | --- | 0.61 | --- | 0.58 | --- |
| 8 | --- | 4.63 | 0.49 | --- | 0.55 | 0.53 |

selective cytotoxicity against RPMI-7951 and KB tumor cell lines (Li et al., 1991; Terada et al., 1993). 7,2',4',5'-Tetramethyoxyisoflavone (**8**) showed potency against RPMI-7951, KB and P388 cell lines. It is rare for an isoflavone to show strong cytotoxicity.

Recently, we have also isolated three flavonols, 5,3'-dihydroxy-3,6,7,8,4'-pentamethoxyflavone (**9**), 5,4'-dihydroxy-3,6,7,8,3'-pentamethoxyflavone (**10**), and quercetin 3-$O$-β-glucopyranosyl-7-$O$-α-L-rhamnopyranoside (**11**), from *Polanisia dodecandia*, a plant native to North America and abundant from Montana to Wisconsin and from Quebec to Mexico (Shi et al., 1995). Compound **9** showed remarkable cytotoxicity *in vitro* against panels of central nervous system cancer (SF-268, SF-593, SNB-75, U-251), non-small cell lung cancer (HOP-62, NCI-H266, NCI-H460, NCI-H522), small cell lung cancer (DMS-114), ovarian cancer (OVCAR-3, SK-OV-3), colon cancer (HCT-116), renal cancer (UO-31), melanoma (SK-MEL-5), and leukemia (HL-60 [TB], SR) cells, with $GI_{50}$ values in the low micromolar concentration range. This compound also inhibited tubulin polymerization ($IC_{50}$=0.83 μM) and the binding of radiolabeled colchicine to tubulin (59% inhibition when present in equimolar concentrations with colchicine). Compound **10** also showed cytotoxicity against medulloblastoma (TE-671) tumor cells with an $ED_{50}$ value of 0.98 μg/ml. Compound **10** appears to be the first example of a flavonol to exhibit potent inhibition of tubulin polymerization and, therefore, warrants further investigation as an antimitotic agent.

Ten known flavonoids were isolated by several groups from the cytotoxic extract of leaves of *Centaurea urvillei* (Ulubelen and Oksuz, 1982). Among them, the major compound, hispidulin (**12**), showed cytotoxic activity at a 0.05 mg/ml dose against L-strain fibroblasts in tissue culture.

Skullcapflavone II (5,2'-dihydroxy-6,7,8,6'-tetramethoxyflavone, **13**) isolated from the root of *Scutellaria baicalensis* exhibited cytotoxic activity with an $ED_{50}$ of 1.5 μg/ml against L1210 cells *in vitro* (Ryu et al., 1985a).

From the whole plant of *Gutierrezia microcephala* (Asteraceae), 3,3'-dimethyl-quercetin (**14**) and 3,7-dimethylquercetin (**15**) were identified as the cytotoxic principles with $IC_{50}$ values of 1.7 and 3.2 μg/ml against P-388 cells *in vitro* respectively (Dong et al., 1987).

Hymenoxin (5,7-dihydroxy-3',4',6,8,-tetramethoxyflavone, **16**) was isolated from the whole plants of *Scoparia dulcis* (Scrophulariaceae), and exhibited cytotoxicity against

Table 2. Cytotoxic flavonols from *Polanisia dodecandia*

**9** : $R_1$ = H, $R_2$ = CH$_3$
**10** : $R_1$ = CH$_3$, $R_2$ = H

|    | $ED_{50}$ (μg/ml) | | | | | |
|----|------|-------|-------|-------|----------|--------|
|    | KB | A-459 | HCT-8 | P-388 | PRMI-7591 | TE-671 |
| 9  | 0.045 | 0.60 | 4.40 | 0.055 | 0.55 | 0.069 |
| 10 | 8.38 | Inactive | Inactive | 6.52 | 5.93 | 0.98 |

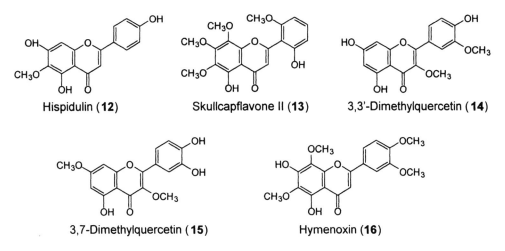

Figure 2. The structures of some cytotoxic flavones.

Table 3. Cytotoxic flavonoids from *Muntingia calabura*

**Flavans**
17 $R_1 = R_2 = H, R_3 = OH$
18 $R_1 = R_3 = OMe, R_2 = H$
19 $R_1 = R_3 = OMe, R_2 = OH$
20 $R_1 = OMe, R_2 = H, R_3 = OH$
21 $R_1 = OH, R_2 = H, R_3 = OMe$
22 $R_1 = R_2 = OH, R_3 = OMe$
23 $R_1 = R_3 = OH, R_3 = H$

24 $R_1 = R_3 = OMe, R_2 = H$
25 $R_1 = OMe, R_2 = H, R_3 = OH$
26 $R_1 = R_3 = OH, R_2 = H$

**Biflavans**
27 R = OMe
28 R = OH

|    | BC1[a] | HT[b] | Lu1[c] | Me12[d] | Co12[e] | KB[f] | KB-V[g] | P-388[h] |
|----|--------|-------|--------|---------|---------|-------|---------|----------|
| 17 | > 20   | > 20  | > 20   | 14.6    | > 20    | 9.4   | 13.3    | 5.9      |
| 18 | > 20   | > 20  | > 20   | 8.9     | 15.8    | 13.3  | 2.1     | 5.4      |
| 19 | 10.9   | 3.3   | 13.5   | 9.7     | 12.0    | 3.4   | 6.2     | 4.9      |
| 20 | > 20   | > 20  | > 20   | 9.0     | > 20    | 15.5  | 12.3    | 2.0      |
| 21 | > 20   | > 20  | > 20   | 9.2     | > 20    | 11.8  | 3.9     | 3.0      |
| 22 | > 20   | > 20  | > 20   | 14.5    | > 20    | 10.2  | > 20    | 2.3      |
| 23 | > 20   | > 20  | > 20   | 10.6    | > 20    | 13.8  | 11.1    | 4.1      |
| 24 | NT     | NT    | NT     | NT      | NT      | NT    | NT      | NT       |
| 25 | > 20   | > 20  | > 20   | > 20    | 15.2    | > 20  | > 20    | 11.9     |
| 26 | > 20   | > 20  | > 20   | > 20    | 5.9     | > 20  | > 20    | 16.7     |
| 27 | 12.0   | 5.5   | 12.4   | 10.2    | 6.2     | 2.2   | 8.3     | 3.7      |
| 28 | 16.0   | 5.0   | 15.6   | 8.7     | 9.0     | 5.2   | 12.6    | 4.8      |

[a] Human breast cancer, [b] Human fibrosarcoma, [c] Human lung cancer, [d] Human melanoma, [e] Human colon cancer, [f] Human nasopharyngeal carcinoma, [g] Vincristine-resistant KB, [h] Murine lymphocytic leukemia.

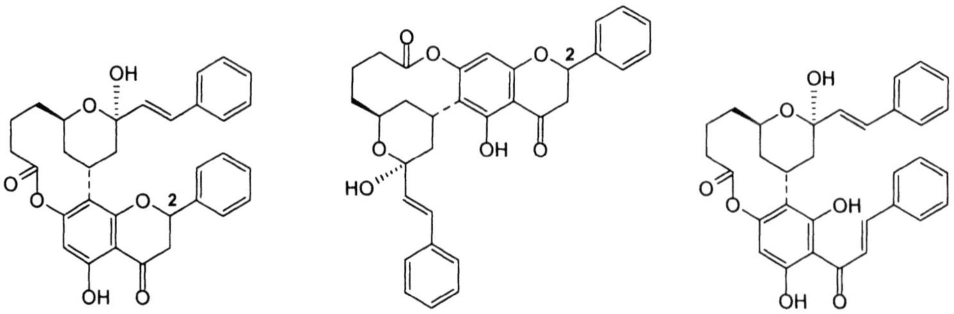

Kurziflavolactone A (**29**), 2R
Kurziflavolactone B (**30**), 2S

Kurziflavolactone C (**31**), 2R
Kurziflavolactone D (**32**), 2S

Kurzichalcolactone D (**33**)

Figure 3. Kurziflavolactones and kurzichalcolactones from *Cryptocarya kurzii*.

human cultured cells such as HeLa 229, HeLa S3, Hep-2, FL, Chang liver and intestine 407 cells with $IC_{50}$ values of 0.097, 0.140, 0.148, 0.283, 0. 510, and 0.264 µg/ml, respectively (Hayashi, 1988).

Twelve new flavonoids, constituting seven flavans (**17–23**), three flavones (**24–26**) and two biflavans (**27–28**), were isolated from the cytotoxic $Et_2O$-soluble extract of *Muntingia calabura* (Elaeocarpaceae) by Kaneda et al (1991). Most of the compounds demonstrated cytotoxic activity against P-388 cells, with the flavans being more active than the flavones.

Four complex flavanones, kurziflavolactones A (**29**), B (**30**), C (**31**), and D (**32**) and a complex kurzichalcolactones (**33**) with an unprecedented carbon side chain on the flavanone or chalcone A ring, have been isolated from a Malaysian plant, *Cryptocarya kurzii* (Lauraceae) (Fu et al., 1993). Compounds **30** and **33** showed slight cytotoxicity against KB cells, with $IC_{50}$ values of 4 and 15 µg/ml, respectively. A biosynthetic pathway for the formation of these compounds has been suggested.

Twenty-seven flavones, including five new flavones (**34–38**), were isolated from Aurantii Nobilis Pericarpium and the fruit peel of *Citrus reticulata* Blanco (Rutaceae) (Sugiyama et al., 1993). Each compound, except for two flavone glucosides, showed dif-

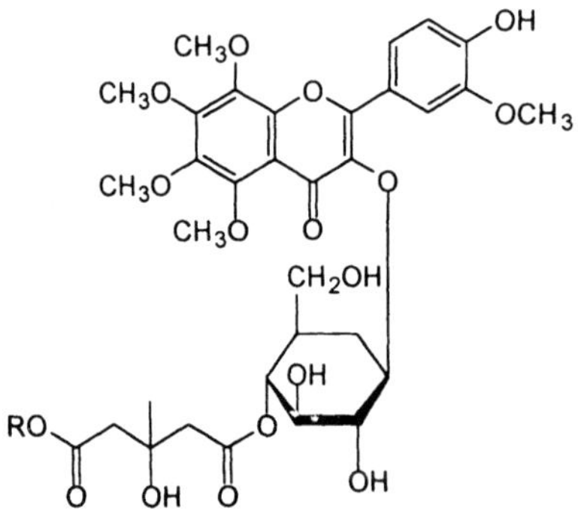

**37** : R = H
**38** : R = Me

Figure 4. New flavone glucosides from Aurantii Nobilis Pericarpium and the fruit peel of *Citrus reticulata* Blanco (Rutaceae).

Table 4. Three cytotoxic biflavonoids from *Calycopteris floribunda*

**Biflavonoids**

Calycopterone (**39**) : R = R$_2$ = Me, R$_1$ = H
Isocalycopterone (**40**) : R = R$_1$ = Me, R$_2$ = H
4-Demethylcalycopterone (**41**) : R = R$_1$ = H, R$_2$ = Me

**42**

| | \multicolumn{9}{c}{Cell lines, ED$_{50}$ (μg/mL)} |
|---|---|---|---|---|---|---|---|---|---|
| | BC1 | HT | Lu1 | Col2 | KB | KB-V1 | P388 | A431[a] | U373[b] |
| 40 | 5.4 | 1.8 | 1.2 | 0.4 | 0.8 | 9.6 | 0.78 | 1.0 | 0.1 |
| 41 | 0.4 | 0.4 | 0.7 | 0.3 | 0.4 | | 0.2 | 0.4 | 0.8 |
| 42 | 1.4 | 0.8 | 1.6 | 0.3 | 1.2 | 8.6 | 0.42 | 0.2 | 0.3 |
| 43 | > 20 | 4.2 | > 20 | > 20 | > 20 | 0.8 | > 20 | > 20 | > 20 |

[a] Epidermoid carcinoma, [b] Glioblastoma.

ferentiation inducing activity toward mouse myeloid leukemia cells (M1), resulting in cellular phagocytic activity. Furthermore, differentiation-inducing activity also was tested using the human acute promyelocytic leukemia cell line (HL-60). The results showed that highly methoxylated flavones exhibited differentiaton-inducing activity toward HL-60 cells at 50 μM. Activity was not affected by the number of substituents on the B-ring, but was affected by the substituent pattern on the A-ring. Flavones containing free hydroxy groups were inactive.

Three biflavonoids, calycopterone (**39**), isocalycopterone (**40**), 4-demethylcalycopterone (**41**), and one flavone 4',5-dihydroxy-3,3',6,7-tetramethoxy flavone (**42**) were isolated as cytotoxic constituents from flowers of *Calycopteris floribunda* Lamk (Combretaceae) (Wall et al., 1994). Compounds **39–41** showed a wide range of activity against a panel of solid tumor cell lines. Calycopterone **39** displayed specific sensitivity in the leukemia cell line panels.

Several *Selaginella* species are used in traditional medicine in various countries to treat a variety of diseases such as cancer, cardiovascular problems, diabetes, gastritis, hepatitis, skin diseases and some urinary tract infections. Bioactivity-guided fractionation of the leaves of *Selaginella willdenowii* (Selaginellaceae) afforded three known biflavones, 4',7''-di-*O*-methylamentoflavone (**43**), isocryptomerin (**44**) and 7''-*O*-methylrobustaflavone (**45**), that were significantly cytotoxic against a panel of human cancer cell lines including breast, lung, colon, and prostate cancer, fibrosarcoma, oral epidermoid carcinoma, glioblastoma, and leukemia (Silva et al., 1995).

Three new dihydroflavonols, gericudranins A-C, were isolated from the stem bark of *Cudrania tricuspidata* (Moraceae) and were identified as 6,8-di-*p*-hydroxybenzyltaxifolin (**46**), 8-*p*-hydroxybenzyltaxifolin (**47**), and 6-*p*-hydroxybenzyltaxifolin (**48**), respectively (Lee et al., 1996). These compounds were cytotoxic to human tumor cell lines, such as CRL 1579 (skin), LOX-IMVI (skin), MOLT-4F (leukemia), KM12 (colon) and UO-31 (renal) in cell culture, with ED$_{50}$ values of 2.7–31.3 μg/ml.

Table 5. Three cytotoxic biflavones from *Selaginella willdenowii*

4',7"-Di-O-methylamentoflavone (**43**)

Isocryptomerin (**44**)

7"-O-Methylrobustaflavone (**45**)

| | \multicolumn{10}{c}{Cell lines, $ED_{50}$ (µg/mL)} | | | | | | | | | |
|---|---|---|---|---|---|---|---|---|---|---|
| | BC1 | HT-1080 | Lu1 | Col2 | KB | KB-V+ | KB-V- | LNCaP | ZR-75-1 | U373 |
| 43 | 5.7 | 11.8 | 12.0 | 2.5 | >20 | 7.0 | >20 | >20 | 16.2 | 3.8 |
| 44 | 1.5 | 0.6 | 0.9 | 1.8 | 1.6 | 1.5 | 2.1 | 2.1 | 0.58 | 3.5 |
| 45 | 3.3 | 0.9 | 0.4 | 6.0 | 3.6 | 8.9 | 7.5 | 3.7 | 1.4 | 0.7 |

[a] Hormone-dependent prostate cancer, [b] Hormone-dependent breast cancer, [c] Glioblastoma

## 2.2. Antitumor-Related Biological Studies

*2.2.1. Cytotoxic and Antitumor Studies of Flavonoids.* Ten flavones with OBn, OH, and OMe substituents were synthesized and evaluated by Ryu et al (1985b) for cytotoxicity against L1210 cells. Among these compounds, 5,2'-dihydroxy-6,7,8,6'-tetramethoxyflavone (**13**) was the most potent.

Table 6. Three cytotoxic benzyldihydroflavonols from *Cudrania tricuspidata*

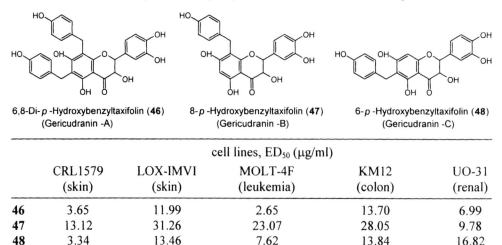

6,8-Di-*p*-Hydroxybenzyltaxifolin (**46**)
(Gericudranin -A)

8-*p*-Hydroxybenzyltaxifolin (**47**)
(Gericudranin -B)

6-*p*-Hydroxybenzyltaxifolin (**48**)
(Gericudranin -C)

| | \multicolumn{5}{c}{cell lines, $ED_{50}$ (µg/ml)} | | | | |
|---|---|---|---|---|---|
| | CRL1579 (skin) | LOX-IMVI (skin) | MOLT-4F (leukemia) | KM12 (colon) | UO-31 (renal) |
| 46 | 3.65 | 11.99 | 2.65 | 13.70 | 6.99 |
| 47 | 13.12 | 31.26 | 23.07 | 28.05 | 9.78 |
| 48 | 3.34 | 13.46 | 7.62 | 13.84 | 16.82 |

A number of representative flavonoids reversibly inhibited human lymphocyte proliferative responses in a concentration-dependent manner. The flavonoids quercetin (**49**) and tangeretin (**50**) were most effective when added during the early phase of lymphocyte exposure to the mitogenic stimuli, but became progressively less effective when added after increasing lengths of time following stimulation, suggesting an early flavonoid-sensitive step(s) in cell activation (Mookerjee et al., 1986).

Fifty-five flavones with a variety of substituents including COOMe, COOH, $NO_2$, $NH_2$, OBn, $OSi(Me)_2$-t-Bu, OH, and OMe were synthesized and evaluated by Cushman and Nagarathnam (Cushman et al., 1991) for cytotoxicity against 5 cancer cell lines: A-549 lung carcinoma, MCF-7 breast carcinoma, HT-29 colon adenocarcinoma, SKMEL-5 melanoma, and MLM melanoma. Of the flavones tested, 14 compounds (**51–64**) exhibited significant activity against at least one of the 5 cancer cell lines. 7,8-Dihydroxy-3',4'-dimethoxyflavone (**59**) was the most potent. It is worth noting that substitution at the 3-position with COOMe or COOH groups generally resulted in noncytotoxic derivatives in all cell lines at 25 µg/ml. After omitting the 3-substituted flavones, five (compounds **51–55**) out of the nine flavones with a 4'-nitro or 4'-amino substituent were significantly active in at least one of the five cancer lines.

Kandaswami et al (1991; 1992) investigated the antiproliferative effect of two polyhydroxylated [quercetin (**49**) and taxifolin (**65**)] and two polymethoxylated [nobiletin (**66**) and tangeretin (**50**)] flavonoids against three cell lines in tissue culture. Tangeretin and nobiletin markedly inhibited the proliferation of a squamous cell carcinoma (HTB 43) and a gliosarcoma (9L) cell line at 2–8 µg/ml concentrations. Quercetin displayed no effect on 9L cell growth at these concentrations; while at 8 µg/ml, it inhibited HTB 43 cell growth. Taxifolin slightly inhibited HTB 43 cell growth at 8 µg/ml, and moderately inhibited HTB

Table 7. The cytotoxicities of some synthetic flavones

| | | | | | | | Cell lines, $ED_{50}$ (µg/ml) | | | | |
|---|---|---|---|---|---|---|---|---|---|---|---|
| | $R_5$ | $R_6$ | $R_7$ | $R_8$ | $R_{3'}$ | $R_{4'}$ | $R_{5'}$ | A-549 | MCF-7 | HT-29[a] | SK-MEL[b] | MALM-3M[c] |
| 51 | OH | H | OH | H | H | $NO_2$ | H | 4.19 | 6.90 | 3.34 | 4.87 | >2 |
| 52 | H | H | OH | H | H | $NO_2$ | H | 2.56 | 4.69 | 3.46 | 3.29 | 3.61 |
| 53 | H | OH | H | H | H | $NH_2$ | H | 2.72 | >25 | 2.65 | 2.03 | >25 |
| 54 | OH | H | OH | H | H | $NH_2$ | H | 2.84 | >25 | 3.44 | 3.86 | >25 |
| 55 | OAc | H | OAc | H | H | NHAc | H | 21.98 | 1.78 | 17.40 | 13.96 | >25 |
| 56 | H | H | OH | OH | OMe | OMe | OMe | 5.27 | 1.11 | 3.66 | >25 | >25 |
| 57 | H | OH | H | H | OMe | $OSi(Me)_2$-t-Bu | OMe | 3.19 | 2.01 | 2.63 | 2.95 | 5.30 |
| 58 | H | H | OH | H | OMe | $OSi(Me)_2$-t-Bu | OMe | 5.30 | 4.47 | 1.00 | >25 | >25 |
| 59 | H | H | OH | OH | OMe | $OSi(Me)_2$-t-Bu | OMe | 3.31 | 0.31 | 0.34 | 0.11 | 0.20 |
| 60 | OH | H | H | H | OMe | OH | OMe | 0.74 | 0.51 | >25 | >2 | >25 |
| 61 | H | H | OAc | H | OMe | $OSi(Me)_2$-t-Bu | OMe | 2.90 | 1.49 | 1.94 | 0.80 | 2.95 |
| 62 | H | H | OH | H | H | $OSi(Me)_2$-t-Bu | H | 6.45 | 4.66 | 3.61 | 5.64 | >10 |
| 63 | OH | H | H | H | OH | H | H | 3.24 | 6.93 | 3.47 | 3.09 | >10 |
| 64 | H | H | OH | OH | OH | H | H | 4.65 | 3.71 | 4.18 | 3.66 | 4.80 |

[a] Colon adenocarcinoma, [b] melanoma, [c] MLM melanoma.

43 cell growth at 2–8 µg/ml. The proliferation of a human lung fibroblast-like cell line (CCL 135) was relatively insensitive to low concentrations of the above flavonoids.

Antileukemic-cell efficacy of 28 naturally occurring and synthetic flavonoids against the human promyelocytic leukemic cell line HL-60 was examined by Hirano et al (1994) using MTT assay methods. Eight of the 28 flavonoids showed considerable suppressive effects on HL-60 cell growth with $IC_{50}$s ranging from 0.01–0.94 µg/ml. Among these compounds, genistein (**67**) had the strongest effects with $IC_{50}$s less than 0.1 µg/ml, which were almost equivalent to the effects of current anti-cancer agents. Genistein, however, showed little or no cytotoxicity against HL-60 cells as assessed by dye exclusion tests ($LC_{50}$s > 2.9 µg/ml), whereas the typical anti-cancer agents showed potent cytotoxicity. All of the flavonoids were less effective against growth of the human T lymphocytic leukemia cell line MOLT-4. In addition, the flavonoids showed little or no inhibiting activity on mitogen-induced blastogenesis of human peripheral-blood lymphocytes. Genistein was strongly suppressive against incorporation of [$^3$H]thymidine, [$^3$H]uridine, and [$^3$H]leucine into HL-60 cells. These results showed that some naturally occurring flavonoids inhibited HL-60 cell growth through a non-toxic mechanism, possibly via cessation of DNA, RNA, and/or protein synthesis in the leukemic cells. In another report, genistein was demonstrated to inhibit the growth of prostate cancer cells *in vitro* (Naik et al., 1994).

Antitumor effects of nine components of a Chinese herbal formula, 'Sho-saiko-to', were investigated in human hepatoma cell lines (PLC/PRF/5, Hep-G2), human liver cells (Chang) and a human pancreatic cancer cell line (BxPC-3) (Motoo and Sawabu, 1994). The concentration of each component required for 50% inhibition of PLC/PRF/5 cell growth was as follows: saikosaponin-d, baicalin (**74**), 20 µg/ml; saikosaponin-a, baicalein (**69**), 50 µg/ml; saikosaponin-$b_2$, -c, ginsenoside-$Rb_1$, -$Rg_1$, glycyrrhizin, > 1000 µg/ml.

**Table 8.** Antileukemic-cell efficacy of some naturally occurring and synthetic flavonoids

|  | $R_5$ | $R_6$ | $R_7$ | $R_8$ | $R_{3'}$ | $R_{4'}$ | Cell lines, $IC_{50}$ (µg/ml) HL-60 | MOLT-4 |
|---|---|---|---|---|---|---|---|---|
| nobiletin (**66**) | OMe | OMe | OMe | OMe | OMe | OMe | 0.10 | > 10.0 |
| genistein (**67**) | OH | H | OH | H | H | OH | 0.01 | 2.85 |
| wogonin (**68**) | OH | H | OH | OMe | H | H | 0.16 | 2.68 |
| baicalein (**69**) | OH | OH | OH | H | H | H | 0.26 | 2.32 |
| liquiritigenin (**70**) | H | H | OH | H | H | OH | 0.29 | 3.20 |
| 6-hydroxyflavone (**71**) | H | OH | H | H | H | H | 0.67 | 1.50 |
| 7-hydroxyflavone (**72**) | H | OH | H | H | H | H | 0.21 | 2.15 |
| 2-cyclohexyl-5-hydroxychromome (**73**) |  |  |  |  |  |  | 0.94 | 5.35 |
| etoposide |  |  |  |  |  |  | 0.01 | 0.02 |

These results indicate that 'Sho-saiko-to' contains antitumor components such as saikosaponin-a, -d, and baicalin with potency against human hepatoma cells as well as other human cell lines.

The cytotoxicity of 21 flavonoids present in *Arnica* species was studied in a human small cell lung carcinoma cell line (GLC4) and in a human colorectal cancer cell line (COLO 320), using the microculture tetrazolium (MTT) assay. Most flavonoids showed moderate to low cytotoxicity; the $IC_{50}$ values varied from 17 to > 200 µM. The most toxic compound was the flavone jaceosidin (**75**) (Woerdenbag et al., 1994).

Certain anti-cancer agents are known to induce apoptosis in human tumor cells. However, these agents are intrinsically cytotoxic against cells of normal tissue origin, including myelocytes and immunocytes. However, Hirano et al (1995) reported that a naturally occurring flavone of citrus origin, tangeretin (5,6,7,8,4'- pentamethoxyflavone, **50**), induced apoptosis in human promyelocytic leukemia HL-60 cells, but showed no cytotoxicity against human peripheral blood mononuclear cells (PBMCs). The growth of HL-60 cells *in vitro* assessed by [$^3$H]thymidine incorporation or tetrazolium crystal formation was strongly suppressed in the presence of tangeretin; the $IC_{50}$ values ranged between 0.062 and 0.173 µM. Apoptosis of HL-60 cells, assessed by cell morphology and DNA fragmentation, was demonstrated in the presence of > 2.7 µM tangeretin. Flow cytometric analysis of tangeretin-treated HL-60 cells also demonstrated apoptotic cells with low DNA content and showed a decrease of G1 cells and a concomitant increase of S and/or G2/M cells. Apoptosis was evident after 24 h of incubation with tangeretin, and the tangeretin effect as assessed by DNA fragmentation or growth inhibition was significantly attenuated in the presence of $Zn^{2+}$, which is known to inhibit $Ca^{2+}$- dependent endonuclease activity. $Ca^{2+}$ and $Mg^{2+}$, in contrast, promoted the effect of tangeretin. Cycloheximide significantly decreased the tangeretin effect on HL-60 cell growth, suggesting that protein synthesis is required for flavonoid-induced apoptosis. Tangeretin showed no cytotoxicity against either HL-60 cells or mitogen-activated PBMCs even at high concentration (27 µM) as determined by a dye exclusion test. Moreover, the flavonoid was less effective on growth of human T-lymphocytic leukemia MOLT-4 cells or on blastogenesis of PBMCs. These results suggest that tangeretin inhibits growth of HL-60 cells *in vitro*, partially through induction of apoptosis, without causing serious side-effects on immune cells.

*2.2.2. Flavonoids as Inhibitors of Tumor Invasion.* Invasion is the hallmark of malignant tumors and generally leads to metastasis, which is the major cause of death of cancer patients. Anti-invasive agents are being studied for both the development of new therapeutic rationales in cancer treatment and the analysis of tumor invasion mechanisms (Parmar et al., 1994). Some flavonoids such as (+)-catechin (**76**), 3,7-dimethoxyflavone (**77**), and tangeretin (**50**), an inhibitor of tumor cell motility, have shown anti-invasive effects *in vitro*.

Bracke et al (1984; 1986; 1987) found that (+)-catechin (**76**) inhibits the invasion of malignant MO4 fibrosarcoma cells into embryonic chick heart fragments *in vitro*. This inhibition is maximal at a drug concentration of 0.5 mM. The anti-invasion activity of (+)-catechin may be related to its binding to laminin. (+)-Catechin abrogates cell adhesion to and spreading on laminin substrates, and could in this way inhibit invasion (Bracke et al., 1988).

Tangeretin (**50**), a flavonoid from citrus plants, was found to inhibit the invasion of MO4 cells (Kirsten murine sarcoma virus transformed fetal mouse cells) into embryonic chick heart fragments *in vitro* (Bracke et al., 1989). The flavonoid appeared to be chemically stable in tissue culture medium, and the anti-invasive effect was reversible on exclusion of the molecule from the medium. Unlike (+)-catechin, tangeretin bound poorly to extracellular matrix. It did not alter fucosylated surface glycopeptides of MO4 cells. Tangeretin seemed not to act as a microtubule inhibitor, as immunocytochemistry revealed no disturbance of the cytoplasmic microtubule complex. However, at anti-invasive concentrations of tangeretin, cell proliferation and thymidine incorporation appeared to be inhibited. When cultured on an artificial substrate, treated MO4 cells were less elongated, covered a larger surface area, and exhibited a slower directional migration than untreated cells. From the decrease in ATP content in MO4 cells after tangeretin treatment, this flavonoid appeared to inhibit several intracellular processes, leading to an inhibition of cell motility and, hence, of invasion.

The effect of 31 polyphenolic compounds, belonging to the flavonoids, chalcones, or coumarins, was tested in an invasion assay (Parmar et al., 1994). Invasion of MCF-7/6 human mammary carcinoma cells into embryonic chick heart fragments was studied in organ culture over 8 days. Anti-invasive activity of 3,7-dimethoxyflavone (**77**) was found at concentrations ranging from 1 to 100 µM. At these anti-invasive concentrations, no cytotoxic effects could be detected; the anti-invasive effect was reversible upon exclusion of the molecule from the medium, and treatment of MCF-7/6 cells or heart fragments did not affect subsequent outgrowth from explants on tissue culture plastic. The compound did not inhibit growth of MCF-7/6 cell aggregates nor of heart fragments kept in suspension culture. Therefore, the mechanism of action of 3,7-dimethoxyflavone is still unclear.

*2.2.3. Flavonoids as Inhibitors of Protein Kinase C.* Fifteen flavonoids were found to inhibit protein kinase C (PKC) in a concentration-dependent manner depending on flavonoid structure (Middleton, 1988a; 1988b; Ferriola et al., 1989). Fisetin (**78**), quercetin (**49**), and luteolin (**79**) were the most potent, while hesperetin (**80**), taxifolin (**65**), and rutin (**81**) were among the least potent. Fisetin was almost 100% inhibitory at a concentration of 100 µM. The X-ray crystal structure analysis of hesperetin monohydrate showed that the molecule is essentially planar despite the sofa conformation of the γ-pyran ring and the 27 degree twist of the 2-phenyl ring. Comparison of the 3-D structure of this inac-

Table 9. Protein kinase C (PKC) inhibitory activities of some flanonoids

Quercetin (**49**)    Fisetin (**78**)    Luteolin (**79**)

| | PKC activity nmol product formed·min-1·(mg protein)-1 | % inhibition |
|---|---|---|
| control | 10.4 ± 1.3 | |
| rutin (**81**) | 11.7 ± 1.8 | 0 |
| quercetin (**49**) | 4.1 ± 1.0 | 61 |
| luteolin (**79**) | | 70 |
| fisetin (**78**) | 1.7 ± 1.0 | 84 |

tive flavanone with those of the active flavones showed that, although hesperetin can adopt a planar profile similar to those of fisetin and quercetin, the 4'-methoxy substituent blocks an essential structural feature required for inhibitory activity. Analysis of these structure-activity data revealed a model of the minimal essential features required for PKC inhibition by flavonoids: a coplanar flavone structure with free hydroxy substituents at the 3', 4' and 7- positions.

*2.2.4. Flavonoids as Inhibitors of Tumor Promotion.* Some flavonoids have shown inhibition of tumor promotion, and early investigations have been reviewed by Fujiki et al (1986). Kato et al (1983) reported quercetin (**49**, 30 μM/mouse) markedly suppressed the effect of 12-*O*-tetradecanoylphorbol-13-acetate (TPA, 20 nM/mouse) on skin tumor formation in CD-1 mice initiated by 7,12- dimethylbenz[a]anthracene (200 nmol/mouse). Quercetin (10–30 μM/mouse) also inhibited TPA (20 nM/mouse)-induced epidermal ornithine decarboxylase (ODC) activity, but failed to inhibit the stimulation of epidermal DNA synthesis by TPA. Fifty percent inhibition of lipoxygenase was observed with quercetin at 1.3 μM. These results suggest that the inhibition of lipoxygenase by quercetin is one of the major mechanisms by which the above agent inhibits tumor promotion and TPA-induced ODC activity. Nakadate et al (1984) reported that morin (**82**), fisetin (**78**), and kaempferol (**83**) inhibit epidermal lipoxygenase activity. Morin inhibited the tumor promoting activity of TPA in a two-stage carcinogenesis experiment.

Two biflavnanones, neochamaejasmin A (**84**) and isochamaejasmin (**85**), isolated from the root of *Stellera chamaejasme* (Thymelaeaceae), inhibit the *in vitro* and *in vivo* effects induced by tumor promotors. Neochamaejasmin A and isochamaejasmin at 11.4 nM caused 60–70% inhibition of the induction of ODC with a given dose of teleocidin in mouse skin 4 h after their concomitant application with teleocidin. Both biflavanones caused 50% inhibition at a dose of only 1.14 nM (Fujiki et al., 1986). These results suggested that flavonoids, and possibly biflavanones, prevent initiation as well as promotion in the process of chemical carcinogenesis and may be useful as new chemopreventive agents in human carcinogenesis.

The tumor promoters teleocidin and aplysiatoxin stimulated human basophil histamine release. The flavonoids luteolin (**79**), nobiletin (**66**), quercetin (**49**), fisetin (**78**), apigenin (**86**), and chalcone (**87**) inhibited histamine release in a concentration-dependent manner, but taxifolin (**65**) and its dihydrochalcone derivative, phloretin (**88**), lacked activ-

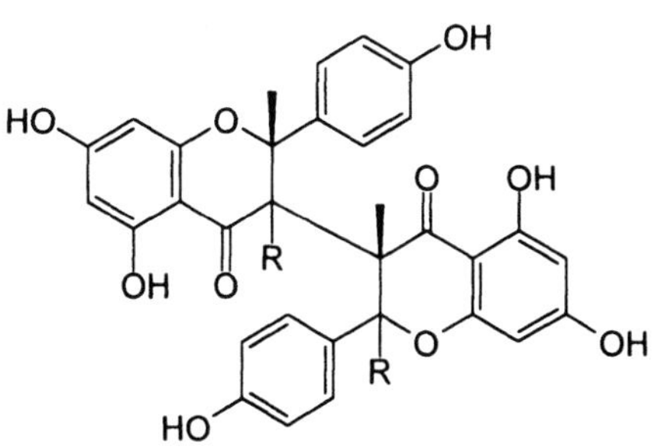

Neochamaejasmin A (**84**) : R = β-H
Isochamaejasmin (**85**) : R = α-H

Figure 5. Two flavonoids from *Stellera chamaejasme* as inhibitors of tumor promotion.

ity, indicating that the flavonoid effects were structure dependent and stereoselective (Middleton et al., 1987).

Mulberry trees (Moraceae) have been widely cultivated in China and Japan for their leaves, which are an indispensable food of silkworms. The root bark of *Morus alba* L. is a traditional Chinese medicine called San Bai Pi and is used for cough, asthma and other diseases. Many flavonoids including thirteen new prenyl (3-methyl-2-butenyl) flavonoids were isolated from the root bark of cultivated mulberry trees, and their chemistry and antitumor promoting activity have been reviewed by Nomura et al (1988).

The two-stage carcinogenesis caused by 7,12-dimethylbenz[a]anthracene and TPA in mice was inhibited by kaempferol (**83**) and flavonol glycosides, whereas naringenin (**89**), a flavanone, had no effect (Yasukawa et al., 1990). The induction of epidermal ODC activity by TPA was also inhibited by kaempferol, whereas mauritianin (**90**), a kaempferol glycoside, failed to inhibit this process. In addition, the effect of the flavonol glycosides on cell-mediated immunosuppression in the two-stage carcinogenesis, as observed in terms of initiation after 14 weeks, was antagonized by mauritianin (**90**) and myricitrin (**91**). Cell-mediated immunosuppression in the two-stage carcinogenesis was unaffected by kaempferol and naringenin (**89**). These results suggest that the inhibitory effects of flavonol glycosides may have been partly due to activation of immune responses against tumors.

Fourteen flavones obtained from the root of *Scutellaria baicalensis* were examined for their inhibitory effects on Epstein-Barr virus early antigen (EBV-EA) activation. A short-term *in vitro* assay designed to discover possible anti-tumor-promoters was used (Konoshima et al., 1992). Among these flavones, 5,7,2'-trihydroxy- (**92**) (63.7% inhibition at $5 \times 10^2$ mol ratio of inhibitor/TPA) and 5,7,2',3'- tetrahydroxyflavone (**93**) (100% inhibition at $5 \times 10^2$ mol ratio of inhibitor/TPA) showed remarkable inhibitory effects on the EBV-EA activation. The potency of **93** was similar to those of retinoic and glycyrrhetinic acid which are known as strong anti-tumor promotors. The effect of **93** on the Raji cell cycle was also examined by flow cytometry. The results showed **93** accumulated in Raji cells in the S phase, and that consequently, the percentages of the $G_2$ and M phases were restored to normal values. These two flavones exhibited remarkable inhibitory effects on mouse skin tumor promotion in an *in vivo* two-stage carcinogenesis test.

Rotenoids amorphispironone (**7**) and tephrosin (**6**) isolated from *Amorpha fruticosa* exhibited remarkable inhibitory effects on EBV-EA activation induced by TPA. Further, compounds **6** and **7** exhibited significant anti-tumor-promotion effects on mouse skin tumor promotion in an *in vivo* two-stage carcinogenesis test (Konoshima et al., 1993).

Table 10. Relative ratio of EBV-EA activation with respect to positive control (100 %) in presence of rotenoids 6 and 7

| | Concentration[a] | | | |
|---|---|---|---|---|
| | 1000 | 500 | 100 | 10 |
| 6 | 8.0 (40)[b] | 26.1 (60) | 60.4 (70) | 89.3 (80) |
| 7 | 0.0 (40) | 31.8 (800) | 70.4 (800) | 100.0 (80) |

[a] Mol ratio/TPA (20 ng = 32 pmol/ml). Values represent percentages relative to the positive control value (100%). [b] Values in parentheses are viability percentages of Raji cell.

*2.2.5. Flavonoids as Inhibitors of Mutagenicity.* Myricetin (**94**), robinetin (**95**), and luteolin (**79**) inhibited the mutagenic activity resulting from the metabolic activation of benzo[a]-pyrene and (+/-)-trans-7,8-dihydroxy-7,8-dihydrobenzo[a]-pyrene by rat liver microsomes. These naturally occurring plant flavonoids and seventeen additional flavonoids and related derivatives with phenolic hydroxy groups inhibited the mutagenic activity (Table 11) of (+/-)-7β,8α-dihydroxy-9α,10α-epoxy- 7,8,9,10- tetrahydrobenzo[a]pyrene (B[a]P 7,8-diol-9,10-epoxide-2), which is an ultimate mutagenic and carcinogenic metabolite of benzo[a]pyrene. Several flavonoids without phenolic hydroxy groups or with methylated phenolic hydroxy groups were inactive. The mutagenic activity of 0.05 nM of BP 7,8-diol-9,10-epoxide-2 towards strain TA 100 of *S. typhimurium* was inhibited 50% by incubation of the bacteria and the diol-epoxide with myricetin (**94**, 2 nM), robinetin (**95**, 2.5 nM), luteolin (**79**, 5 nM), 7-methoxyquercetin (**96**, 5 nM), rutin (**81**, 5 nM), quercetin (**49**, 5 nM), delphinidin (**97**) chloride (5 nM), morin (**82**, 10 nM), myricitrin (**91**, 10 nM), kaempferol (**83**, 10 nM), diosmetin (**98**, 10 nM), fisetin (**78**, 10 nM), or apigenin (**86**, 10 nM). Considerably less antimutagenic activity was observed for dihydroquercetin (**99**), naringenin (**89**), robinin (**100**), D-catechin (**76**), genistein (**67**), kaempferid (**101**) and chrysin (**102**). Pentamethoxyquercetin (**103**), tangeretin (**50**), nobiletin (**66**), 7,8-benzoflavone (**104**), 5,6-benzoflavone (**105**), and flavone (**106**), which lack free phenolic groups, were inactive. The antimutagenic activity of hydroxylated flavonoids results from their direct interaction with B[a]P 7,8- diol-9,10-epoxide-2, since the rate of disappearance of the diol-epoxide from cell-free solutions in 1:9 dioxane:water was markedly stimulated by

Table 11. Inhibitory activities of mutagenicity of some flavonoids

| Compound | IC$_{50}$, μM | 3 | 5 | 6 | 7 | 2' | 3' | 4' | 5' |
|---|---|---|---|---|---|---|---|---|---|
| Myricetin (**94**) | 2 | OH | OH | H | OH | H | OH | OH | OH |
| Robinetin (**95**) | 2.5 | OH | H | H | OH | H | OH | OH | OH |
| Quercetin (**49**) | 5 | OH | OH | H | OH | H | OH | OH | H |
| Luteolin (**79**) | 5 | H | OH | H | OH | H | OH | OH | H |
| Rutin (**81**) | 5 | O-GR$^1$ | OH | H | OH | H | OH | OH | H |
| 7-Methoxyquercetin (**96**) | 5 | OH | OH | H | OMe | H | OH | OH | H |
| Delphinidin (**97**) chloride | 5 | | | | | | | | |
| Fisetin (**78**) | 10 | OH | H | H | OH | H | OH | OH | H |
| Morin (**82**) | 10 | OH | OH | H | OH | OH | H | OH | H |
| Kaempferol (**83**) | 10 | OH | OH | H | OH | H | H | OH | H |
| Apigenin (**86**) | 10 | H | OH | H | OH | H | H | OH | H |
| Myricitrin (**91**) | 10 | O-Rha$^2$ | OH | H | OH | H | OH | OH | OH |
| Diosmetin (**98**) | 10 | H | OH | OMe | OH | | OH | OH | H |

1. GR = 6-α-L-rhamnopyranosyl-β-D-glucopyranosyl.
2. Rha = α-L-rhamnopyranosyl.

myricetin (**94**), robinetin (**95**) and quercetin (**49**). Myricetin was a highly potent inhibitor of the mutagenic activity of bay-region diol-epoxides of benzo[a]pyrene, dibenzo[a,h]pyrene and dibenzo[a,i]pyrene, but higher concentrations of myricetin were needed to inhibit the mutagenicity of the chemically less reactive benzo[a]pyrene 4,5-oxide and bay region diol-epoxides of benz[a]anthracene, chrysene and benzo[c]phenanthrene (Yasukawa et al., 1988).

Axillarin (**107**) isolated from *Pulicaria crispa* (Compositae) was active at 25 mg/ml and decreased the metabolism of benzo(a)pyrene by an average of 61.3% over DMSO controls using benzo(a)pyrene metabolism as a mutagen in cultured hamster embryo cells (Al-Yahya et al., 1988).

Two new isoflavones, fremontin [(5'-α,α-dimethylallyl)-5,7,2',4'-tetrahydroxy-isoflavone, **108**] and fremontone [3'-(γ,γ-dimethylallyl)-(5'-α,α-dimethylallyl)-5,7,2',4'-tetrahydroxyisoflavone, **109**], were isolated from roots of the desert plant *Psorothamnus fremontii* (Fabaceae) (Manikumar et al., 1989). The α,α-dimethylallyl substituent is rarely observed, and the combination of the α,α- and γ,γ-dimethylallyl substituents is unprecedented. Both **108** and **109** were nontoxic toward *Salmonella typhimurium*, and inhibited the mutagenicity of ethyl methanesulfonate (EMS) at all concentrations tested. Compound **109** was more active than **108** in the inhibiting the mutagenicity of 2-aminoanthracene (2AN) and acetylaminofluorene (AAF) toward *S. typhimurium*.

Intricatin (7,4'-dimethoxy-8-hydroxyhomoisoflavone, **110**) and intricatinol (4'-methoxy-7,8- dihydroxyhomoisoflavone, **111**) are new homoisoflavonoids isolated from the desert plant *Hoffmanosseggia intricata*, which was collected in Baja California, Mexico (Wall et al., 1989). Both compounds **110** and **111** inhibited the mutagenicity of 2AN toward *Salmonella typhimurium* (T98). Compound **111** was more active than **110** in inhibiting the mutagenicity of AAF toward *S. typhimurium* (T98) and of EMS towards *S. typhimurium* (T100). Compounds **110** and **111** are the first examples of antimutagenic activity in homoisoflavones.

Table 12. Percentage inhibition of various mutagens by flavonoids **108–111**

Fremontin (**108**) : R = H
Fremontone (**109**) : R = CH$_2$-CH=C(CH$_3$)$_2$

Intricatin (**110**) : R = Me
Intricatinol (**111**) : R = H

| | Mutagen[a] and Dose (μg/plate) | | | | | | | | | | | |
|---|---|---|---|---|---|---|---|---|---|---|---|---|
| | 2AN (2-aminoanthracene) | | | | AAF (acetylaminofluorene) | | | | EMS (ethyl methanesulfonate) | | | |
| μg | *300* | *150* | *75* | *37.5* | *300* | *150* | *75* | *37.5* | *300* | *150* | *75* | *37.5* |
| **108** | 93 | 24 | 0 | 65 | 0 | 0 | 0 | 0 | 87 | 81 | 90 | 58 |
| **109** | 93 | 17 | 78 | 89 | 70 | 27 | 3 | 0 | 94 | 86 | 81 | 63 |
| **110** | 62 | 84 | 92 | 70 | 7 | 0 | 0 | 8 | 45 | 27 | 4 | 9 |
| **111** | 87 | 57 | 36 | 8 | 70 | 54 | 48 | 26 | 74 | 82 | 74 | 30 |

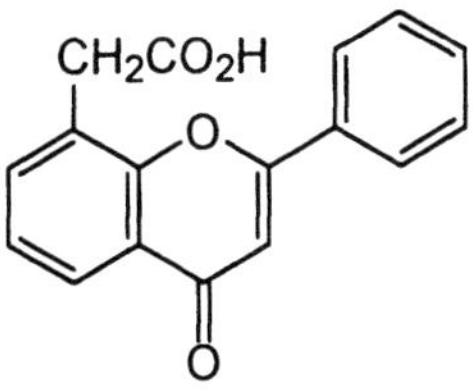

FAA (112)

Figure 6. Flavone 8-acetic acid (FAA).

*2.2.6. Mutagenic and Carcinogenic Effects of Flavonoids.* The first report regarding the mutagenic effects of flavonoids appeared as early as 1977. Since then, extensive investigations on the mutagenic and carcinogenic effects of flavonoids have been conducted due to concerns regarding the widespread existence of flavonoids in human foods. Quercetin (**49**) is probably the most frequently investigated flavone in the past two decades, but its carcinogenicity is still controversial (Gaspar et al., 1994). Some reviews have covered the mutagenic and carcinogenic effects of flavonoids (MacGregor, 1986; Glusker and Rossi, 1986; Das et al., 1994); however, detailed discussion of these studies would exceed the scope of this review.

## 2.3. Flavone Acetic Acid as an Anticancer drug in Clinical Trials

Flavone 8-acetic acid (**112**, FAA) is one of a series of flavonoids synthesized as potential antiinflammatory agents by Lyonnaise Industrielle Pharmacentique in France. FAA is the second of a series of flavonoids to undergo clinical evaluation in malignant diseases. The parent compound (flavone acetic acid ester, LM985) was not recommended for phase II assessment because of drug-associated acute hypotension; it also appeared to function as a prodrug with rapid hydrolysis *in vivo* to FAA (Kerr and Kaye, 1989). Preclinical studies with FAA indicate that it is active against a broad spectrum of murine transplantable solid tumors that tend to be refractory to conventional cytotoxic agents, including a range of colon adenocarcinoma and Glasgow's osteosarcoma (Weiss et al., 1988; O'Dwyer et al., 1987; Humber, 1987). In addition, a soft agar colony formation assay showed that FAA was selectively cytotoxic *in vitro* for solid tumors compared to the L1210 and P-388 leukemias. Human lung tumor xenografts have also been shown to respond to FAA. Thus, the phase I trials (Weiss et al., 1988; Pratt et al., 1991; Havlin et al., 1991; Olver et al., 1992; Holmund et al., 1995) have been performed in patients with various malignant solid tumors. FAA was used safely in cancer patients in phase I clinical trial, and the dose recommended (Olver et al., 1992) for further phase II studies is 10 g/m$^2$. The phase II trials of FAA in patients with non-small cell lung cancer (Siegenthaler et al., 1992) and metastatic melanoma (Maughan et al., 1992) have been ongoing in Switzerland and Great Britain, respectively.

## 3. ANTI-HIV FLAVONOIDS AND THEIR ANALOGS

The antiviral property of flavones was first discovered in the 1940s when quercetin (**49**) was found to have a prophylactic effect when administered in the diet of mice infected intracerebrally with attenuated rabies and other viruses (Kato et al., 1983). Sub-

**Table 13.** Structures of antitumor or antiviral flavonoids cited in the article

| Compound No. | Skeleton | Substituents (R) | | | | | | | | |
|---|---|---|---|---|---|---|---|---|---|---|
| | | 3 | 5 | 6 | 7 | 8 | 2' | 3' | 4' | 5' |
| 34 | A | OMe | OMe | OMe | OMe | OMe | H | H | OMe | H |
| 35 | A | OH | OMe | OMe | OMe | OMe | H | OMe | OMe | H |
| 36 | A | H | OMe | OMe | OH | OMe | H | OMe | OMe | H |
| 49 | A | OH | OH | H | OH | H | H | OH | OH | H |
| 50 | A | H | OMe | OMe | OMe | OMe | H | H | OMe | H |
| 65 | B | OH | OH | H | OH | H | H | OH | OH | H |
| 74 | A | H | OH | OH | O-Glr[1] | H | H | H | H | H |
| 75 | A | H | OH | OMe | OH | H | H | OMe | OH | H |
| 76 | E | OH | OH | H | OH | H | H | OH | OH | H |
| 77 | A | OMe | H | H | OMe | H | H | H | H | H |
| 80 | B | H | OH | H | OH | H | H | OH | OMe | H |
| 86 | A | H | OH | H | OH | H | H | H | OH | H |
| 87 | D1 | H | H | H | H | H | H | H | H | H |
| 88 | D2 | H | OH | H | OH | H | H | H | OH | H |
| 89 | B | H | OH | H | OH | H | H | H | OH | H |
| 90 | A | O-GRR[2] | OH | H | OH | H | H | OH | OH | H |
| 92 | A | H | OH | H | OH | H | OH | H | H | H |
| 93 | A | H | OH | H | OH | H | OH | OH | H | H |
| 99 | B | OH | OH | H | OH | H | H | OH | OH | H |
| 100 | A | O-GR[3] | OH | H | O-Rha[4] | H | H | H | OH | H |
| 101 | A | OH | OH | H | OH | H | H | H | OMe | H |
| 103 | A | OMe | OMe | H | OMe | H | H | OMe | OMe | H |
| 104 | A | H | H | H | $C_4H_4$ | H | H | H | H | H |
| 105 | A | H | $C_4H_4$ | H | H | H | H | H | H | H |
| 106 | A | H | H | H | H | H | H | H | H | H |
| 107 | A | OMe | OH | OMe | OH | H | H | OH | OH | H |
| 113 | A | H | OH | H | OH | H | H | OMe | OH | H |
| 114 | A | OMe | OH | OMe | OMe | H | H | OMe | OH | H |
| 115 | A | OMe | OH | OH | OMe | H | H | OMe | OH | H |
| 116 | A | OMe | OH | OMe | OMe | H | H | OH | OH | H |
| 117 | A | OMe | OH | H | OMe | H | O-Glu[5] | H | OMe | OH |
| 118 | A | OMe | OH | OMe | OMe | H | H | OMe | O-Glu | H |
| 119 | B | OH | H | H | OH | H | H | OH | OH | H |
| 120 | A | OMe | OMe | OMe | OMe | H | H | OMe | OH | H |
| 121 | A | OH | OH | H | OH | H | H | H | H | H |
| 122 | A | OH | OH | H | OH | H | H | OMe | OH | H |
| 123 | E (R₄=OH) | OH | OH | H | OH | H | H | OH | OH | H |
| 124 | A | OMe | H | H | H | H | H | H | H | H |
| 125 | C | H | OMe | H | O-Glu | H | H | H | OH | H |
| 126 | B | H | OH | H | O-Neo[6] | H | H | H | OH | H |
| 127 | A | OMe | OH | H | OMe | H | OH | H | OMe | OH |
| 128 | A | H | OH | OMe | OH | H | H | H | OMe | H |
| 129 | F | OH | OH | H | OH | H | H | H | OH | H |

1. Glr = β-D-glucuropyranosyl.
2. GRR = (6-α-L-rhamnopyranosyl)-(2-α-L-rhamnopyranosyl)-β-D-glucopyranosyl.
3. GR = 6-α-L-rhamnopyranosyl-β-D-glucopyranosyl.
4. Rha = α-L-rhamnopyranosyl.
5. Glu = β-D-glucopyranosyl.
6. Neo = 2-α-L-rhamnopyranosyl-β-D-glucopyranosyl.

sequently, many flavonoids such as quercetin (**49**), catechin (**76**), luteolin (**79**), hesperetin (**80**), rutin (**81**), morin (**82**), kaempferol (**83**), apigenin (**86**), diosmetin (**98**), dihydroquercetin (**99**), chrysin (**102**), axillarin (**107**), chrysoeriol (**113**), chrysoplenol -B (**114**), -C (**115**), -D (**116**), chrysoplenoside -A (**117**), -B (**118**), dihydrofisetin (**119**), 4'-hydroxy-3,5,6,7,3'-pentamethoxy flavone (**120**), galangin (**121**), isorhamnetin (**122**), leucicyanidin (**123**), 3-methoxyflavones (**124**) (Vlietinck et al., 1986; Van Hoof et al., 1984), 5-O-methylgenistein-7-O-β-D-glucopyranoside (**125**) (De Rodriguez et al., 1990), naringin (**126**), oxyayanin A (**127**), pectolinarigenin (**128**) (Perry and Foster, 1994), pelargonidin (**129**) chloride, and many synthetic flavonoids (Burnham et al., 1972; De Meyer et al., 1991) have been found to have antiviral activity against various viruses including herpes simplex (Amoros et al., 1992), respiratory syncytial virus, parainfluenza, adenovirus, pseudorabies/aujeszky's virus, poliovirus, rhinovirus, Sindbis virus, or potato virus, respectively (Kaul et al., 1985; Vlietinck et al., 1988). Structure-activity relationship studies indicted that the presence of the 3-methoxy, the carbonyl, and the $C_2$-$C_3$ double bond are necessary for antiviral activity (Vlietinck et al., 1986). Research conducted before 1985 has been reviewed by Selway (Selway, 1986). Due to the worldwide spread of the AIDS problem, since the 1980s, the investigations regarding antiviral activity of flavonoids have focused on their anti-HIV activity. Much progress has been achieved and will be discussed herein.

## 3.1. Flavonoids as Anti-HIV Principles

The medicinal herb, *Kummerowia striata (Thunb.)* Schindle (Papilionaceae), a plant widely distributed in Guangxi Province, China, is known as "Renzi-Cao" and has been used folklorically for treatment of hepatitis and some other diseases. *K. striata* is a hitherto phytochemically uninvestigated species. We have now isolated and characterized an anti-HIV principle, apigenin-7-O-β-D-glucopyranoside (**130**), from this plant. The anti-HIV bioassay results indicated that compound **130** showed good anti-HIV activity ($EC_{50}$ = 1.8 µg/ml) (Tang et al., 1994).

The flowering heads of *Chrysantbemum morifolium* Ramar (Compositae) have been used as an herbal tea in Chinese folklore and are known as "Ju Hua". They have been found to possess antibacterial, antifungal, antiviral, antispirochetal, and anti-inflammatory

Table 14. Some naturally occurring flavonoids as anti-HIV principles

**49** : $R_1$ = $R_4$ = H, $R_2$ = $R_3$ = OH
**86** : $R_1$ = $R_2$ = $R_3$ = $R_4$ = H
**130** : $R_1$ = glucopyranosyl, $R_2$ = $R_3$ = $R_4$ = H
**131** : $R_1$ = galatopyranosyl, $R_2$ = $R_3$ = H, $R_4$ = $CH_3$

|  | $IC_{50}$ (µg/ml) | $EC_{50}$ (µg/ml) | Therapeutic Index |
|---|---|---|---|
| Quercetin (**49**) | < 4 | 1.6 | < 2.5 |
| Apigenin (**86**) | 35 | 9 | 4 |
| Apigenin-7-O-β-D-glucopyranoside (**130**) | > 100 | 1.8 | > 55.6 |
| Acacetin-7-O-β-D-galactopyranoside (**131**) | 37 | 8 | 4.6 |

activities. The MeOH extract of C. *morifolium* was found to show significant anti-HIV activity. Acacetin-7-*O*-β-D-galactopyranoside (**131**) has been identified as a new active anti-HIV principle in our laboratory and showed potent activity ($EC_{50}$ = 8 μM) with relatively low toxicity ($IC_{50}$ = 37 μM) (Hu et al., 1994).

From the dried flower buds of *Egletes viscosa* (Compositae), ternatin (4',5-dihydroxy-3,3',7,8-tetramethoxyflavone, **132**), was isolated and showed anti-inflammatory, hepatoprotection and gastroprotection properties. According to the NCI protocols, this compound also showed moderate activity against HIV (Lima et al., 1996).

Three anti-HIV flavonoids, morusin (**133**), morusin-4'-glucoside (**134**), and kuwanon H (**135**), were isolated from the Chinese herb San Bai Pi (the root bark of *Morus alba*). The site between the free 2-hydroxyl and the 3-γ,γ-dimethylallylic group of these flavonoids was found to be sensitive to photo-oxidation, and contributed to anti-HIV activity (Luo et al., 1994). Kuwanon G (**136**) and H (**135**) also inhibited the specific binding of [$^{125}$I]gastrin-releasing peptide (GRP) to GRP-preferring receptors in murine Swiss 3T3 fibroblasts with Ki values of 470 and 290 nM, respectively. Kuwanon H was one order of magnitude less potent in inhibiting [$^{125}$I]bombesin binding to neuromedin B (NMB)-preferring receptors in rat esophagus membranes. This compound antagonized bombesin-induced increases in the cytosolic free calcium concentration and GRP-induced DNA synthesis in Swiss 3T3 cells. Thus, kuwanon H, and possibly kuwanon G, are specific antagonists for the GRP-preferring receptor and can be useful for studying the physiological and pathological role of GRP (Mihara et al., 1995).

The first flavone-xanthone C-glucoside, swertifrancheside (**137**), was isolated from *Swertia franchetiana* (Gentianaceae), a Chinese herb used to reduce fever and to treat hepatogenous jaundice and cholecystitis. Its structure was elucidated as 1,5,8- trihydroxy-3-methoxy-7-(5',7',3",4"- tetrahydroxy-6'-C-β–D- glucopyranosyl-4'-oxy-8'-flavyl)-xanthone (Wang et al., 1994). Swertifrancheside was found to be a potent inhibitor of the HIV-1 reverse transcriptase ($ED_{50}$ = 30.9 μg/ml).

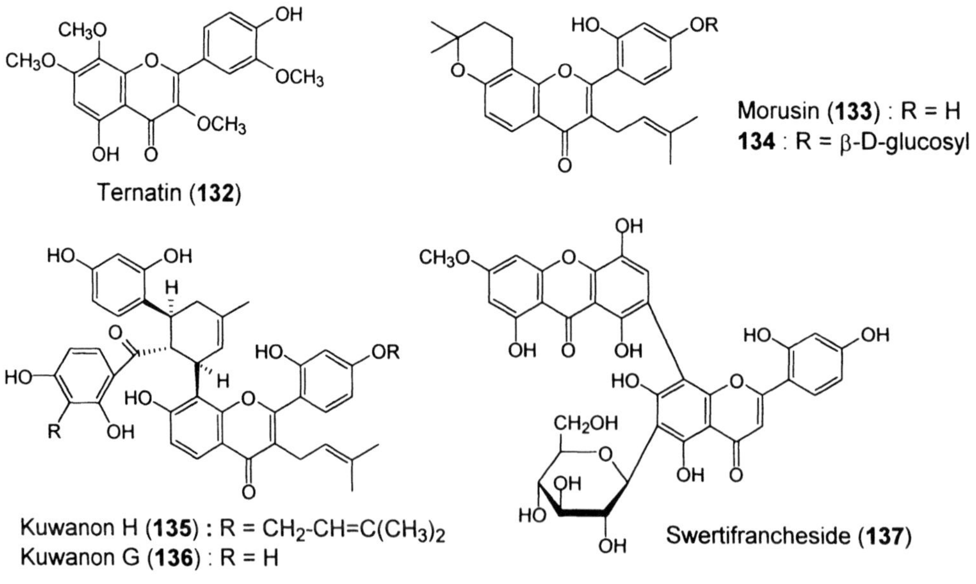

**Figure 7.** Flavonoids from various Chinese herbs as anti-HIV principles.

Table 15. Inhibitory efficacy of some flavones and flavans against HIV-1 infection

**A (Flavones)**     **B (Flavans)**

| | $EC_{50}$ (μg/ml)[a] | $TC_{50}$ (μg/ml)[b] | Skeleton | 3 | 7 | 3' | 4' | 5' |
|---|---|---|---|---|---|---|---|---|
| Chrysin (**102**) | 20 | 50 | A | H | OH | H | H | H |
| Myricetin (**94**) | 2 | 40 | A | OH | OH | OH | OH | OH |
| Myricetrin (**91**) | 100 | > 200 | A | O-Rha | OH | OH | OH | OH |
| Quercetin-3-rhamnoside (**138**) | 50 | > 100 | A | O-Rha | OH | OH | OH | H |
| Kaempferol-3-glucoside (**2**) | 10 | 100 | A | O-Glc | OH | H | OH | H |
| (+)-Gallocatechin (**139**) | 5 | > 80 | B | (+)-OH | OH | OH | OH | OH |
| (-)-Epicatechin | 2 | > 100 | B | (-)-OH | OH | OH | OH | H |
| (+)-Epicatechin | 4 | > 100 | B | (+)-OH | OH | OH | OH | H |
| (-)-Catechin-3-gallate (**140**) | 1 | > 100 | B | (-)-O-gallate | OH | OH | OH | H |
| (+)-Catechin-7-gallate (**141**) | 10 | > 100 | B | (+)-OH | O-gallate | OH | OH | H |

[a] Concentration which reduces by 50 % the production of gp120 in infected C8166 cells.
[b] Concentration which causes 50 % cytotoxicity to uninfected C8166 cells.

## 3.2. Biological Studies of Anti-HIV Flavonoids

*3.2.1. Flavonoids as Inhibitors of HIV Infection.* Mahmood et al examined the antiviral activity of a variety of flavanoids including 17 flavones, 6 flavans, and 5 flavanones (Mahmood et al., 1993). The results showed that the flavans were generally more effective than the flavones and flavanones in selective inhibition of HIV-1, HIV-2 or SIV infection. The galloyl derivative, (-)-epicatechin-3-*O*-gallate (**140**), consistently exhibited the greatest activity with an $EC_{50}$ of 1 μg/ml and selectivity index >100, and was somewhat more active than (-)-epicatechin. Studies of their effects on the binding of sCD4 and antibody to gp120 indicated that the effective compounds interact irreversibly with gp120 to inactive virus infectivity and block infection.

*3.2.2. Flavonoids as Inhibitors of HIV-1 Reverse Transcriptase.* Eighteen naturally occurring flavonoids were investigated by Spedding et al (1989) for their inhibition of three reverse transcriptases (RT): avian myeloblastosis (AMV) RT, Rous-associated virus-2 (RAV-2) RT and Maloney murine leukemia virus (MMLV) RT. Amentoflavone (**142**), scutellarein (**143**) and quercetin (**49**) were the most active compounds, and their effect was concentration-dependent. The enzymes exhibited differential sensitivity to the inhibitory effects of the flavonoids. The compounds also inhibited rabbit globin mRNA-directed, MMLV RT- catalyzed DNA synthesis. Amentoflavone and scutellarein inhibited ongoing new DNA synthesis catalyzed by RAV-2 RT.

Four flavonoids, baicalein (**69**), quercetin (**49**), quercetagetin (**144**), and myricetin (**94**), were found to be potent inhibitors of reverse transcriptases from Rauscher murine leukemia virus (RLV) and HIV (Ono et al., 1989; 1990). Under the reaction conditions

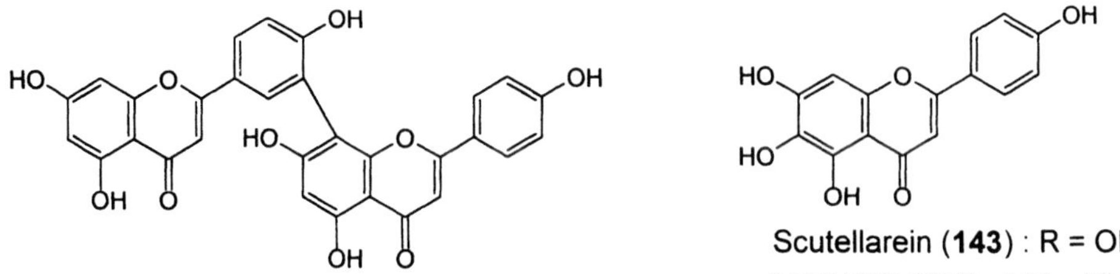

**Amentoflavone (142)**
*AMV RT (51% at 50 µM)

**Scutellarein (143)** : R = OH
*AMV RT (37% at 50 µM)

* Avian myeloblastosis

**Figure 8.** Two flavonoids showing potent HIV-1 reverse transcriptase inhibitory activity.

employed, any one of these flavonoids almost completely inhibited the activity of RLV reverse transcriptase at a concentration of 1 µg/ml. HIV reverse transcriptase was inhibited by 100%, 100%, 90% and 70% in the presence of 2 µg/ml quercetin, myricetin, quercetagetin and baicalein, respectively. The mode of inhibition of these flavonoids was competitive (RLV reverse transcriptase) or partially competitive (HIV reverse transcriptase) with respect to the template.primer complex (rA)n.(dT) and noncompetitive with respect to the triphosphate substrate dTTP. The $K_i$ values for RLV reverse transcriptase were found to be 0.37 µM and 0.08 µM for baicalein and quercetin, respectively and those for HIV reverse transcriptase were 2.52 µM, 0.52 µM, 0.46 µM and 0.08 µM for baicalein, quercetin, quercetagetin and myricetin, respectively. Comparative studies with other flavonoids (hydroxyflavones, dihydroxy- and polyhydroxy-flavones and flavanones), carried out to clarify the structure/activity relationships, revealed that the presence of both the unsaturated double bond between positions 2 and 3 of the flavonoid pyrone ring and the three hydroxy groups introduced on positions 5, 6 and 7, (i.e. baicalein) were a prerequisite for inhibition of reverse transcriptase activity. Removal of the 6-hydroxy group of baicalein required the introduction of three additional hydroxy groups at positions 3, 3' and 4' (quercetin) to afford a compound still capable of inhibiting the reverse transcriptase activity. Quercetagetin, which contains the structures of both baicalein and quercetin, and myricetin, which has the structure of quercetin with an additional hydroxy group on the 5' position, also proved to be strong inhibitors of reverse transcriptase activity. Baicalein's inhibition of reverse transcriptase was highly specific, whereas quercetin and quercetagetin were also strong inhibitors of DNA polymerase beta and DNA polymerase I, respectively. Myricetin was also a potent inhibitor of both DNA polymerase alpha and DNA polymerase I.

Swertifrancheside (**137**), mentioned above previously, was found to be a potent inhibitor of the DNA polymerase activity of human immunodeficiency virus-1 reverse transcriptase (HIV-1 RT), with an $IC_{50}$ value of 43 µM (Pengsuparp et al., 1995). It was not cytotoxic to cultured mammalian cells. The kinetic mechanism by which this compound inhibited HIV-1 RT was studied, as was its potential to inhibit other nucleic acid polymerases. Swertifrancheside bound to DNA and was shown to be a competitive inhibitor with respect to template-primer, but a mixed-type competitive inhibitor with respect to TTP.

*3.2.3. Flavonoids as Inhibitors of HIV-1 Protease.* Six substituted γ-chromones and 17 flavones were investigated by Brinkworth et al. (1992) for their inhibition of HIV-1

Table 16. Inhibitory efficacy of some flavones against HIV-1 protease

|  | IC$_{50}$ (μM) | \multicolumn{9}{c}{Substituents (R)} | | | | | | | | |
|---|---|---|---|---|---|---|---|---|---|---|
|  |  | 3 | 5 | 6 | 7 | 8 | 2' | 3' | 4' | 5' |
| Gardenin A (**145**) | 11 | H | OH | OH | OH | OH | H | OH | OH | OH |
| 3,2'-Dihydroxyflavone | 12 | OH | H | H | H | H | OH | H | H | H |
| Myricetin (**94**) | 22 | OH | OH | H | OH | H | H | OH | OH | OH |
| Morin (**82**) | 24 | OH | OH | H | OH | H | OH | H | OH | H |
| Quercetin (**49**) | 36 | OH | OH | H | OH | H | H | OH | OH | H |
| Fisetin (**78**) | 50 | OH | H | H | OH | H | H | OH | OH | H |
| Apigenin (**86**) | 60 | H | OH | H | OH | H | H | H | OH | H |

protease, an important enzyme in the replication and processing of the AIDS virus. Hydroxy flavones were chosen as potential inhibitors with the rationale that the 2-phenyl substituent might be able to fit into the $P_1$ or $P_{1'}$ binding pockets that flank either side of the active site Asp25 and 125 side chains of the viral protease. Substituted γ-chromones were found to weakly inhibit HIV-1 protease, while flavones and chromones bearing hydroxy substituents were the most active compounds. The 2-phenyl substituent is best hydroxylated, although adjacent hydoxyls (e.g. 2',3' or 3',4') usually confer toxicity to flavones and are best avoided. Removing the hydroxyl from the 3-position increases the inhibition by ~4 fold. Replacing the 3-OH with OMe or with O-rutinose dramatically reduces activity (> 10 fold). The 3 position tolerates phenyl substituents on the pyrone ring (genistein & biochanin A). The carbonyl in the 4-position appears critical for activity based on data for catechin (**76**) and epi-catechin. The 5-OH substituent has little effect according to the comparison between fisetin (**78**) and quercetin. Hydroxyl substitution in the 6-position [chrysin vs. baicalein (**69**)] reduces inhibition, while the effect of 7-and 8- substitution by hydroxyls appears to increase inhibitor potency.

*3.2.4. Flavonoids as Inhibitors of HIV-1 Integrase.* Efficient replication of HIV-1 requires integration of a DNA copy of the viral genome into a chromosome of the host cell; this process is catalyzed by the viral integrase. Integrase presents an attractive possibility as an antiviral target because host cells do not make or require such enzymes. The inhibition of HIV-1 integrase by flavones and related compounds was investigated biochemically and by means of structure-activity relationships (Fesen et al., 1994). Inhibition by flavones generally occurred in parallel and usually required the presence of at least one ortho pair of phenolic hydroxy groups and at least one or two additional hydroxy groups. Potency was enhanced by the presence of additional hydroxy groups, especially when present in ortho pairs or in adjacent groups of three. Inhibitory activity was reduced or eliminated by methoxy or glycosidic substitutions or by saturation of the 2,3 double bond.

Table 17. Inhibitory efficacy of some flavones against HIV-1 integrase

|  | $IC_{50}$ (µM) Cleavage | $IC_{50}$ (µM) Integration | Substituents (R) | | | | | |
|---|---|---|---|---|---|---|---|---|
|  |  |  | 3 | 5 | 6 | 3' | 4' | 5' |
| Quercetagetin (**144**) | 0.8 | 0.1 | OH | OH | OH | OH | OH |  |
| Baicalein (**69**) | 1.2 | 4.3 |  | OH | OH |  |  |  |
| Robinetin (**95**) | 5.9 | 1.6 | OH |  |  | OH | OH | OH |
| Myricetin (**94**) | 7.6 | 2.5 | OH | OH |  | OH | OH | OH |
| Quercetin (**49**) | 23.6 | 13.6 | OH | OH |  | OH | OH |  |
| Fisetin (**78**) | 28.4 | 8.5 | OH |  |  | OH | OH |  |
| Luteolin (**79**) | 32.9 | 25.0 |  | OH |  | OH | OH |  |
| Myricetrin (**91**) | 39.6 | 10.3 | O-Rha | OH |  | OH | OH | OH |

These structure-activity results for flavones were generally concordant with those previously reported for inhibition of reverse transcriptase and topoisomerase II.

Recently, Raghavan et al (1995) published results from a comparative molecular field analysis (CoMFA) of a set of 15 flavones that inhibit HIV-1 integrase-mediated cleavage and integration *in vitro* as tested by Fesen et al (1994). The results indicate a strong correlation between the inhibitory activity of these flavones and the steric and electrostatic fields around them. CoMFA quantitative structure-activity relationship models with considerable predictive ability (cross-validated $r^2$ as high as 0.8) were obtained.

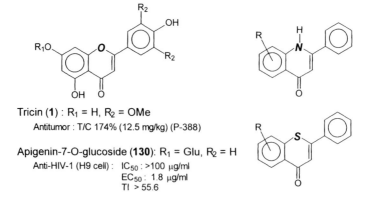

Tricin (**1**) : $R_1$ = H, $R_2$ = OMe
  Antitumor : T/C 174% (12.5 mg/kg) (P-388)

Apigenin-7-O-glucoside (**130**): $R_1$ = Glu, $R_2$ = H
  Anti-HIV-1 (H9 cell) : $IC_{50}$ : >100 µg/ml
                        $EC_{50}$ : 1.8 µg/ml
                        TI > 55.6

Figure 9. The structural similarity of flavones, 2-phenyl-4-quinolones, and 2-phenylthiochromen-4-ones.

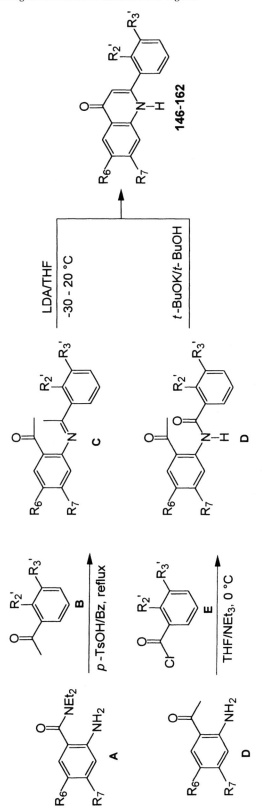

Figure 10. Synthesis of substituted 2-phenyl-4-quinolones.

## 4. FLAVONOID ANALOGS SYNTHESIZED IN OUR LABORATORY AS ANTITUMOR OR ANTI-HIV AGENTS

Comparison of the structures of flavones, 2-phenyl-4-quinolones, and 2-phenylthio-chromen-4-ones reveals that these three compound types share the same tricyclic skeleton except for the identity of the hetero atom in position 1 of the C ring. Therefore, similar or even new biological activities could been anticipated when the oxygen of bioactive flavonoids is replaced by another atom such as nitrogen or sulfur, which line up closely with oxygen in the Periodic Table.

### 4.1. Synthesis and Cytotoxicity of 2-Phenyl-4-Quinolones and Related Compounds as Antimitotic Agents Interacting with Tubulin

A series of substituted 2-phenyl-4-quinolones and related compounds has been synthesized (Scheme 1) in our laboratory and evaluated as cytotoxic compounds and antimitotic agents interacting with tubulin (Kuo et al., 1993; Li et al., 1994a; 1994b). Some of the most active compounds have been summarized in Table 2. These compounds were potent inhibitors of tubulin polymerization with activities comparable to those of the antimitotic natural products colchicine, podophyllotoxin, and combretastatin A-4. For example, compound **161** totally inhibited the growth of about half of the NCI tumor cell lines at subnanomolar concentrations (log TGI<-9.00) and was also a potent inhibitor of tubulin polymerization with an $IC_{50}$ value of 0.44 µM.

### 4.2. Synthesis and Biological Evaluation of Substituted 2-Phenylthio-Chromen-4-Ones

A series of 5,6,7,8-substituted-2-phenylthio-chromen-4-ones has also been synthesized (Schemes 10 and 11) in our laboratory and evaluated for cytotoxic, topoisomerase I and II inhibitory, and anti-HIV activities (Tables 3 and 4) (Wang et al., 1996). Among them, compounds **169**, **172**, **174**, and **175** displayed significant growth inhibitory action against a panel of tumor cell lines including human ileocecal carcinoma (HCT-8), murine leukemia (P-388), human melanoma (RPMI), and human central nervous system tumor (TE6671) cells. Compounds **172**, **174**, and **177** displayed DNA topoisomerase I inhibitory activity *in vitro*, and compound **173** was an *in vitro* inhibitor of DNA topoisomerase II. Compound **173** was the most active compound ($ED_{50}$ value, 0.65 µM) against HIV in acutely infected H9 lymphocytes and had a therapeutic index of about 5 (Table 18).

## 5. CONCLUSION

Due to the wide existence of flavonoids, all plant-eating animals are influenced by these natural products through their food. Moreover, flavonoids are used in more purified forms as drugs and food supplements. Indeed, the average daily human intake of these compounds in western countries has been estimated to be 1 g or more (Brandi, 1992). However, current safety studies are not adequate and are sometimes controversial or even contradictory; therefore, more in-depth investigations are required.

Substantial progress has been made recently in discovering flavonoids having antitumor and anti-HIV activities, with one compound, FAA, now in clinical trial as a chemotherapeutic

Table 18. Biological activities of selected substituted 2-phenyl-4-quinolones and related compounds[a]

| No. | $R_6$ | $R_7$ | $R_{2'}$ | $R_{3'}$ | ITP[b] $IC_{50}(\mu m) \pm SD$ | ICB[c] % inhibition | Cytotoxicity log $GI_{50}$ (μm)[d] HCT-116 | OVCAR-3 | SK-Mel-5 | SF-295 | RXF-393 |
|---|---|---|---|---|---|---|---|---|---|---|---|
| 146 | OCH$_2$O | | | | 0.63±0.3 | 26±10 | -7.15 | -6.79 | -7.45 | -6.72 | -6.95 |
| 147 | OCH$_2$O | | | OMe | 0.57±0.1 | 39±8 | -6.76 | -7.34 | -7.43 | -7.04 | -7.34 |
| 148 | OCH$_2$O | | | N(Me)$_2$ | 0.7±0.03 | 29±7 | -6.43 | -6.61 | -6.71 | -6.62 | -6.62 |
| 149 | OCH$_2$O | | F | | 0.85±0.1 | 37±3 | -7.22 | -7.09 | -7.68 | -7.26 | — [e] |
| 150 | OCH$_2$O | | Cl | | 0.89±0.1 | 24±4 | -7.06 | -7.11 | -7.44 | — | — |
| 151 | OCH$_2$O | | | NH$_2$ | 0.73±0.2 | 16±7 | -6.32 | — | -6.03 | -5.62 | -6.22 |
| 152 | OCH$_2$O | | | NHCH$_3$ | 0.65±0.1 | 17±11 | -6.61 | -6.47 | -6.35 | -6.49 | -6.44 |
| 153 | OCH$_2$O | | | F | 0.53±0.1 | 51±3 | -6.66 | -6.54 | -6.57 | -6.85 | — |
| 154 | OCH$_2$O | | | Cl | 0.37±0.03 | 45±7 | -6.47 | -6.74 | -6.68 | -6.65 | -6.79 |
| 155 | OCH$_2$O | | | OCF$_3$ | 0.50±0.1 | 40±5 | -7.85 | -7.20 | -6.77 | -7.59 | -7.39 |
| 156 | OCH$_2$O | | | OCH$_2$CH$_3$ | 0.47±0.1 | 40±2 | -6.76 | -6.78 | -6.72 | -7.19 | -6.56 |
| 157 | OCH$_2$O | | | OCH$_2$C$_6$H$_5$ | 0.45±0.2 | 26±5 | -6.41 | -6.61 | -6.55 | -6.35 | -6.39 |
| 158 | OMe | H | | N(Me)$_2$ | 0.84±0.2 | 18±6 | -6.71 | -6.77 | -6.71 | -6.69 | -6.92 |
| 159 | OMe | H | | OMe | 0.74±0.1 | 13±2 | -6.81 | — | -6.93 | -6.82 | -6.72 |
| 160 | N(Me)$_2$ | | | OMe | 0.38±0.1 | 14±8 | — | -7.82 | -7.59 | -7.81 | -7.73 |
| 161 | morpholinyl | H | | OMe | 0.44±0.2 | 84±6 | <-9.00 | <-9.00 | <-9.00 | <-9.00 | <-9.00 |
| 162 | morpholinyl | H | | OMe | 0.36±0.1 | 43±10 | -6.36 | -7.70 | -7.45 | -7.31 | -7.91 |
| Colchicine | | | | | 0.80±0.07 | | | | | | |
| Podophyllotoxin | | | | | 0.46±0.02 | 86±1 | | | | | |
| Combretastatin A-4 | | | | | 0.53±0.05 | 94±2 | | | | | |
| Dihydro combretastatin A-4 | | | | | 0.63±0.05 | 65±6 | | | | | |

[a] Data was summarized from references 2 and 3. [b] ITP = inhibition of tubulin polymerization. [c] ICB = inhibition of colchicine binding. [d] log concentrations which reduced cell growth to 50% of level at start of experiment; cancer cell lines: HCT-116, colon cancer; OVCAR-3, ovarian cancer; SK-Mel-5, melanoma; SF-295, CNS tumor; RXF-393, renal cancer. [e] "—" means not tested.

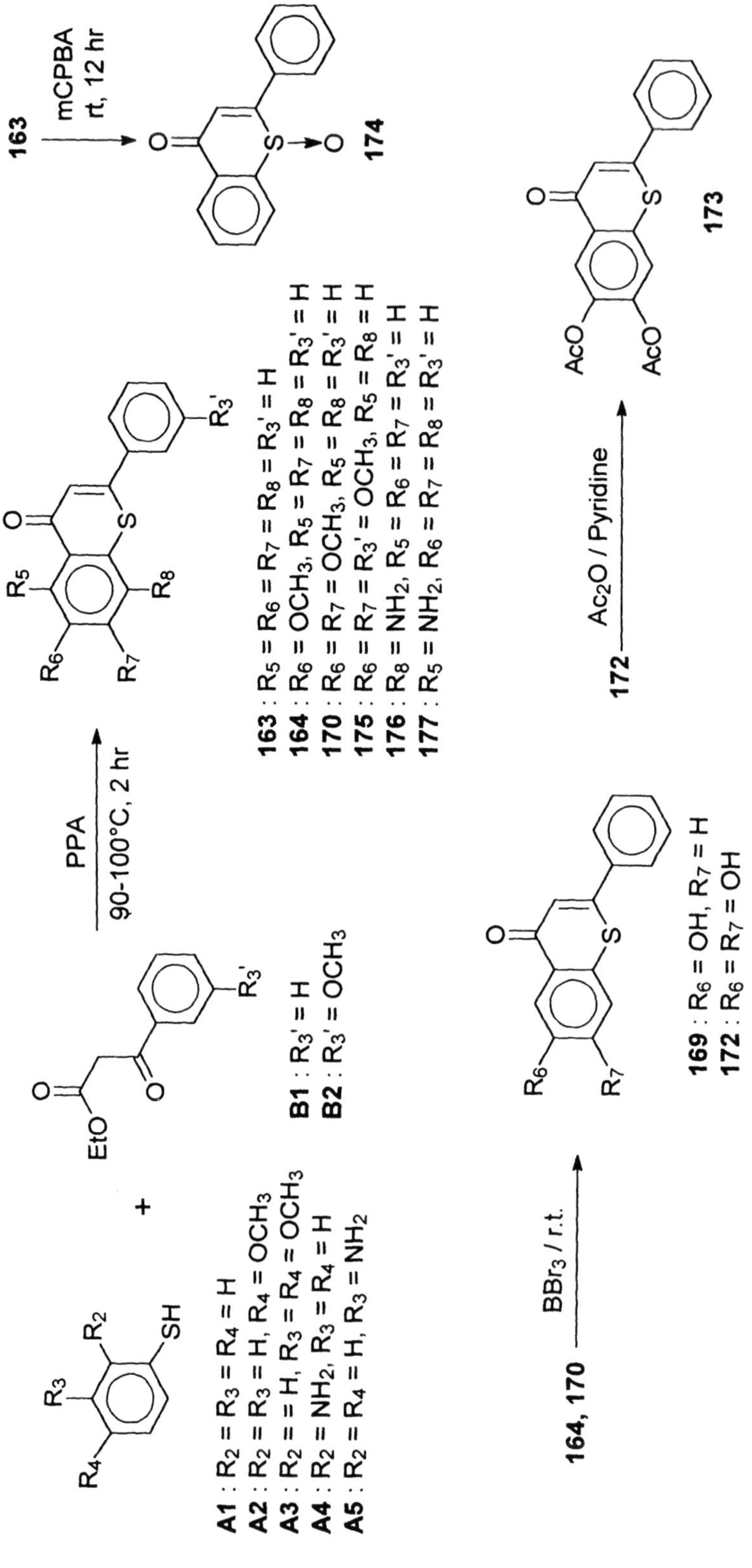

**Figure 11.** Synthesis of 2-phenyl-thiochromen-4-one derivatives (1).

PPA: polyphosphoric acid; mCPBA: m-Chloro perbenzoic acid.

Table 19. Biological activities of substituted 2-phenyl thiochromen-4-ones in various tumor cells

| No. | R$_5$ | R$_6$ | R$_7$ | R$_8$ | R$_{3'}$ | KB[b] | HCT-8[b] | P-388[b] | RPMI[b] | TE671[b] | Topo I[d] | Topo II[e] |
|---|---|---|---|---|---|---|---|---|---|---|---|---|
| 163 | | | | | | [c] | — | — | — | — | ---[f] | --- |
| 164 | | OCH$_3$ | | | | — | — | — | — | — | --- | --- |
| 165 | | | OCH$_3$ | | | 5.5 | 9.16 | — | — | — | --- | --- |
| 166 | OCH$_3$ | | | | | — | — | — | 5.5 | — | --- | --- |
| 167 | OH | | | | | — | — | — | — | — | --- | --- |
| 168 | | | OH | | | — | — | — | — | — | --- | --- |
| 169 | | OH | | | | — | — | — | 0.55 | 2.87 | --- | --- |
| 170 | | OCH$_3$ | OCH$_3$ | | | 4.23 | 9.33 | — | 8.89 | — | --- | --- |
| 171 | | | OAc | | | — | — | — | — | — | --- | --- |
| 172 | | OH | OH | | | 5.5 | 1.45 | 4.51 | 3.92 | 5.50 | 100 | 0 |
| 173 | | OAc | OAc | | | 5.5 | — | 4.83 | 5.5 | 4.58 | 0 | 100 |
| 174 | | | | | | 5.5 | 5.79 | 1.18 | 5.5 | 3.66 | 100 | 0 |
| 175 | | OCH$_3$ | OCH$_3$ | | OCH$_3$ | 5.5 | 6.35 | 0.61 | 5.5 | — | --- | --- |
| 176 | | | | NH$_2$ | | — | — | 7.46 | 5.5 | — | --- | --- |
| 177 | NH$_2$ | | | | | — | — | 9.31 | — | — | 100 | 0 |

[a] EC$_{50}$ was the concentration of drug which affords 50% reduction in cell number after a 3-day incubation. For significant activity of the pure compound, an EC$_{50}$ ≤ 4.0 μg/mL is required. [b] Human epidermoid carcinoma of the nasopharynx (KB), human lung carcinoma (A-549), human ileocecal carcinoma (HCT-8), murine leukimia (P-388), human melanoma (RPMI), human CNS tumor (TE671). [c] "—" means inactive. [d] Measured as ATP-independent relaxation of supercoiled plasmid DNA compared to enzyme and DNA controlled reactions. [e] Measured as ATP-dependent unknotting of P4 DNA compared to enzyme and DNA controlled reactions. [f] "---" means data for compounds (157-165, and 169-170) was not available due to their insolubilities at the testing concentration.

Table 20. HIV-inhibitory effects of substituted 2-phenyl thiochromen-4-ones on infected H9 lymphocytes

| Compd. No. | $R_5$ | $R_6$ | $R_7$ | $R_8$ | $IC_{50}$ (μg/mL)[a] | $ED_{50}$ (μg/mL)[b] | Therapeutic Index |
|---|---|---|---|---|---|---|---|
| 166[c] | $OCH_3$ | | | | 7 | 7 | 1.0 |
| 168 | | | OH | | 40 | 7 | 5.7 |
| 169 | | OH | | | 40 | 9.5 | 4.2 |
| 171 | | | OAc | | 45 | 80 | 0.6 |
| 172 | | OH | OH | | 1.8 | 0.8 | 2.3 |
| 173 | | OAc | OAc | | 3 | 0.65 | 4.6 |
| 176 | | | | $NH_2$ | 15 | 4 | 3.8 |
| 177 | $NH_2$ | | | | 50 | 20 | 2.5 |

[a] Concentration which inhibits uninfected growth by 50%; AZT has an $IC_{50}$ value of 2000 μm.
[b] Concentration which inhibits virus replication growth by 50%; AZT has an $EC_{50}$ value of 0.04 μm.
[c] Crystals of the test agents were observed at both 100 and 20 μg/mL

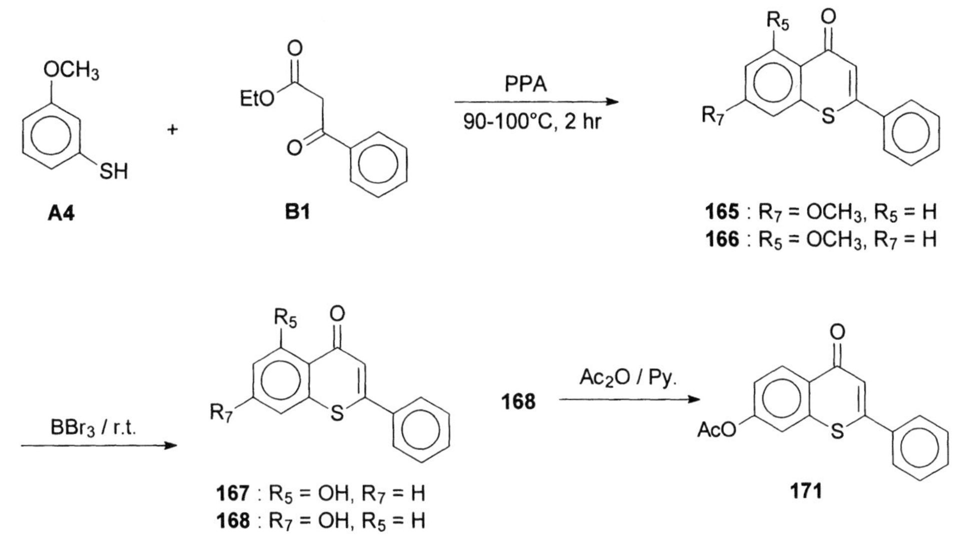

**Figure 12.** Synthesis of 2-phenyl-thiochromen-4-one derivatives (2).

agent. Thus, the future must be visualized as an optimistic one. The isolation and structural modification of such plant-derived active principles provide a continuing source of potential antitumor and anti-HIV agents.

## ACKNOWLEDGMENTS

This investigation was supported by grants from the National Cancer Institute (CA-17625) and National Institute of Allergies and Infectious Diseases (AI-33066) awarded to K. H. Lee.

## REFERENCES

Al-Yahya, M. A.; El-Sayed, A. M.; Mossa, J. S.; Kozlowski, J. F.; Antoun, M. D.; Ferin, M.; Baird, W. M.; Cassady, J. M. Potential cancer chemopreventive and cytotoxic agents from *Pulicaria crispa*. *J. Nat. Prod.* **1988**, *51*, 621–624.

Amoros, M.; Simoes, C. M.; Girre, L.; Sauvager, F.; Cormier, M. Synergistic effect of flavones and flavonols against herpes simplex virus type 1 in cell culture. Comparison with the antiviral activity of propolis. *J. Nat. Prod.* **1992**, *55*, 1732–1740.

Bracke, M. E.; Van Cauwenberge, R. M.; Mareel, M. M. (+)-Catechin inhibits the invasion of malignant fibrosarcoma cells into chick heart in vitro. *Clin. & Experi. Metastasis* **1984**, *2*, 161–170.

Bracke, M. E.; Van Cauwenberge, R. M.; Mareel, M. M.; Castronovo, V.; Foidart, J. M. Flavonoids: tools for the study of tumor invasion in vitro. *Prog. Clin. & Biol. Res.* **1986**, *213*, 441–444.

Bracke, M. E.; Castronovo, V.; Van Cauwenberge, R. M.; Coopman, P.; Vakaet, L. ,Jr.; Strojny, P.; Foidart, J. M.; Mareel, M. M. The anti-invasive flavonoid (+)-catechin binds to laminin and abrogates the effect of laminin on cell morphology and adhesion. *Experi. Cell Res.* **1987**, *173*, 193–205.

Bracke, M. E.; De Pestel, G.; Castronovo, V.; Vyncke, B.; Foidart, J. M.; Vakaet, L. C.; Mareel, M. M. Flavonoids inhibit malignant tumor invasion in vitro. *Prog. Clin. & Biol. Res.* **1988**, *280*, 219–233.

Bracke, M. E.; Vyncke, B. M.; Van Larebeke, N. A.; Bruyneel, E. A.; De Bruyne, G. K.; De Pestel, G. H.; De Coster, W. J.; Espeel, M. F.; Mareel, M. M. The flavonoid tangeretin inhibits invasion of MO4 mouse cells into embryonic chick heart in vitro. *Clin. & Experi. Metastasis* **1989**, *7*, 283–300.

Brandi, M. L. Flavonoids: biochemical effects and therapeutic applications. *Bone & Mineral* **1992**, *19*, S3–14.

Brinkworth, R. I.; Stoermer, M. J.; Fairlie, D. P. Flavones are inhibitors of HIV-1 proteinase. *Biochem. & Biophys. Res. Commun.* **1992**, *188*, 631–637.

Burnham, W. S.; Sidwell, R. W.; Tolman, R. L.; Stout, M. G. Synthesis and antiviral activity of 4'-hydroxy-5,6,7,8-tetramethoxyflavone. *J. Med. Chem.* **1972**, *15*, 1075–1076.

Cushman, M.; Nagarathnam, D. Cytotoxicities of some flavonoid analogs. *J. Nat. Prod.* **1991**, *54*, 1656–1660.

Das, A.; Wang, J. H.; Lien, E. J. Carcinogenicity, mutagenicity and cancer preventing activities of flavonoids: a structure-system-activity relationship (SSAR) analysis. *Prog. Drug Res.* **1994**, *42*, 133–166.

De Meyer, N.; Haemers, A.; Mishra, L.; Pandey, H. K.; Pieters, L. A.; Vanden Berghe, D. A.; Vlietinck, A. J. 4'-Hydroxy-3-methoxyflavones with potent antipicornavirus activity [published erratum appears in *J. Med. Chem.* **1992**, *35*, 4923]. *J. Med. Chem.* **1991**, *34*, 736–746.

De Rodriguez, D. J.; Chulia, J.; Simoes, C. M.; Amoros, M.; Mariotte, A. M.; Girre, L. Search for "in vitro" antiviral activity of a new isoflavonic glycoside from *Ulex europaeus*. *Planta Med.* **1990**, *56*, 59–62.

Dong, X. P.; Che, C. T.; Farnsworth, N. R. Cytotoxic flavonols from *Gutierrezia microcephala*. *J. Nat. Prod.* **1987**, *50*, 337–338.

Edwards, J. M.; Raffauf, R. F.; Le Quesne, P. W. Antineoplastic activity and cytotoxicity of flavones, isoflavones, and flavanones. *J. Nat. Prod.* **1979**, *42*, 85–91.

Fesen, M. R.; Pommier, Y.; Leteurtre, F.; Hiroguchi, S.; Yung, J.; Kohn, K. W. Inhibition of HIV-1 integrase by flavones, caffeic acid phenethyl ester (CAPE) and related compounds. *Biochem. Pharmacol.* **1994**, *48*, 595–608.

Fu, X.; Sevenet, T.; Remy, F.; Pais, M.; Hamid, A.; Hadi, A.; Zeng, L. M. Flavonone and chalcone derivatives from *Cryptocarya kurzii*. *J. Nat. Prod.* **1993**, *56*, 1153–1163.

Ferriola, P. C.; Cody, V.; Middleton, E. Jr. Protein kinase C inhibition by plant flavonoids. Kinetic mechanisms and structure-activity relationships. *Biochem. Pharmacol.* **1989**, *38*, 1617–1624.

Fujiki, H.; Horiuchi, T.; Yamashita, K.; Hakii, H.; Suganuma, M.; Nishino, H.; Iwashima, A.; Hirata, Y.; Sugimura, T. Inhibition of tumor promotion by flavonoids. *Prog. Clin. & Biol. Res.* **1986**, *213*, 429–440.

Harborne, J. B.; Mabry, T. J.; Mabry, H. eds., *The Flavonoids*, Academic Press, New York, 1975.

Harborne, J. B. Nature, distribution and function of plant flavonoids. *Prog. Clin. & Biol. Res.* **1986**, *213*, 15–24.

Havlin, K. A.; Kuhn, J. G.; Craig, J. B.; Boldt, D. H.; Weiss, G. R.; Koeller, J.; Harman, G.; Schwartz, R.; Clark, G. N.; Von Hoff, D. D. Phase I clinical and pharmacokinetic trial of flavone acetic acid. *J. the National Cancer Institute.* **1991**, *83*, 124–128.

Hayashi, T.; Uchida, K.; Hayashi, K.; Niwayama, S.; Morita, N. A cytotoxic flavone from *Scoparia dulcis* L. *Chem. & Pharm. Bull.* **1988**, *36*, 4849–4851.

Hirano, T.; Abe, K.; Gotoh, M.; Oka, K. Citrus flavone tangeretin inhibits leukaemic HL-60 cell growth partially through induction of apoptosis with less cytotoxicity on normal lymphocytes. *British. J. Cancer* **1995**, *72*, 1380–1388.

Horowitz, R. M. Taste effects of flavonoids. *Prog. Clin. & Biol. Res.* **1986**, *213*, 163–175.

Hu, C. Q.; Chen, K.; Shi, Q.; Kilkuskie, R. E.; Cheng, Y. C.; Lee, K. H. Anti-AIDS agents, 10. Acacetin-7-O-beta-D-galactopyranoside, an anti-HIV principle from Chrysanthemum morifolium and a structure-activity correlation with some related flavonoids. *J. Nat. Prod.* **1994**, *57*, 42–51.

Gabor, M. Anti-inflammatory and anti-allergic properties of flavonoids. *Prog. Clin. & Biol. Res.* **1986**, *213*, 471–480.

Gaspar, J.; Laires, A.; Rueff, J. Genotoxic flavonoids and red wine: a possible role in stomach carcinogenesis. *European. J. Cancer Prevention.* **1994**, *3*, Suppl 2, 13–17.

Glusker, J. P.; Rossi, M. Molecular aspects of chemical carcinogens and bioflavonoids. *Prog. Clin. & Biol. Res.* **1986**, *213*, 395–410.

Hirano, T.; Gotoh, M.; Oka, K. Natural flavonoids and lignans are potent cytostatic agents against human leukemic HL-60 cells. *Life Sci.* **1994**, *55*, 1061–1069.

Holmund, J. T.; Kopp, W. C.; Wiltrout, R. H.; Longo, D. L.; Urba, W. J.; Janik, J. E.; Sznol, M.; Conlon, K. C.; Fenton, R. G.; Hornung, R.; de Forni, M.; Chabot, G. G.; Armand, J. P.; Gouyette, A.; Klink-Alak, M.; Recondo, G. A phase I clinical trial of flavone-8-acetic acid in combination with interleukin 2 Phase I and pharmacology study of flavone acetic acid administered two or three times weekly without alkalinization. *Cancer Chemother. & Pharmacol.* **1995**, *35*, 219–224.

Humber, L. G. The medicinal chemistry of aldose reductase inhibitors. *Prog. Med. Chem.* **1987**, *24*, 299–343.

Kato, R.; Nakadate, T.; Yamamoto, S.; Sugimura, T. Inhibition of 12-O-tetradecanoylphorbol-13-acetate-induced tumor promotion and ornithine decarboxylase activity by quercetin: possible involvement of lipoxygenase inhibition. *Carcinogenesis* **1983**, *4*, 1301–1305.

Kandaswami, C.; Perkins, E.; Soloniuk, D. S.; Drzewiecki, G.; Middleton, E. Jr. Antiproliferative effects of citrus flavonoids on a human squamous cell carcinoma in vitro. *Cancer Lett.* **1991**, *56*, 147–152.

Kandaswami, C.; Perkins, E.; Drzewiecki, G.; Soloniuk, D. S.; Middleton, E. Jr. Differential inhibition of proliferation of human squamous cell carcinoma, gliosarcoma and embryonic fibroblast-like lung cells in culture by plant flavonoids. *Anti-Cancer Drugs* **1992**, *3*, 525–530.

Kandaswami, C.; Middleton, E. Jr. Free radical scavenging and antioxidant activity of plant flavonoids. *Adv. in Experi. Med. & Biol.* **1994**, *366*, 351–376.

Kaneda, N.; Pezzuto, J. M.; Soejarto, D. D.; Kinghorn, A. D.; Reutrakul, V.; Plant anticancer agents. Part 48. New cytotoxic flavonoids from *Muntingia calabura* roots. *J. Nat. Prod.* **1991**, *54*, 196–206.

Kaul, T. N.; Middleton, E. Jr.; Ogra, P. L. Antiviral effect of flavonoids on human viruses. *J. Med. Virology* **1985**, *15*, 71–79.

Kerr, D. J.; Kaye, S. B. Flavone acetic acid-preclinical and clinical activity. *European. J. Cancer & Clin. Oncol.* **1989**, *25*, 1271–1272.

Konoshima, T.; Kokumai, M.; Kozuka, M.; Iinuma, M.; Mizuno, M.; Tanaka, T.; Tokuda, H.; Nishino, H.; Iwashima, A. Studies on inhibitors of skin tumor promotion. XI. Inhibitory effects of flavonoids from Scutellaria baicalensis on Epstein-Barr virus activation and their anti-tumor-promoting activities. *Chem. & Pharm. Bull.* **1992**, *40*, 531–533.

Konoshima, T.; Terada, H.; Kokumai, M.; Kozuka, M.; Tokuda, H.; Estes, J. R.; Li, L.; Wang, H. K.; Lee, K. H. Studies on inhibitors of skin tumor promotion, XII. Rotenoids from *Amorpha fruticosa*. *J. Nat. Prod.* **1993**, *56*, 843–848.

Kuo, S. C.; Lee, H. Z.; Juang, J. P.; Lin, Y. T.; Wu, T. S.; Chang, J. J.; Lednicer, D.; Paull, K. D.; Lin, C. M.; Hamel, E. Synthesis and cytotoxicity of 1,6,7,8-substituted 2-(4'-substituted phenyl)-4-quinolones and related compounds: identification as antimitotic agents interacting with tubulin. *J. Med. Chem.* **1993**, *36*, 1146–1156.

Lee, I. K.; Kim, C. J.; Song, K. S.; Kim, H. M.; Koshino, H.; Uramoto, M.; Yoo, I. D. Cytotoxic benzyl dihydroflavonols from *Cudrania tricuspidata*. *Phytochem.* **1996**, *41*, 213–216.

Lee, K. H.; Tagahara, K.; Suzuki, H.; Wu, R. Y.; Haruna, M.; Hall, I. H.; Huang, H. C.; Ito, K.; Iida, T.; Lai, J. S. Antitumor agents. 49 Tricin, kaempferol-3-O-β-D- glucopyranoside and (+)-nortrachelogenin, antileukemic principles from *Wikstroemia indica*. *J. Nat. Prod.* **1981**, *44*, 530–535.

Li, L.; Wang, H. K.; Fujioka, T.; Chang, J. J.; Kozuka, M.; Konoshima, T.; Estes, J. R.; McPhail, D. R.; McPhail, A. T.; Lee, K. H. Structure and sterochemistry of amorphispironone. *J. Chem. Soc., Chem. Commu.* **1991**, 1652–1653.

Li, L.; Wang, H. K.; Chang, J. J.; McPhail, A. T.; McPhail, D. R.; Terada, H.; Konoshima, T.; Kokumai, M.; Kozuka, M.; Estes, J. R.; Lee, K. H. Antitumor agents, 138. Rotenoids and isoflavones as cytotoxic constitutents from *Amorpha fruticosa*. *J. Nat. Prod.* **1993**, *56*, 690–698.

Li, L.; Wang, H. K.; Kuo, S. C.; Wu, T. S.; Lednicer, D.; Lin, C. M.; Hamel, E.; Lee, K. H. Antitumor agents. 150. 2',3',4',5',5,6,7-substituted 2-phenyl-4-quinolones and related compounds: their synthesis, cytotoxicity, and inhibition of tubulin polymerization. *J. Med. Chem.* **1994**, *37*, 1126–1135.

Li, L.; Wang, H. K.; Kuo, S. C.; Wu, T. S.; Mauger, A.; Lin, C. M.; Hamel, E.; Lee, K. H. Antitumor agents. 155. Synthesis and biological evaluation of 3',6,7-substituted 2-phenyl-4-quinolones as antimicrotubule agents. *J. Med. Chem.* **1994**, *37*, 3400–3407.

Lima, M. A.; Silveira, E. R.; Marques, M. S.; Santos, R. H.; Gambardela, M. T. Biologically active flavonoids and terpenoids from *Egletes viscosa*. *Phytochem.* **1996**, *41*, 217–223.

Lin, Y. M.; Chen, F. C.; Lee, K. H. Hinokiflavone, a cytotoxic principle from *Rhus succedanea* and the cytotoxicity of the related biflavonoids. *Planta Med.* **1989**, *55*, 166–168.

Luo, S. D.; Nemec, J.; Ning, B. M.; Li, Q. X. Anti-HIV flavonoids from *Morus alba* L. *Int. Conf. AIDS* **1994**, *10*, 203.

MacGregor, J. T. Mutagenic carcinogenic effects of flavonoids. *Prog. Clin. & Biol. Res.* **1986**, *213*, 411–424.

Mahmood, N.; Pizza, C.; Aquino, R.; De Tommasi, N.; Piacente, S.; Colman, A.; Burke, A.; Hay, A. J. Inhibition of HIV infection by flavanoids. *Antiviral Res.* **1993**, *22*, 189–199.

Manikumar, G.; Gaetano, K.; Wani, M. C.; Taylor, H.; Hughes, T. J.; Warner, J.; McGivney, R.; Wall, M. E. Plant antimutagenic agents, 5. Isolation and structure of two new isoflavones, fremontin and fremontone from *Psorothamnus fremontii*. *J. Nat. Prod.* **1989**, *52*, 769–773.

Maughan, T. S.; Ward, R.; Dennis, I.; Honess, D. J.; Workman, P.; Bleehen, N. M. Tumour concentrations of flavone acetic acid (FAA) in human melanoma: comparison with mouse data. *British. J. Cancer* **1992**, *66*, 579–582.

Middleton, E. ,Jr.; Fujiki, H.; Savliwala, M.; Drzewiecki, G. Tumor promoter-induced basophil histamine release: effect of selected flavonoids. *Biochem. Pharmacol.* **1987**, *36*, 2048–2052.

Middleton, E. Jr.; Ferriola, P. Effect of flavonoids on protein kinase C: relationship to inhibition of human basophil histamine release. *Prog. Clin. & Biol. Res.* **1988a**, *280*, 251–266.

Middleton, E. ,Jr.; Ferriola, P.; Drzewiecki, G. Effect of flavonoids on tumor promoter-induced basophil histamine release and protein kinase C. *Transactions of the Am. Clin. & Climatological Assoc.* **1988b**, *100*, 47–56.

Mihara, S.; Hara, M.; Nakamura, M.; Sakurawi, K.; Tokura, K.; Fujimoto, M.; Fukai, T.; Nomura, T. Non-peptide bombesin receptor antagonists, kuwanon G and H, isolated from mulberry. *Biochem. & Biophys. Res. Commun.* **1995**, *213*, 594–599.

Mookerjee, B. K.; Lee, T. P.; Lippes, H. A.; Middleton, E. Jr. Some effects of flavonoids on lymphocyte proliferative responses. *J. Immunopharmacol.* **1986**, *8*, 371–392.

Motoo, Y.; Sawabu, N. Antitumor effects of saikosaponins, baicalin and baicalein on human hepatoma cell lines. *Cancer Lett.* **1994**, *86*, 91–95.

Naik, H. R.; Lehr, J. E.; Pienta, K. J. An in vitro and in vivo study of antitumor effects of genistein on hormone refractory prostate cancer. *Anticancer Res.* **1994**, *14*, 2617–2619.

Nakadate, T.; Yamamoto, S.; Aizu, E.; Kato, R. Effects of flavonoids and antioxidants on 12-O-tetradecanoyl-phorbol-13-acetate-caused epidermal ornithine decarboxylase induction and tumor promotion in relation to lipoxygenase inhibition by these compounds. *Gann.* **1984**, *75*, 214–222.

Noble, A. C. Bitterness in wine. *Physiology & Behavior* **1994**, *56*, 1251–1255.

Nomura, T.; Fukai, T.; Hano, Y.; Yoshizawa, S.; Suganuma, M.; Fujiki, H. Chemistry and anti-tumor promoting activity of Morus flavonoids. *Prog. Clin. & Biol. Res.* **1988**, *280*, 267–281.

O'Dwyer, P. J.; Shoemaker, D.; Zaharko, D. S.; Grieshaber, C.; Plowman, J.; Corbett, T.; Valeriote, F.; King, S. A.; Cradock, J.; Hoth, D. F. Flavone acetic acid (LM 975, NSC 347512). A novel antitumor agent. *Cancer Chemother. & Pharmacol.* **1987**, *19*, 6–10.

Olver, I. N.; Webster, L. K.; Bishop, J. F.; Stokes, K. H. A phase I and pharmacokinetic study of 12-h infusion of flavone acetic acid. *Cancer Chemother. & Pharmacol.* **1992**, *29*, 354–360.

Ono, K.; Nakane, H.; Fukushima, M.; Chermann, J. C.; Barre-Sinoussi, F. Inhibition of reverse transcriptase activity by a flavonoid compound, 5,6,7-trihydroxyflavone. *Biochem. & Biophys. Res. Commun.* **1989**, *160*, 982–987.

Ono, K.; Nakane, H.; Fukushima, M.; Chermann, J. C.; Barre-Sinoussi, F. Differential inhibitory effects of various flavonoids on the activities of reverse transcriptase and cellular DNA and RNA polymerases [published erratum appears in Eur. J. Biochem. **1991**, *199*, 769]. *Eur. J. Biochem.* **1990**, *190*, 469–476.

Parmar, V. S.; Jain, R.; Sharma, S. K.; Vardhan, A.; Jha, A.; Taneja, P.; Singh, S.; Vyncke, B. M.; Bracke, M. E.; Mareel, M. M. Anti-invasive activity of 3,7-dimethoxyflavone in vitro. *J. Pharm. Sci.* **1994**, *83*, 1217–1221.

Pengsuparp, T.; Cai, L.; Constant, H.; Fong, H. H.; Lin, L. Z.; Kinghorn, A. D.; Pezzuto, J. M.; Cordell, G. A.; Ingolfsdottir, K.; Wagner, H. Mechanistic evaluation of new plant-derived compounds that inhibit HIV-1 reverse transcriptase. *J. Nat. Prod.* **1995**, *58*, 1024–1031.

Perry, N. B.; Foster, L. M. Antiviral and antifungal flavonoids, plus a triterpene, from *Hebe cupressoides*. *Planta Med.* **1994**, *60*, 491–492.

Pratt, C. B.; Relling, M. V.; Meyer, W. H.; Douglass, E. C.; Kellie, S. J.; Avery, L. Phase I study of flavone acetic acid (NSC 347512, LM975) in patients with pediatric malignant solid tumors. *Am. J. Clin. Oncol.* **1991**, *14*, 483–486.

Raghavan, K.; Buolamwini, J. K.; Fesen, M. R.; Pommier, Y.; Kohn, K. W.; Weinstein, J. N. Three-dimensional quantitative structure-activity relationship (QSAR) of HIV integrase inhibitors: a comparative molecular field analysis (CoMFA) study. *J. Med. Chem.* **1995**, *38*, 890–897.

Ryu, S. H.; Ahn, B. Z.; Pack, M. Y. The cytotoxic principle of Scutellariae radix against L1210 cell. *Planta Med.* **1985a**, *51*, 462–462.

Ryu, S. H.; Yoo, B. T.; Ahn, B. Z.; Pack, M. Y. Synthesis of flavones which are cytotoxic against L1210 cells. *Archiv. der. Pharmazie.* **1985b**, *318*, 659–661.

Selway, J. W. Antiviral activity of flavones and flavans. *Prog. Clin. & Biol. Res.* **1986**, *213*, 521–536.

Shi, Q.; Chen, K.; Li, L.; Chang, J. J.; Autry, C.; Kozuka, M.; Konoshima, T.; Estes, J. R.; Lin, C. M.; Hamel, E.; Lee, K. H. Antitumor agents, 154. Cytotoxic and antimitotic flavonols from *Polanisia dodecandra*. *J. Nat. Prod.* **1995**, *58*, 475–482.

Siegenthaler, P.; Kaye, S. B.; Monfardini, S.; Renard, J. Phase II trial with Flavone Acetic Acid (NSC-347512, LM975) in patients with non-small cell lung cancer. *Ann. of Oncol.* **1992**, *3*, 169–170.

Silva, G. L.; Chai, H.; Gupta, M. P.; Farnsworth, N. R.; Cordell, G. A.; Pezzuto, J. M.; Beecher, C. W.; Kinghorn, A. D. Cytotoxic biflavonoids from *Selaginella willdenowii*. *Phytochem.* **1995**, *40*, 129–134.

Spedding, G.; Ratty, A.; Middleton, E. Jr. Inhibition of reverse transcriptases by flavonoids. *Antiviral Res.* **1989**, *12*, 99–110.

Sugi, M.; Nagashio, Y. In *Cancer Therapy in Modern China*; Kondo, K., ed.; Shizen Sha, Tokyo, Japan, 1977.

Sugiyama, S.; Umehara, K.; Kuroyanagi, M.; Ueno, A.; Taki, T. Studies on the differentiation inducers of myeloid leukemic cells from *Citrus* species. *Chem. & Pharm. Bull.* **1993**, *41*, 714–719.

Tang, R. J.; Chen, K.; Cosentino, L. M.; Lee, K. H. Apigenin-7-O-β-D-glucopyranoside, an anti-HIV principle from *Kummerowia striata*. *Bioorg. & Med. Chem. Lett.* **1994**, *4*, 455–458.
Terada, H.; Kokumai, M.; Konoshima, T.; Kozuka, M.; Haruna, M.; Ito, K.; Estes, J. R.; Li, L.; Wang, H. K.; Lee, K. H. Structural Elucidation and chmical conversion of amorphispironone, a novel spironone from *Amorpha fruticosa*, to rotenoids. *Chem. & Pharm. Bull.* **1993**, *41*, 187–190.
Ulubelen, A.; Oksuz, S. Cytotoxic flavones from *Centaurea urvillei*. *J. Nat. Prod.* **1982**, *45*, 373.
Van Hoof, L.; Vanden Berghe, D. A.; Hatfield, G. M.; Vlietinck, A. J. Plant antiviral agents; V. 3-Methoxyflavones as potent inhibitors of viral-induced block of cell synthesis. *Planta Med.* **1984**, *50*, 513–517.
Vlietinck, A. J.; Vanden Berghe, D. A.; Van Hoof, L. M.; Vrijsen, R.; Boeye, A. Antiviral activity of 3-methoxyflavones. *Prog. Clin. & Biol. Res.* **1986**, *213*, 537–540.
Vlietinck, A. J.; Vanden Berghe, D. A.; Haemers, A. Present status and prospects of flavonoids as anti-viral agents. *Prog. Clin. & Biol. Res.* **1988**, *280*, 283–299.
Wagner, H. Mechanistic evaluation of new plant-derived compounds that inhibit HIV-1 reverse transcriptase. *Prog. Clin. & Biol. Res.* **1986**, *213*, 545–558.
Wall, M. E.; Wani, M. C.; Manikumar, G.; Taylor, H.; McGivney, R. Plant antimutagens, 6. Intricatin and intricatinol, new antimutagenic homoisoflavonoids from *Hoffmanosseggia intricata*. *J. Nat. Prod.* **1989**, *52*, 774–778.
Wall, M. E.; Wani, M. C.; Fullas, F.; Oswald, J. B.; Brown, D. M.; Santisuk, T.; Reutrakul, V.; McPhail, A. T.; Farnsworth, N. R.; Pezzuto, J. M. Plant antitumor agents. 31. The calycopterones, a new class of biflavonoids with novel cytotoxicity in a diverse panel of human tumor cell lines. *J. Med. Chem.* **1994**, *37*, 1465–1470.
Wang, H. K.; Bastow, K. F.; Cosentino, L. M.; Lee, K. H. Antitumor agents. 166. Synthesis and biological evaluation of 5,6,7,8-substituted-2-phenylthiochromen-4-ones. *J. Med. Chem.* **1996**, *39*, 1975–1980.
Wang, J. N.; Hou, C. Y.; Liu, Y. L.; Lin, L. Z.; Gil, R. R.; Cordell, G. A. Swertifrancheside, an HIV-reverse transcriptase inhibitor and the first flavone-xanthone dimer, from *Swertia franchetiana*. *J. Nat. Prod.* **1994**, *57*, 211–217.
Weiss, R. B.; Greene, R. F.; Knight, R. D.; Collins, J. M.; Pelosi, J. J.; Sulkes, A.; Curt, G. A. New anticancer agents. *Cancer Res.* **1988**, *48*, 5878–5882.
Woerdenbag, H. J.; Merfort, I.; Passreiter, C. M.; Schmidt, T. J.; Willuhn, G.; van Uden, W.; Pras, N.; Kampinga, H. H.; Konings, A. W. Cytotoxicity of flavonoids and sesquiterpene lactones from Arnica species against the GLC4 and the COLO 320 cell lines. *Planta Med.* **1994**, *60*, 434–437.
Yasukawa, K.; Takido, M.; Takeuchi, M.; Nitta, K. Effect of flavonoids on tumor promoter's activity. *Prog. Clin. Biol. Res.* **1988**, *280*, 247–250.
Yasukawa, K.; Takido, M.; Takeuchi, M.; Sato, Y.; Nitta, K.; Nakagawa, S. Inhibitory effects of flavonol glycosides on 12-O-tetradecanoylphorbol-13-acetate-induced tumor promotion. *Chem. & Pharm. Bull.* **1990**, *38*, 774–776.

# 16

# INHIBITION OF MAMMARY CANCER BY CITRUS FLAVONOIDS

N. Guthrie and K. K. Carroll

Centre for Human Nutrition
Department of Biochemistry
The University of Western Ontario
London, Ontario, N6A 5C1, Canada

## 1. INTRODUCTION

Much of the emphasis in research on diet and cancer has been focused on the promotion of cancer by dietary fat (Carroll, 1994). However, the lack of support for the fat hypothesis by case-control and cohort studies has led people to question its validity (Land, 1994; Hunter et al., 1996). Epidemiological studies on diet and cancer have provided leads in the search for naturally-occurring anticancer agents (Steinmetz and Potter, 1991a; Block et al., 1992). There is general agreement that plant-based diets, rich in whole grains, legumes, fruits and vegetables reduce the risk of various types of cancer, including breast cancer. A variety of compounds produced by plants have been investigated (Steinmetz and Potter, 1991b; Wattenburg, 1992). These include the flavonoids, which are an integral part of the human diet.

Flavonoids are a class of chemically-related compounds that are ubiquitous in nature. Over 4000 different naturally-occurring flavonoids have been described (Middleton and Kandaswami, 1994a). They are present in many plant products, including fruits, vegetables, nuts, seeds, grains, tea and wine (Hertog et al., 1993; Kuhnau, 1976; Adlercreutz et al., 1991). They occur as aglycones, glycosides and methylated derivatives (Kuhnau, 1976). Reports of their estimated daily intake range from 20 mg/day (Hertog et al., 1993) to 1 g/day (expressed as glycosides) (Kuhnau, 1976). Citrus flavonoids are a major class of secondary metabolites that have important biological activity (Middleton and Kandaswami, 1994b) including anti-cancer activity (So et al., 1996a; Guthrie et al., 1997a). These include naringin from grapefruit and pummelo, hesperidin from oranges, tangeretin and nobiletin from tangerines, lemons and limes.

Many studies have reported that flavonoids inhibited, and sometimes induced, a large variety of mammalian enzyme systems (Middleton and Kandaswami, 1994b). Some

*Flavonoids in the Living System*, edited by Manthey and Buslig
Plenum Press, New York, 1998.

of these enzymes affect cell division and proliferation, platelet aggregation, detoxification, and inflammatory and immune responses. Flavonoids have also been investigated for their anti-cancer activities (Edwards et al., 1988; Middleton and Kandaswami, 1994b). Cytotoxic effects on Hela cells and Raji lymphoma cells (Ramanathan et al., 1992) and antiproliferative effects on human lymphocytes (Mookerjee et al., 1986) and human breast cancer cells have been reported (So et al., 1996a; Guthrie et al., 1997a).

Genistein, an isoflavone found in soybeans, has been extensively studied as a possible anti-cancer agent (Messina et al., 1994; Barnes et al., 1995; Peterson and Barnes, 1991; Messina, 1995). It was found to inhibit the growth of human breast cancer cells in culture and soybean preparations inhibited chemically-induced mammary carcinogenesis in rats.

Quercetin, another flavonoid found in many fruits and vegetables, has also been investigated for anti-cancer activity. It was shown to have growth inhibitory activity *in vitro* in human breast cancer cells (Singhal et al., 1995) and to reduce the incidence of chemically-induced mammary tumors in rats (Verma et al., 1988). Quercetin was been found to be a potent inhibitor of aromatase activity (Pelissero et al., 1996) and topoisomerase I-catalyzed DNA religation (Boege et al., 1996).

Our interest in dietary flavonoids developed as a result of studies by colleagues at The University of Western Ontario. They observed that drugs, such as felodipine and nifedipine, used for treatment of hypertension, were more effective when given in grapefruit juice and suggested that the effect might be due to naringenin, a flavonoid present in grapefruit juice (Bailey et al., 1991). The structure of naringenin is somewhat analogous to that of genistein (Fig. 1). When tested for its ability to inhibit proliferation of human breast cancer cells *in vitro*, naringenin was found to be effective at a substantially lower concentration than genistein (Guthrie et al., 1993).

## 2. ANIMAL STUDIES

This observation led us to carry out an animal experiment. Rats were given a 5 mg dose of 7,12-dimethylbenz(a)anthracene (DMBA) and fed a semipurified diet containing 5% corn oil. After one week, the animals were started on the experimental diets. One group was given double strength orange juice and another double strength grapefruit juice in place of drinking water. For these groups, the carbohydrate component of the semipurified diets was reduced to compensate for the carbohydrate in the fruit juices. A third group was given naringin and a fourth naringenin, mixed in the food in amounts comparable to those obtained by drinking the double strength grapefruit juice. A fifth group was fed the semipurified diet with plain drinking water and served as a control. The rats were weighed, examined and palpated for tumors weekly. After 16 weeks, they were sacrificed and the tumors were excised, weighed and sent for histological examination.

The grapefruit juice and naringenin had little effect on tumor development. Naringin delayed the development of tumors, but this may have been a non-specific effect, since it also inhibited growth of the rats. Orange juice, which was used as a control, also tended to inhibit tumorigenesis. In this case the animals gained more weight than those in any of the other groups, indicating that the inhibition of tumorigenesis was a more specific growth inhibitory effect (So et al., 1996a).

Because of this unexpected finding, this experiment was repeated, with the exception of using a semipurified diet containing 20% corn oil. Similar results were obtained in this experiment (So et al., 1996a). The group fed naringin showed the smallest weight gain whereas the rats in the orange juice group showed the best weight gain. Double strength or-

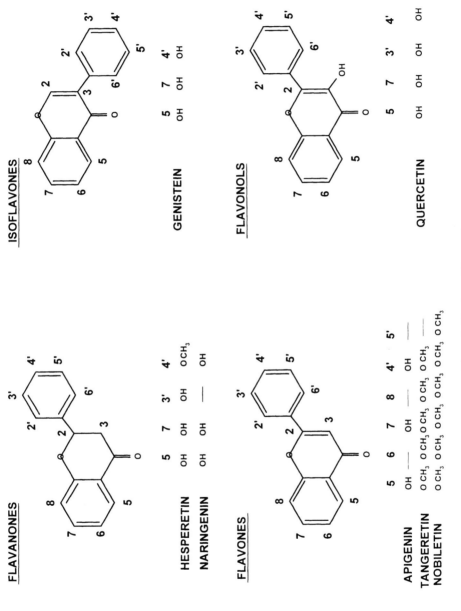

Figure 1. Structure of flavonoids.

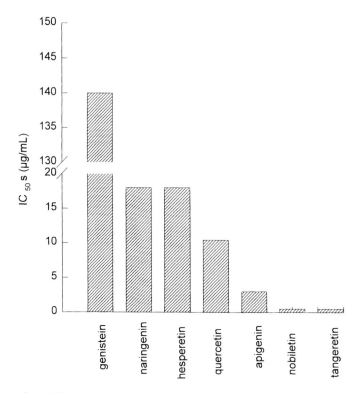

**Figure 2.** Inhibition of estrogen receptor-negative human breast cancer cells by flavonoids.

ange juice given to the rats in place of drinking water inhibited tumorigenesis more effectively than double strength grapefruit juice. Tumor development was delayed in the rats eating the naringin-supplemented diet and in the rats drinking orange juice (So et al., 1996a).

## 3. STUDIES WITH HUMAN BREAST CANCER CELLS

### 3.1. Inhibition of Estrogen Receptor-Negative (ER-) Cells

A number of naturally-occurring flavonoids were investigated for their effects on proliferation of a human breast cancer cell line, MDA-MB-435 (So et al., 1996a; Guthrie et al., 1997a). The $IC_{50}$s are presented in Fig. 2. Hesperetin, the aglycone of the flavonoid present in oranges, was found to inhibit human breast cancer cells as effectively as naringenin ($IC_{50}$s for hesperetin and naringenin were 18 µg/ml compared to 140 µg/ml for genistein) (So et al., 1996a). Two other citrus flavonoids, tangeretin and nobiletin, found in tangerines, had even lower $IC_{50}$s (0.5 µg/ml for both compounds) (Guthrie et al., 1997a).

The ability of flavonoids to inhibit cell growth was also investigated by treating the cells at their $IC_{50}$ concentrations and following the growth of the cells over a ten-day period (So et al., 1996a; Guthrie et al., 1997a). They inhibited the growth of these cells and the effect was apparent after two days of treatment. Cytotoxic effects of the flavonoids were also investigated using the MTT assay (Hansen et al., 1989). Most cells were viable at the $IC_{50}$ concentrations of the flavonoids (So et al., 1996a; Guthrie et al., 1997a), indicating that the antiproliferative activity of the compounds was not due to non-specific cytotoxicity.

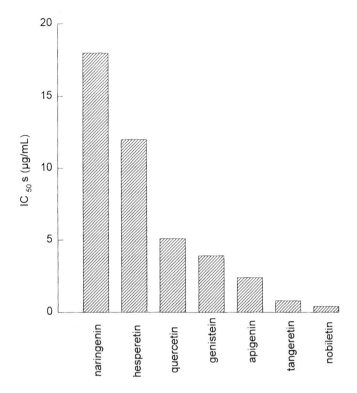

**Figure 3.** Inhibition of estrogen receptor-positive human breast cancer cells by flavonoids.

## 3.2. Inhibition of Estrogen Receptor-Positive (ER+) Cells

The effects of various flavonoids on the proliferation, growth and viability of MCF-7 estrogen receptor-positive human breast cancer were also investigated (So et al., 1997; Guthrie et al., 1997b). The $IC_{50}$s are presented in Fig. 3. Tangeretin and nobiletin were again the most effective inhibitors, with $IC_{50}$s of 0.8 and 0.4 µg/mL respectively (Fig. 3).

Further studies were performed to investigate whether this inhibition was acting on the estrogen receptor. MCF-7 cells were depleted of steroids and treated with flavonoids or tamoxifen in the presence or absence of 100 nM estradiol. The results (Table 1) show that inhibition by all of these flavonoids, except for genistein, was unaffected by estradiol, in contrast to the results with tamoxifen. Thus, unlike genistein, the citrus flavonoids do not appear to be acting as weak estrogens (So et al., 1997b) and this may be why they are more effective inhibitors of the estrogen receptor-negative MDA-MB-435 human breast cancer cells in the above experiments.

## 4. SYNERGISTIC EFFECTS

In other studies we have shown that proliferation of ER- and ER+ cells is inhibited by tocotrienols from palm oil (Nesaretnam et al., 1995; Carroll et al., 1995; Guthrie et al., 1997b). These compounds are a form of vitamin E having an unsaturated phytyl side chain instead of the saturated side chain of the more common tocopherols (Nesaretnam et al., 1995). We have tested the tocotrienol-rich fraction (TRF) and the individual tocotrienols

**Table 1.** Inhibition of proliferation of MCF-7 human breast cancer cells by tamoxifen and by various flavonoids in the presence and absence of excess estrogen

| Compound | % of Control | % of Control (+ 100nM estradiol) |
|---|---|---|
| Naringenin | 53 | 52 |
| Hesperetin | 60 | 63 |
| Quercetin | 71 | 67 |
| Baicalein | 63 | 59 |
| Genistein | 67 | 91 |
| Tamoxifen | 49 | 88 |

($\alpha$-, $\gamma$-, $\delta$-) on proliferation of both ER- and ER+ human breast cancer cells in culture. Tocotrienols were effective inhibitors of proliferation of both cell types, with $IC_{50}$s of 30–180 µg/mL in ER- cells and 2–6 µg/mL in ER+ cells, whereas $\alpha$-tocopherol was ineffective (Nesaretnam et al., 1995; Carroll et al., 1995; Guthrie et al., 1997b).

In other studies with ER- and ER+ human breast cancer cells *in vitro*, we have observed that 1:1 combinations of flavonoids with tocotrienols or tamoxifen and 1:1 combinations of tocotrienols with tamoxifen inhibit proliferation of the cells more effectively than the individual compounds by themselves. The most effective combination with ER- cells was tangeretin and $\gamma$-tocotrienol ($IC_{50}$ 0.05 µg/mL) (Guthrie et al., 1997a). With ER+ cells, the best results were obtained with tangeretin and $\gamma$-tocotrienol ($IC_{50}$ 0.02 µg/mL), nobiletin + tamoxifen ($IC_{50}$ 0.004 µg/mL) and $\delta$-tocotrienol + tamoxifen ($IC_{50}$ 0.003 µg/mL) (Guthrie et al., 1997a). When combined in 1:1:1 combinations of flavonoids, tocotrienols and tamoxifen, tangeretin + $\gamma$-tocotrienol + tamoxifen was the most effective in ER- cells ($IC_{50}$ 0.01 µg/mL) (Fig. 4) and hesperetin + $\delta$-tocotrienol + tamoxifen was the most effective in ER+ cells ($IC_{50}$ 0.0005 µg/mL) (Fig. 5). The synergism observed suggests that these compounds may be acting by different mechanisms.

## 5. MECHANISM OF ACTION

Tamoxifen is widely used in the treatment of hormone-responsive breast cancer (Jordan, 1994, 1995), and acts mainly by competing with estrogen for its receptor. Our data indicate that tocotrienols (Fig. 6) (Guthrie et al., 1997b) and flavonoids (Table 1) (So et al., 1997b) act via an estrogen receptor-independent pathway, since they inhibit both ER+ and ER- cell lines.

Although the mechanisms for the observed inhibition of cellular proliferation are not clear, genistein is a specific inhibitor of epidermal growth factor receptor tyrosine kinase (Akiyama et al., 1988). Quercetin has also been shown to inhibit the activity of tyrosine protein kinase activity and phosphoinositide phosphorylation, both of which are necessary for cell growth and development (Levy et al., 1984; Sharoni et al., 1986). Protein kinase C is believed to be involved in the regulation of cellular proliferation and quercetin has been shown to inhibit this enzyme as well (Ferriola, et al., 1989). Preliminary studies on the mechanism by which flavonoids inhibit the proliferation of these cells indicated that hesperetin inhibited protein kinase C activity to about the same extent as quercetin and genistein (Table 2) (So et al., 1996b; Carroll et al., 1998). Tocotrienols completely abol-

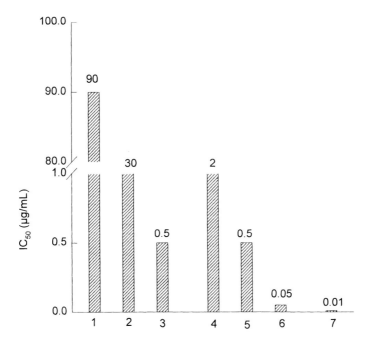

**Figure 4.** Effect of gamma tocotrienol, tangeretin and tamoxifen on proliferation of MDA-MB-435 human breast cancer cells alone and in combination. 1-tamoxifen; 2-gamma tocotrienol; 3-tangeretin; 4-gama tocotrienol + tamoxifen (1:1); 5-tangeretin + tamoxifen (1:1); 6-gama tocotrienol + tangeretin (1:1); 7-tangeretin + gama tocotrienol + tamoxifen (1:1:1).

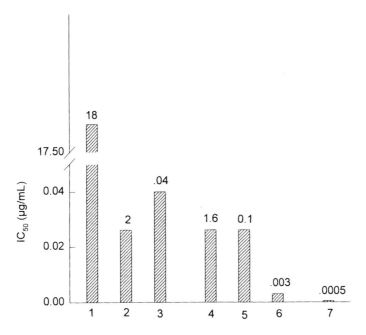

**Figure 5.** Effect of delta tocotrienol, hesperetin and tamoxifen on proliferation of MCF-7 human breast cancer cells alone and in combination. 1-hesperetin; 2-delta tocotrienol; 3-tamoxifen; 4-hesperetin + tamoxifen (1:1); 5-delta tocotrienol + hesperetin (1:1); 6-delta tocotrienol + tamoxifen (1:1); 7-hesperetin + delta tocotrienol + tamoxifen (1:1:1).

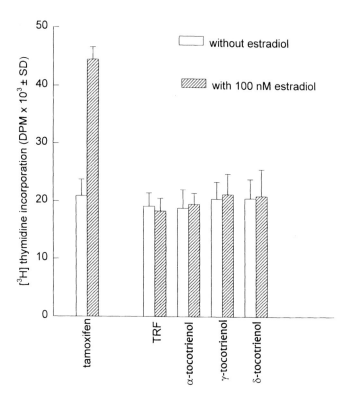

**Figure 6.** Inhibition of proliferation of MCF-7 human breast cancer cells by tamoxifen, and tocotrienols in the presence and absence of excess estrogen.

ished protein kinase C activity in MDA-MB-435 cells at these $IC_{50}$ concentrations (Guthrie et al., 1996).

## 6. SUMMARY AND CONCLUSIONS

Double strength orange juice given to the rats in place of drinking water inhibited mammary tumorigenesis induced in female Sprague-Dawley rats by DMBA more effectively than double strength grapefruit juice. This may mean that hesperetin retains its ef-

Table 2. Effect of flavonoids on protein kinase C activity in MDA-MB-435 human breast cancer cells

| Flavonoid | % Inhibition of Protein Kinase C Activity |
|---|---|
| Naringenin | 16 |
| Genistein | 21 |
| Quercetin | 22 |
| Hesperetin | 23 |
| Baicalein | 38 |

fectiveness *in vivo* better than naringenin, since the flavonoids are present in the juices at similar levels. It is also possible that orange juice contains other compounds that have anti-cancer activity and that may act synergistically with hesperetin.

Citrus flavonoids are effective inhibitors of both estrogen receptor-negative MDA-MB-435 and estrogen receptor-positive MCF-7 human breast cancer cells *in vitro*. Furthermore, 1:1 combinations of flavonoids with tocotrienols and/or tamoxifen inhibit proliferation of the cells more effectively than the individual compounds. This synergism may be due to the fact that the compounds are exerting their inhibitory effects by different mechanisms.

## ACKNOWLEDGMENTS

This work was supported by the State of Florida Department of Citrus. The authors would like to thank Hung Daotran, Josephine Ho, and Juliet Ho for excellent technical assistance and Charlotte Harman for assistance in typing the manuscript.

## REFERENCES

Akiyama, T.; Ishida, J.; Nakagawa, S.; Ogawara, H.; Watanabe, S.; Itoh, N.; Shibuya, M.; Fukami, Y. Genistein, a specific inhibitor of tyrosine-specific protein kinases. *J. Biol. Chem.* **1987**, *262*, 5592–5595.

Adlercreutz, H; Honjo, H.; Higashi, A.; Fotsis, T.; Hamalainen, E.; Hasegawa, T.; Okada, H. Urinary excretion of lignans and isoflavonoid phytoestrogens in Japanese men and women consuming a traditional Japanese diet. *Am. J. Clin. Nutr.* **1991**, *54*, 1093–1100.

Bailey, D.G.; Spence, J.D.; Munoz, C.; Arnold, J.M.O. Interaction of citrus juices with felodipine and nifedipine. *Lancet* **1991**, *337*, 268–269.

Barnes, S. Effect of genistein on *in vitro* and *in vivo* models of cancer. *J. Nutr.* **1995**, *125*, 777S-783S.

Block, G.; Patterson, B.; Subar, A. Fruits, vegetables and cancer prevention: A review of the epidemiological evidence. *Nutr. Cancer* **1992**, *18*, 1–29.

Boege, F.; Straub, T.; Kehr, A.; Boesenberg, C.; Christiansen, K.; Anderson, A.; Jakob, F.; Kohrle, J. Selected novel flavones inhibit the DNA binding or the DNA religation step of eukaryotic topoisomerase I. *J. Biol. Chem.* **1996**, *271(4)*, 2262–2270.

Carroll, K.K. Lipids and cancer. In *Nutrition and Disease Update. Cancer*; Carroll, K.K.; Kritchevsky, D., Eds.; AOCS Press: Champaign, IL, **1994**; pp 235–296.

Carroll, K.K.; Guthrie, N.; Nesaretnam, K.; Gapor, A.; Chambers, A.F. Anti-cancer properties of tocotrienols from palm oil. In *Nutrition, Lipids, Health, and Disease;* Ong, A.S.H.; Niki, E.; Packer, L., Eds.; AOCS Press, Champaign, IL, **1995**; pp. 117–121.

Carroll, K.K.; Guthrie, N.; So, F.V.; Chambers, A.F. Anti-cancer properties of flavonoids, with emphasis on citrus flavonoids. In *Flavonoids in Health and Disease;* Rice-Evans, C.; Packer, L. Eds.; Marcel Dekker, Inc., New York, **1998**; pp. 437–446.

Edwards, J.M.; Mori, A.; Nishino, C.; Enoki, N.; Tawata, S. Cytotoxicity of plant flavonoids against Hela cells. *Phytochemistry* **1988**, 27, 1017–1020.

Ferriola, P.C.; Cody, V.; Middleton, E. Jr. Protein kinase C inhibition by plant flavonoids. Kinetic mechanisms and structure-activity relationships. *Biochem. Pharmacol.*, **1989**, 38, 1617–1624.

Guthrie, N.; Gapor, A.; Chambers, A.F.; Carroll, K.K. Palm oil tocotrienols and plant flavonoids act synergistically with each other and with tamoxifen in inhibiting proliferation and growth of estrogen receptor-negative MDA-MB-435 and -positive MCF-7 human breast cancer cells in culture. *Asia Pacific J. Clin. Nutr.* **1997a**, 6, 41–45.

Guthrie, N.; Gabor, A.; Chambers, A.F.; Carroll, K.K. Inhibition of proliferation of estrogen receptor-negative MDA-MB-435 and -positive MCF-7 human breast cancer cells by palm oil tocotrienols and tamoxifen, alone and in combination. *J. Nutr.* **1997b**, 127, 544S-548S.

Guthrie, N.; Moffatt, M.; Chambers, A.F.; Spence, J.D.; Carroll, K.K. Inhibition of proliferation of human breast cancer cells by naringenin, a flavonoid in grapefruit. *Abst. Natl. Forum Breast Cancer*, Montreal, PQ, **1993**, p 118.

Guthrie, N.; Gapor, A.; Chambers, A.F.; Carroll, K.K. Inhibition of protein kinase C in MDA-MB-435 human breast cancer cells by palm oil tocotrienols. *Proc. Can. Fed. Biol. Soc.* **1996**, 39, Abstract 262.

Hansen, M.B.; Nielsen, S.E.; Berg, K. Re-examination and further development of a precise and rapid dye method for measuring cell growth/cell kill. *J. Immunol. Methods* **1989**, *119*, 203–210.

Hertog, M.G.L.; Hollman, P.C.H.; Katan, M.B.; Kromhout, D. Intake of potentially anticarcinogenic flavonoids and their determinants in adults in the Netherlands. *Nutr. Cancer* **1993**, 20, 21–29.

Hunter, D.J.; Spiegelman, D.; Adami, H.-O. et al. Cohort studies of fat intake and risk of breast cancer: a pooled analysis. *N. Engl. J. Med.* **1996**, *334*, 356–361.

Jordan, V.C. Tamoxifen and tumorigenicity: a predictable concern. *J. Natl. Cancer Inst.* **1995**, *87*, 623–626.

Jordan, V.C. Long-term tamoxifen treatment for breast cancer. *J. Natl. Cancer Inst.* **1995**, 87, 1176–1177.

Kuhnau, J. The flavonoids. A class of semi-essential food components: Their role in human nutrition. *World Rev. Nutr. Diet* **1976**, *24*, 117–191.

Levy, J.; Teuerstein, I.; Marbach, M.; Radian, S.; Sharoni, Y. Tyrosine protein kinase Activity in the DMBA-induced rat mammary tumor:inhibition by quercetin. *Biochem. Biophys. Res. Commun.* **1984**, 123, 1227–1233.

Lund, E. The research tide ebbs for the dietary fat hypothesis in breast cancer. *Epidemiology* **1994**, 5, 387–388.

Messina, M.J.; Persky, V.; Setchell, K.D.R.; Barnes, S. Soy intake and cancer risk:a review of the *in vitro* and *in vivo* data. *Nutr. Cancer* **1994**, 21, 113–131.

Messina, M.; Erdman, J.W. Jr., Eds. First international symposium on the role of soy in preventing and treating chronic disease. *J. Nutr.* **1995**, 125, 567s-569s.

Middleton, E. Jr.; Kandaswami, C. Potential health-promoting properties of citrus flavonoids. *Food Technol.* **1994a**, *48(11)*, 115–119.

Middleton, E.; Kandaswami, C. The impact of plant flavonoids on mammalian biology:implications for immunity, inflammation and cancer. In *The Flavonoids: Advances in Research Since 1986*; Harborne, J.B., Ed.; Chapman and Hall, London, **1994b**; pp. 619–652.

Mookerjee, B.K.; Lee, T.-P.; Lippes, H.A.; Middleton, E. Jr. Some effects of flavonoids on lymphocyte proliferative responses. *J. Immun. Pharmacol.* **1986**, 8, 371–392.

Nesaretnam, K.; Guthrie, N.; Chambers, A.F.; Carroll, K.K. Effect of tocotrienols on the growth of a human breast cancer cell line in culture. *Lipids* **1995**, 30, 1139–1143.

Peterson, G.; Barnes, S. Genistein inhibition of the growth of human breast cancer cells: independence from the estrogen receptors and the multi-drug resistance gene. *Biochem. Biophys. Res. Commun.* **1991**, *179*, 661–667.

Pelissero, C.; Lenczowski, M.J.; Chinzi, D.; Davail-Cuisset, B.; Sumpter, J.P.; Fostier, A. Effects of flavonoids on aromatase activity, an in vitro study. *J. Steroid Biochem. Mol. Biol.* **1996**, *57*, 215–223.

Raffauf, R.F.; Le Quesne, P.W. Antineoplastic activity and cytotoxicity of flavones, soflavones, and flavonones. *J. Nat. Prod.* **1979**, *42*, 85–91.

Ramanathan, R.; Tan, C.H.; Das, N.P. Cytotoxic effect of plant polyphenols and fat-soluble vitamins on malignant human cultured cells. *Cancer Lett.* **1992**, *62*, 217–224.

Sharoni, Y.; Teuerstein, I.; Levy, J. Phosphoinositide phosphorylation precedes growth in rat mammary tumors. *Biochem. Biophys. Res. Commun.* **1998**, *134*, 876–882.

Singhal, R.L.; Yeh, Y.A.; Prajda, N.; Olah, E.; Sledge, G.W. Jr.; Weber, G. Quercetin down regulates signal transduction in human breast carcinoma. *Biochem. Biophys. Res. Commun.* **1995**, *208*, 425–431.

So, F.V.; Guthrie, N.; Chambers, A.F.; Moussa, M.; Carroll, K.K. Inhibition of human breast cancer cell proliferation and delay of mammary tumorigenesis by flavonoids and citrus juices. *Nutr. Cancer* **1996a**, *26*, 167–181.

So, F.V. Dietary flavonoids and breast cancer: *in vitro, in vivo* and mechanism studies. M.Sc. Thesis, **1996b**; The Univ. of Western Ontario, London, Ontario.

So, F.V.; Guthrie, N.; Chambers, A.F.; Carroll, K.K. Inhibition of proliferation of estrogen receptor-positive MCF-7 human breast cancer cells by flavonoids in the presence and absence of excess estrogen. *Cancer Lett.* **1997**, *112*, 127–133.

Steinmetz, K.A.; Potter, J.D. Vegetables, fruit, and cancer. I. Epidemiology. *Cancer Causes Control* **1991a**, *2*, 325–357.

Steinmetz, K.A.; Potter, J.D. Vegetables, fruit, and cancer. II. Mechanisms. *Cancer Causes Control* **1991b**, *2*, 427–442.

Verma, A.K.; Johnson, J.A.; Gould, M.N.; Tanner, M.A. Inhibition of 7,12-dimethylbenz[a]anthracene- and n-nitrosomethylurea-induced rat mammary cancer by the dietary flavonol quercetin. *Cancer Res.* **1988**, *48*, 5754–5758.

Wattenburg, L.W. Inhibition of carcinogenesis by minor dietary constituents. *Cancer Res.* **1992**, 52, 2085s-2091s.

# 17

# INHIBITION OF NEOPLASTIC TRANSFORMATION AND BIOAVAILABILITY OF DIETARY FLAVONOID AGENTS

Adrian A. Franke, Robert V. Cooney, Laurie J. Custer, Lawrence J. Mordan, and Yuichiro Tanaka

Cancer Research Center of Hawaii
1236 Lauhala Street
Honolulu, Hawaii 96813

## 1. ABSTRACT

Evaluation of unknown biological effects of chemicals including food plant products requires the assessment of bioactivity and bioavailability. Epidemiologic studies show consistently a cancer protective effect of fruit and vegetable consumption, but there is little understanding of which phytochemicals account for this observation. Commonly studied antioxidant micronutrients are less consistently correlated with cancer protection relative to the food groups themselves, suggesting that other phytochemicals or a combination of food products play key roles in preventing cancer. We investigated the effects of the predominant dietary flavonoids and isoflavonoids at inhibiting neoplastic transformation induced by 3-methylcholanthrene in C3H 10T1/2 murine fibroblasts. We found that most phenolic agents tested were equal to or superior to known chemopreventive agents such as carotenoids or vitamins in effectiveness. Hesperetin, hesperidin and catechin were the most potent agents among the flavonoids tested, inhibiting transformation completely when applied at 1.0 µM after exposure to the carcinogen. Structure-activity comparison revealed that among the compounds tested, flavonoids with a vicinal diphenol structure in ring 'B' and a saturated 'C' ring exhibited the strongest effects. Most agents tested showed dose-dependent patterns. Interestingly, the soy isoflavonoids were weakly active except when applied in combination, suggesting a synergistic effect. In addition, HPLC techniques were developed for determining the bioavailability of isoflavonoids in human biological fluids including urine, plasma and breast milk. We observed a relatively fast absorption, distribution and elimination of isoflavonoids including a biphasic pattern probably due to enterohepatic circulation. Total peak isoflavone levels in urine, plasma and in breast milk were found to be 60 µM, 2 µM and 0.2 µM, respectively and were reached 8–12 hours after consumption of soy foods. Levels detected in human body fluids were found to be highly active at inhibiting neoplastic transformation, especially considering synergistic effects observed for combinations of daidzein and genistein, the predominant isoflavonoids occurring in soy foods.

## 2. INTRODUCTION

Epidemiologic studies show consistently a negative correlation between consumption of fruits and vegetables and the incidence of a variety of cancers (Block et al., 1992). However, it is little understood which phytochemical(s) or mechanism(s) account for this effect, since commonly studied chemopreventive micronutrients are not as consistently correlated with cancer preventive effects compared to the food groups themselves. A case-control study in Hawaii (Le Marchand et al., 1989) showed that intake of ß-carotene and food groups rich in particular carotenoids were negatively associated with lung cancer. Subsequently, using recently published food composition data for carotenoids (Mangels et al., 1993), dose-dependent inverse associations for α-carotene, ß-carotene and lutein were observed (Le Marchand et al., 1993). However, the inverse association for all vegetables as a group was stronger than those for the three carotenoids or for total carotenoid intake. Although flavonoids, isoflavonoids and other phenolic agents are known to occur in high levels in many fruits and vegetables, limited knowledge is available regarding their role in preventing cancer through dietary means (reviewed in Middleton and Kandaswami, 1994). Flavonoids have been reported to interfere with various stages of cancer development, for example, by scavenging free radicals and thus detoxifying carcinogens (Stavric, 1995; Das, 1990; Huang and Ferraro, 1992) and by inhibiting tumor growth (Middleton and Kandaswami, 1994). However, only a limited number of phenolic dietary agents were tested regarding antiproliferative effects (Hirano et al., 1994; Peterson and Barnes, 1991; Peterson and Barnes, 1993; Hirano et al., 1989; Fotsis et al., 1993; Schweigerer et al., 1992) and regarding interfering with cancer-initiation processes including cellular mechanisms of action (reviewed in Middleton and Kandaswami, 1994). Carcinogenesis is believed to be a multistep process involving repeated DNA damage, initiation, proliferation, clonal selection and progression (Arnold et al., 1995). For selecting and ranking of potential chemopreventive chemicals, inhibition of transformation has been recommended due to relatively low costs, high speed, inclusion of the various stages of carcinogenesis and most importantly, due to the good prediction of in vivo activity (Arnold et al., 1995). Only selected phenolic agents have been used in previous studies to determine effects at inhibition of transformation (Arnold et al., 1995) and only concentrations greater than 10 µM were applied (Leighton et al., 1992); these are levels greater than those reached by dietary exposure (Herrmann, 1988). In order to identify candidate dietary agents with cancer protective effects, we applied a transformation assay system using C3H 10T1/2 mouse fibroblasts and 3-methyl cholanthrene as carcinogen and screened flavonoids occurring predominantly in foods. We tested those flavonoids known to be found in humans systemically (Franke et al., 1997; Coward et al., 1996; Adlercreutz et al., 1993; Xu et al., 1994; Fuhr and Kummert, 1995; Hollman et al., 1996) and also those occurring in non-detectable or trace levels systemically (Lee et al., 1995). Although these latter agents are marginally or not absorbed by humans, they may still exhibit a beneficial effect in the digestive system and therefore, deserve consideration. We tested the selected agents at physiological concentrations (0.1–10.0 µM) as occurring in humans after dietary exposure. We hypothesize that the system applied is a valid tool for preliminary evaluation of anticancer effects reflecting in vivo situations.

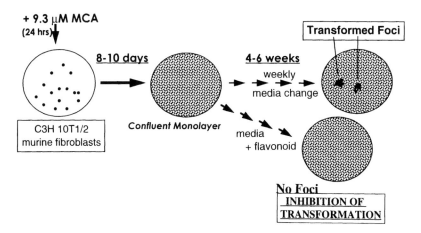

**Figure 1.** Performance of transformation assay. Mouse fibroblasts were seeded at a density of 1000 cells per dish and treated the next day with 9.3 μM 3-methyl cholanthrene (MCA) for 24 hours. A negative control group lacking MCA treatment was also maintained. The medium with the carcinogen was then replaced with fresh medium, and the cells were grown to confluence during 8–14 days. Thereafter cells were incubated for 5 weeks with weekly replenishment of medium containing either 0.1–10 μM flavonoids, or carcinogen alone (positive control). Four to six weeks after confluence, all cultures were fixed, stained, and scored for the presence of types II and III foci according to the guidelines described by Reznikoff et al., (1973).

## 3. RESULTS AND DISCUSSION

### 3.1. Performance of Transformation Assay

Inhibition of neoplastic transformation was performed as shown in Figure 1 and described in detail previously (Cooney et al., 1993). C3H 10T1/2 murine fibroblasts were seeded at a density of 1000 cells per 60 mm culture dish followed by treating the next day with 3-methyl cholanthrene (MCA) at a final concentration of 9.3 μM for 24 hours. The medium with the carcinogen was then replaced weekly with fresh medium and the cells were grown to confluence during 8–14 days. Thereafter, cells were incubated for four to six weeks with medium containing 0.1–10 μM flavonoids as listed in Figure 2. The flavonoid-containing culture medium was replenished weekly. Negative control cultures were treated with vehicle alone whereas positive controls were treated once with the carcinogen only. Four to six weeks after confluence, all cultures were fixed, stained, and scored for the presence of types II and III foci according to the guidelines described by Reznikoff et al., (1973). An average of 12 dishes per compound and concentration were used and final results are reported as the mean # of transformed foci per dish. A total of four batches of this assay was performed and replicate results are reported in Figure 3 as mean of all assays with error bars indicating standard deviation.

### 3.2. Transformation Assay Results

The transformation assay involved exposure and removal of the carcinogen prior to flavonoid exposure and revealed that most phenolic agents tested were equal to or superior to known chemopreventive agents in effectiveness. This includes retinoids (data in Mordan et al., 1985) and vitamin E compounds (Cooney et al., 1993) which we used for

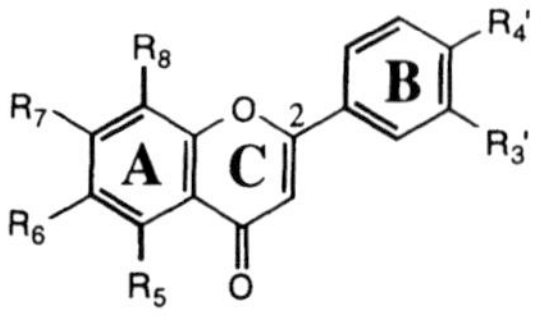

| FLAVONES | code | R3' | R4' | R5 | R6 | R7 | R8 |
|---|---|---|---|---|---|---|---|
| CHRYSIN | CHR | H | H | H | H | H | H |
| APIGENIN | A | H | OH | H | H | H | H |
| TANGERITIN | TAN | H | OCH$_3$ | OCH$_3$ | OCH$_3$ | OCH$_3$ | OCH$_3$ |
| NOBILETIN | NOB | OCH$_3$ | OCH$_3$ | OCH$_3$ | OCH$_3$ | OCH$_3$ | OCH$_3$ |

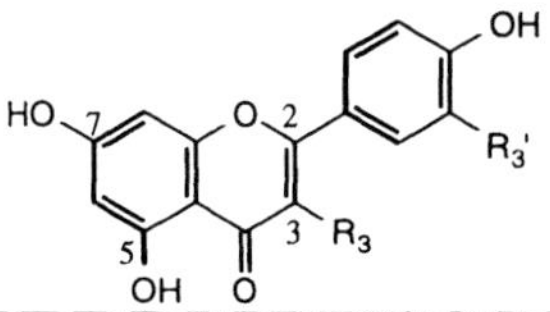

| FLAVONOLS | code | R3 | R3' |
|---|---|---|---|
| KÄMPFEROL | K | OH | H |
| QUERCETIN | Q | OH | OH |
| RUTIN | RUT | O-rutinosyl | OH |

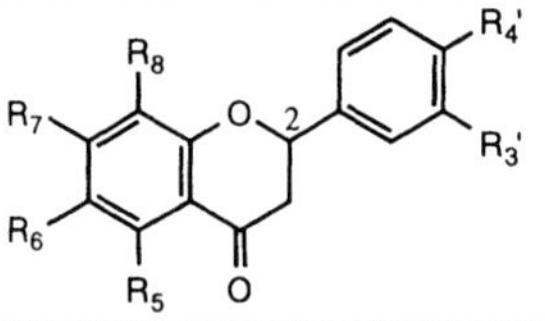

| FLAVANONES | code | R3' | R4' | R5 | R6 | R7 | R8 |
|---|---|---|---|---|---|---|---|
| NARINGENIN | NE | H | OH | OH | H | OH | H |
| NARINGIN | NI | H | OH | OH | H | O-neo-hesperi-dosyl | H |
| ERIODICTYOL | E | OH | OH | OH | H | OH | H |
| HESPERETIN | HT | OH | OCH$_3$ | OH | H | OH | H |
| HESPERIDIN | HD | OH | OCH$_3$ | OH | H | O-rutinosyl | H |

Figure 2. Molecular structure of compounds according to flavonoid class.

comparison in our model (Figure 3). Hesperetin, hesperidin and catechin were observed to be the most effective agents tested inhibiting transformation almost completely (≥98%). Twelve among the 19 flavonoids tested as well as caffeic acid (Figure 3A) showed higher activity than vitamin E when applied at 1.0 µM. Only 4 flavonoids showed weaker activity than vitamin E. Chrysin was the only agent tested without any effect on transformation. Structure-activity comparison revealed that among the compounds tested, flavonoids with a vicinal diphenol structure in ring 'B' and a saturated ring 'C' (Figure 2) exhibited the strongest effects. Isoflavones seem to possess weaker activity compared to flavones with the same substitution pattern. Interestingly, the soy isoflavones daidzein and genistein were weakly active except when applied in combination, suggesting a synergistic effect. A synergistic effect was also observed for the citrus flavones nobiletin and tangeritin. Both agents showed already relatively high activity (ca. 80%) when applied individually at 1.0 µM. This activity was not altered when the

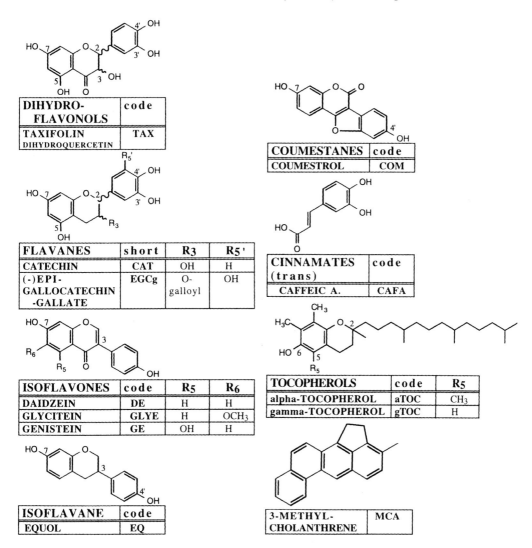

Figure 2. (*Continued*)

combined dose was reduced to 0.3 μM equaling a 3.3-fold dose reduction. Also, the effects at inhibiting transformation described above were found to be dose-dependent for most agents tested (Figure 3B). Only genistein showed toxic effects when applied at 10 μM. Rutin and epi-gallocatechingallate (EGCg) exhibited a biphasic pattern with maximum effects observed at 0.3 μM. Biphasic and/or opposing behavior of flavonoids have been reported in previous studies and was attributed to the variability of the test system utilized, to the different concentrations applied, and/or to the structure of the molecule used (Jung et al., 1983; Li et al., 1994). Further studies elucidating mechanisms, which potentially explain the effects observed, are ongoing in our lab. Currently, we are investigating cellular bioavailability and specific potencies of flavonoids regarding inhibiting enzyme systems involved in carcinogenesis.

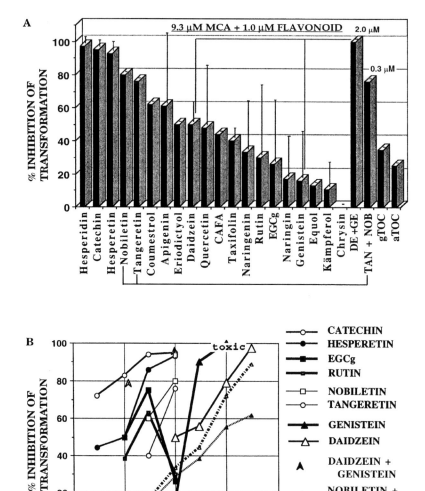

**Figure 3.** Effects of flavonoid agents at inhibiting neoplastic transformation. A: Compounds tested at a final concentration of 1.0 μM except for combinations of daidzein and genistein (DE+GE=2.0 μM) and tangeritin and nobiletin (TAN+NOB=0.3 μM). Error bars indicate standard deviations observed from 2–4 replicate experiments. B: Analytes tested at various concentrations except for combinations of daidzein and genistein (DE+GE=2.0 μM) and tangeritin and nobiletin (TAN+NOB=0.3 μM).

## 3.3. Availability of Isoflavones in Human Urine, Plasma, and Breast Milk after Soy Consumption

In order to determine the availability of cancer preventive dietary flavonoids in humans, we applied HPLC techniques measuring levels of isoflavonoids in foods (Franke et al., 1994) and in human urine, plasma and breast milk after soy consumption (Franke and Custer, 1994; Franke et al., 1998; Franke and Custer, 1996). In the past, GC/MS was commonly utilized for bioavailability studies (Cassidy et al., 1994; Adlercreutz et al., 1991; Kelly et al., 1993; Adlercreutz et al., 1993; Kurzer et al., 1995; Adlercreutz et al., 1995). However, HPLC methods require fewer steps for sample preparation and analysis, and demand less technician time and less expensive instrumentation compared to GC/MS. Although GC/MS techniques result in higher sensitivity, our HPLC system showed detection limits sufficiently low for the desired applications (1.09, 0.53, 3.28 and 1.00 pmoles per 20 µL injection for daidzein, genistein, equol and $O$-desmethylangolensin, respectively). These limits could be reduced three to five fold by coulometric detection and even further, by extended purification and concentration through partitioning with ethyl acetate. Most importantly, HPLC procedures (Franke and Custer, 1994; Xu et al., 1994) allowed rapid quantitation of isoflavonoid aglycones and also their conjugates (glucosides, malonylglucosides, acetylglucosides, glucuronides, sulfates), their human metabolites (equol, $O$-desmethylangolensin) and also a variety of other phytoestrogens including coumestrol, formononetin and biochanin-A in a single chromatographic run (Franke et al., 1998).

Among the vast variety of dietary flavonoid agents (Herrmann, 1988), we chose to focus on the bioavailability of isoflavonoids due to established knowledge regarding isoflavone occurrence in soy foods (Franke et al., 1994; Barnes et al., 1994; Kudou et al., 1991; Wang and Murphy, 1994), due to known isoflavone metabolism in humans (Price and Fenwick, 1985; Setchell, 1985) and due to required support of epidemiologic studies assessing the potential cancer protective role of soya or diets containing isoflavones (Middleton and Kandaswami, 1994). Isoflavonoids were analyzed from urine, plasma and breast milk using the HPLC system described previously after enzymatic hydrolysis and solid phase (C18 material) or solvent extraction (Franke et al., 1998). Peaks were separated using a NovaPak C18 column (150 x 3.9 mm; 4.0 µm), eluted by a linear gradient consisting of $MeOH/MeCN/CH_2Cl_2=10:5:1$ and 10% aqueous acetic acid, and monitored by diode-array and coulometric (+500mV) detection. Signal assignment was also confirmed by co-chromatography with authentic standards and by GC/MS analysis of trimethylsilylated extracts (Franke and Custer, 1996).

Dietary isoflavone exposure was achieved through consumption of roasted soybeans containing 0.79 mg/g daidzein, 0.89 mg/g genistein and 0.17 mg/g glycitein (Franke et al., 1994; Franke and Custer, 1996). Levels in plasma or urine (Franke et al., 1998) and in breast milk or urine (Franke and Custer, 1996) were monitored after exposure to a single dose of 20 g roasted soybeans (Figure 4A) and to consecutive doses of 5, 10 and 20 g of roasted soybeans, respectively (Figure 4B).

Isoflavone metabolites were not observed in analyzed plasma samples probably due to the presence of $O$-desmethylangolensin that was below detection limit and due to the absence of equol. We concluded this because these analytes are concentrated in urine and $O$-desmethylangolensin was detected in urine while equol was not. In contrast, parent isoflavones were detected in these plasma samples. Although timewise slightly shifted, similar response patterns of urinary and plasma isoflavone levels were observed after exposing a male individual to a single dose of 20 g roasted soybeans (Figure 4B). Isoflavone absorption and distribution was fast with peak levels reached 8–12 hours after soy consumption (Figure 4A) and fast elimination resulted in exposure of isoflavonoids for approximately 48 hours in this experiment. Similarly, maximum milk levels were reached 10–14 hours after soy intake followed by baseline levels reached 2–4 days later depending on the dose (Figure 4B). Also, the isoflavone patterns in breast milk were similar to those in urine except, with a slight delay. This is in good agreement with other micronutrients or drugs showing a faster urinary excretion than secretion into milk (Lawrence, 1994). Most importantly, milk levels of genistein conjugates were found to be higher than those of

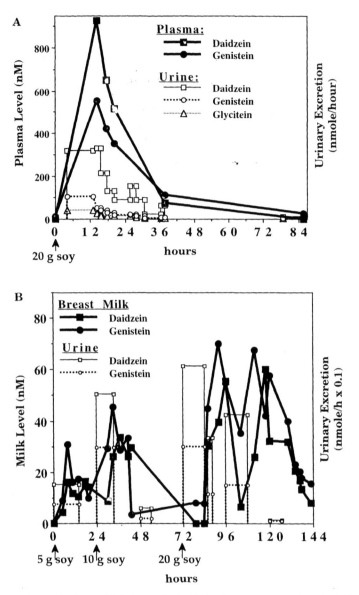

Figure 4. Concentrations of isoflavonoids in human body fluids after soy consumption. A: Isoflavonoid plasma levels and urinary excretion rates after consumption of 20 g roasted soybeans in a male individual. B: Isoflavonoid milk levels and urinary excretion rates after consumption of 5, 10 and 20 g roasted soybeans in a female individual.

daidzein conjugates as was observed for plasma levels for this individual (genistein to daidzein ratio=2:1). In contrast, urinary levels of genistein conjugates were found to be smaller than those of daidzein conjugates which is in agreement with other studies (Adlercreutz et al., 1991; Xu et al., 1994; Hutchins et al., 1995b; Hutchins et al., 1995a; Cruz et al., 1994; Baird et al., 1995; Kelly et al., 1993; Franke and Custer, 1994). This might be due to the higher polarity of daidzein conjugates favoring its urinary excretion and leading to increasing genistein to daidzein ratios in blood. Secretory processes will consequently result in isoflavone patterns in breast milk being similar to those in blood. A biphasic flavonoid elimination pattern observed, especially monitoring milk levels after 10 and 20 g soybean consumption (Figure 4B), has already been reported when animal plasma was analyzed after treatment with the flavane hydroxyfarrerol (Marzo et al., 1990) or the flavone baicalin (Wakui et al., 1992) and most importantly, in human plasma (Barnes, 1994) and urine (Franke and Custer, 1994) after soy intervention. This biphasic phenomenon was suggested to be due to enterohepatic circulation (Xu et al., 1994; Sfakianos et al., 1996), a process flavonoids frequently undergo (Griffith, 1982). Therefore, enterohepatic circulation might also explain the biphasic pattern observed in human milk. The lack of detectable isoflavones after extraction without hydrolysis using sulfatase or glucuronidase suggested that the majority of isoflavones occur as glucuronide and/or sulfate conjugates in human urine, plasma and milk. None of the other soy isoflavones or metabolites were detected in human milk extracts using GC/MS/SIM analysis suggesting their absence in human milk after soybean consumption by this particular individual. A preferential excretion of the metabolites over the parent isoflavones is suggested by the isoflavone to metabolite ratio observed to be much higher in plasma than in urine or feces (Adlercreutz et al., 1993; Adlercreutz et al., 1995). Because milk is produced by secretory processes from blood (Lawrence, 1994), the low plasma concentrations of isoflavonoid metabolites might explain their absence in milk.

Great inter-individual variation of plasma isoflavone recovery (5–10%) was observed by analyzing 6 plasma samples from three individuals (data not shown). Genistein-to-daidzein ratios in plasma were also highly variable (0.5–2.1) and consistent only within an individual similar to earlier findings from urinary patterns (Franke and Custer, 1994). The similarity of urinary and plasma isoflavone patterns suggests that future epidemiological and clinical trials assessing the protective role of isoflavonoids or soy intake at inhibiting chronic diseases would not necessarily require plasma but could design studies using non-invasive protocols posing less risks to participants and facilitating easier subject recruitment, better compliance and faster approval from study reviews.

## 4. CONCLUSION

In summary, we showed that dietary flavonoids inhibit neoplastic transformation. Depending on the dose and structure of the molecule, this inhibition ranges from no effect to complete inhibition when applied at 1.0 μM, a concentration considered to be physiological after dietary exposure. Among the compounds tested, flavonoids with a vicinal diphenol structure in ring 'B' and most importantly, a saturated pyran 'C' ring, exhibited the strongest effects. Isoflavones showed a weaker inhibitory effect on transformation than flavones when comparing compounds with identical substitution pattern. However, isoflavones (and flavones) applied in combination showed extremely potent inhibition of transformation suggesting a synergistic effect.

Furthermore, we established an HPLC method for isoflavonoid analysis from foods and from human body fluids applicable in future epidemiologic and clinical studies for assessing the role of soy or (iso-)flavonoids in preventing chronic diseases. After moderate soy challenge, we found mean total peak isoflavone levels of 60 μM in urine, 2 μM in plasma and 0.2 μM in breast milk. Although demonstrated exclusively in carcinogen or cytokine-induced cell and animal models (Messina et al., 1994), the anticancer properties of genistein and daidzein are particularly intriguing when considered in combination with the results presented. The soy isoflavones daidzein and genistein were shown to inhibit neoplastic transformation completely at concentrations present in human biological fluids after moderate soy intake. In addition, cancer incidence and severity is reported to be significantly reduced when newborn animals are treated with only three single doses of genistein (Lamartiniere et al., 1995). The data presented may suggest that infants breast-feeding on mothers consuming soy foods are protected from cancer due to exposure of genistein and daidzein to the offspring. This effect may take place at a very early and most critical developmental period and might protect the individual throughout life (Lamartiniere et al., 1995). It is conceivable that the isoflavone conjugates obtained from mother's milk are more bioavailable to the newborn child compared to those from soy foods. The young infant might not be able to absorb isoflavones from soy foods as efficiently as from breast milk due to incompletely developed gut flora (Setchell et al., 1987; Cruz et al., 1994), thereby preventing hydrolysis of acylated and non-acylated isoflavone glucosides present in soy foods. Biotransformation and bioavailability studies of soy isoflavones in the baby are required to further explore the cancer preventive effects of these compounds in humans.

## ACKNOWLEDGMENTS

We are very grateful for the most generous support of the participants in the intervention trials. We also thank Dr. Hans Geyer, German Sports University, Köln (Germany) for the skillful performance of the GC/MS measurements and Dr. Bill Widmer, Department of Citrus, Lake Alfred, FL, and Dr. Y. Hara, Mitsui Norin Co. Ltd., Fujieda, Japan, for providing authentic flavonoid standards.

## REFERENCES

Adlercreutz, H.; Fotsis, T.; Bannwart, C.; Wahala, K.; Brunow, G.; Hase, T. Isotope dilution gas chromatographic-mass spectrometric method for the determination of lignans and isoflavonoids in human urine, including identification of genistein. Clin Chim Acta 1991, 199, 263–278.

Adlercreutz, H.; Fotsis, T.; Lampe, J.; Wahala, T.; Makela, T.; Brunow, G.; Hase, T. Quantitative determination of lignans and isoflavonoids in plasma of omnivorous and vegetarian women by isotope dilution gas chromatography-mass spectrometry. Scand J Clin Lab Invest 1993, 53, 5–18.

Adlercreutz, H.; Fotsis, T.; Kurzer, M. S.; Wähälä, K.; Mäkelä, T.; Hase, T. Isotope dilution gas chromatographic-mass spectrometric method for the determination of unconjugated lignans and isoflavonoids in human feces, with preliminary results in omnivorous and vegetarian women. Anal. Biochem. 1995, 225, 101–108.

Arnold, J. T.; Wilkinson, B. P.; Sharma, S.; Steele, V. E. Evaluation of chemopreventive agents in different mechanistic classes using a rat tracheal epithelial cell culture transformation assay. Cancer Research 1995, 55, 537–543.

Baird, D. D.; Umbach, D. M.; Lansdell, L.; Hughes, C. L.; Setchell, K. D. R.; Weinberg, C. R.; Haney, A. F.; Wilcox, A. J.; McLachlan, J. A. Dietary intervention study to assess estrogenicity of dietary soy among postmenopausal women. J. Clin. Endocrinol. Metab. 1995, 80, 1685–1690.

Barnes, S. Serum conjugated and unconjugated isoflavones by capillary reversed-phase HPLC-electrospray ionization-mass Spectrometry. NCI Meeting on "Dietary Phytoestrogens: Cancer Cause or Prevention?"; Sept.21–23, 1994; Herndon, VA; Abstract #1, 1994.

Barnes, S.; Kirk, M.; Coward, L. Isoflavones and their conjugates in soy foods: extraction conditions and analysis by hplc-mass spectrometry. J. Agr. Food Chem. 1994, 42, 2466–2474.

Block, G.; Patterson, B.; Subar, A. Fruit, Vegetables, and Cancer Prevention: A Review of the Epidemiological Evidence. Nutr. Cancer 1992, 18, 1–29.

Cassidy, A.; Bingham, S.; Setchell, K. D. R. Biological effect of a diet of soy protein rich in isoflavones on the menstrual cycle of premenopausal women. Am J Clin Nutr 1994, 60, 333–340.

Cooney, R. V.; Franke, A. A.; Harwood, P. J.; Hatch-Pigott, V.; Custer, L. J.; Mordan, L. J. Gamma-tocopherol detoxification of nitrogen dioxide: Superiority to alpha-tocopherol. Proc. Natl. Acad. Sci. USA 1993, 90, 1771–1775.

Coward, L.; Kirk, M.; Albin, N.; Barnes, S. Analysis of plasma isoflavones by reversed-phase HPLC-multiple reaction ion monitoring-mass spectrometry. Clin. Chim. Acta 1996, 247, 121–142.

Cruz, M. L. A.; Wong, W. W.; Mimouni, F.; Hachey, D. L.; Setchell, K. D. R.; Klein, P. D.; Tsang, R. C. Effects of infant nutrition on cholesterol synthesis rates. Pediatric Research 1994, 35, 135–140.

Das, N.P. Plant flavonoids in biology and medicine III, NY:A.R. Liss, 1990.

Fotsis, T.; Pepper, M.; Adlercreutz, H.; Fleischmann, G.; Hase, T.; Montesano, T.; Schweigerer, L. Genistein, a dietary-derived inhibitor of in vitro aniogenesis. Proc. Natl. Acad. Sci. 1993, 90, 2690–2694.

Franke, A. A.; Custer, L. J. High-performance liquid chromatography assay of isoflavonoids and coumestrol from human urine. J Chromatogr. B 1994, 662, 47–60.

Franke, A. A.; Custer, L. J.; Cerna, C. M.; Narala, K. K. Quantitation of phytoestrogens in legumes by HPLC. J Agric Food Chem 1994, 42, 1905–1913.

Franke, A. A.; Custer, L. J. Daidzein and genistein concentrations in human milk after soy consumption. Clin. Chem. 1996, 42, 955–964.

Franke, A. A.; Custer, L. J.; Wang, W.; Shi, S. J. HPLC analysis of isoflavonoids and other phenolic agents from foods and from human fluids. Proc Soc Exp Biol Med 1998, 217, 263–273.

Fuhr, U.; Kummert, A. The fate of naringenin in humans: a key to grapefruit juice-drug interactions? Clin. Pharmacol. Ther. 1995, 58, 365–373.

Griffith, L.A. Mammalian metabolism of flavonoids. In: The Flavonoids: Advances in Research, edited by Harborne, J.B. and Mabry, T.J. London/New York: Chapman and Hall, 1982, p. 681–718.

Herrmann, K. On the occurrence of flavonol and flavone glycosides in vegetables. Z. Lebensm. Unters. Forsch. 1988, 186, 1–5.

Hirano, T.; Oka, K.; Akiba, M. Antiproliferative effects of synthetic and naturally occurring flavonoids on tumor cells of the human breast carcinoma cell line, ZR-75-1. Res Comm Chem Path Pharm 1989, 64, 69–78.

Hirano, T.; Gotoh, M.; Oka, K. Natural flavonoids and lignans are potent cytostatic agents against human leukemic HL-60 cells. Life Sci. 1994, 55, 1061–1069.

Hollman, P. C. H.; Gaag, M.; Mengelers, M. J. B.; Van Trijp, J. M. P.; de Vries, J. H. M.; Katan, M. B. Absorption and disposition kinetics of the dietary antioxidant quercetin in man. Free Rad. Biol. Med. 1996, 21, 703–707.

Huang, M.-T. and Ferraro, T. Phenolic compounds and their effects on Health II. In: Phenolic Compounds and Cancer Prevention, edited by Huang, M.-T., Ho, C.-T. and Lee, C.Y. Washington, D.C.: American Cancer Society Press, 1992, p. 9–34.

Hutchins, A. M.; Lampe, J. W.; Martini, M. C.; Campbell, D. R.; Slavin, J. L. Vegetables, fruits, and legumes: Effect on urinary isoflavonoid phytoestrogen and lignan excretion. J. Am. Diet Assoc. 1995a, 95, 769–774.

Hutchins, A. M.; Slavin, J. L.; Lampe, J. W. Urinary isoflavonoid phytoestrogen and lignan excretion after consumption of fermented and unfermented soy products. J. Am. Diet Assoc. 1995b, 95, 545–551.

Jung, G.; Hennings, G.; Pfeifer, M.; Bessler, W. G. Interaction of metal-complexing compounds with lymphocytes and lymphoid cell lines. Molecular Pharmacology 1983, 23, 698–702.

Kelly, G. E.; Nelson, C.; Waring, M. A.; Joannou, G. E.; Reeder, A. Y. Metabolites of dietary (soya) isoflavones in human urine. Clinica Chimica Acta 1993, 223, 9–22.

Kudou, S.; Fluery, Y.; Welti, D.; Magnolato, D.; Uchida, T.; Kitamura, K.; Okubo, M. Malonyl isoflavone glycosides in soybean seeds (*Glycine max* MERRILL). Agric Biol Chem 1991, 55, 2227–2233.

Kurzer, M. S.; Lampe, J. W.; Martini, M. C.; Adlercreutz, H. Fecal lignan and isoflavonoid excretion in premenopausal women consuming flaxseed power. Cancer Epidemiology, Biomarkers & Prevention 1995, 4, 353–358.

Lamartiniere, C. A.; Moore, J.; Holland, M.; Barnes, S. Neonatal genistein chemoprevents mammary cancer. Proc. Soc. Exp. Bio. Med. 1995, 208, 120–123.

Lawrence, R.A. Breastfeeding, St. Louis:Mosby, 1994. Ed. 4

Le Marchand, L.; Yoshizawa, C. N.; Kolonel, L. N.; Hankin, J. H.; Goodman, M. T. Vegetable consumption and lung cancer risk: A population-based case-control study in Hawaii. J. Natl. Cancer Institute 1989, 81, 1158–1164.

Le Marchand, L.; Hankin, J. H.; Kolonel, L. N.; Beecher, G. R.; Wilkens, L. R.; Zhao, L. P. Intake of Specific Carotenoids and Lung Cancer Risk. Cancer Epidemiol., Biomarkers, & Prev. 1993, 2, 183–187.

Lee, M. -J.; Wang, Z. -Y.; Li, H.; Chen, L.; Sun, Y.; Gobbo, S.; Balentine, D. A.; Yang, C. S. Analysis of plasma and urinary tea polyphenols in human subjects. Cancer Epidemiology, Biomarkers & Prevention 1995, 4, 393–399.

Leighton, T., Ginther, C., Fluss, L., Harter, W.K., Cansado, J. and Notario, V. Molecular characterization of quercetin and quercetin glycosides in Allium vegetables. In: Phenolic compounds in foods and their effects on Health II, edited by Huang, M.-T., Ho, C.-T. and Lee, C.Y. Washington, DC: American Chemical Society, 1992, p. 220–238.

Li, Y.; Wang, E.; Patten, C. J.; Chen, L.; Yang, C. S. Effects of flavonoids on cytochrome P450-dependent acetaminophen metabolism in rats and human liver microsomes. Drug Metab. Dispo. 1994, 22, 566–571.

Mangels, A. R.; Holden, J. M.; Beecher, G. R.; Forman, M. R.; Lanza, E. Carotenoid Content of Fruits and Vegetables: An Evaluation of Analytic Data. Am. Diet. Assoc. 1993, 93, 284–296.

Marzo, A.; Arrigoni Martelli, E.; Bruno, G. Assay of hydroxyfarrerol in biological fluids. J. Chromatography 1990, 535, 255–261.

Messina, M.; Persky, V.; Setchell, K. D. R.; Barnes, S. Soy intake and cancer risk: A review of in vitro and in vivo data. Nutrition and Cancer 1994, 21, 113–131.

Middleton, E., Jr. and Kandaswami, C. The impact of plant flavonoids on mammalian biology: Implications for immunity, inflammation and cancer. In: The Flavonoids: Advances in Research Since 1986, edited by Harborne, J.B. London: Chapman & Hall, 1994, p. 619–652.

Mordan, L. J.; Bergin, L. M.; Budnick, J. E. L.; Meegan, R. R.; Bertram, J. S. Isolation of methylcholanthrene-initiated C3H/10T1/2 cells by inhibiting neoplastic progression with retinyl acetate. Carcinogenesis 1985, 3, 279–285.

Peterson, G.; Barnes, S. Genistein and biochanin A inhibit the growth of human prostate cancer cells but not epidermal growth factor receptor tyrosine autophosphorylation. The Prostate 1993, 22, 335–345.

Peterson, T. G.; Barnes, S. Genistein inhibition of the growth of human breast cancer cells: indepdnence from estrogen receptors and the multi-drug resistance gene. Biochem. Biophys. Res. Comm. 1991, 179, 661–667.

Price, K. R.; Fenwick, G. R. Naturally occurring oestrogens in foods-A review. Food Additives & Contaminants 1985, 2, 73–106.

Reznikoff, C. A.; Bertram, J. S.; Brankow, D. W.; Heidelberger, C. Quantitative and qualitative studies on chemical transformation of clones C3H mouse embryo cells sensitive to postconfluence inhibiton of cell division. Cancer Res. 1973, 33, 3239–3249.

Schweigerer, L.; Christeleit, K.; Fleischmann, G.; Adlercreutz, H.; Wahala, K.; Hasa, T.; Schwab, M.; Ludwig, R.; Fotsis, T. Identification in human urine of a natural growth inhibitor for cells derived from solid paediatric tumours. Eur J Clin Invest 1992, 22, 260–264.

Setchell, K.D.R. in McLachlan, J.A. (editor) Estrogens in the environment, New York:Elsevier, 1985. pp. 69–83.

Setchell, K. D. R.; Welch, M. B.; Lim, C. K. HPLC analysis of phytoestrogens in soy protein preparations with ultraviolet, electrochemical and thermospray mass spectrometric detection. J Chromatogr 1987, 386, 315–323.

Sfakianos, J.; Coward, L.; Kirk, M.; Barnes, S. Intestinal uptake and biliary excretion of the isoflavone genistein in the rat: evidence for its enterohepatic circulation. J. Nutr. 1996 (in press).

Stavric, B. The role of polyphenols as chemopreventers. Polyphenols Actualites 1995, 13, 19–21.

Wakui, Y.; Yanagisawa, E.; Ishibashi, E.; Matsuzaki, Y.; Takeda, S.; Sasaki, H.; Aburada, M.; Oyama, T. Determination of baicalin and baicalein in rat plasma by high- performance liquid chromatography with electrochemical detection. J. Chromatogr. 1992, 575, 131–136.

Wang, H.; Murphy, P. A. Isoflavone content in commercial soybean foods. J Agric Food Chem 1994, 42, 1666–1673.

Xu, X.; Wang, H. -J.; Murphy, P. A.; Cook, L.; Hendrich, S. Daidzein is a more bioavailable soymilk isoflavone than is genistein in adult women. J. Nutr. 1994, 124, 825–832.

# 18

# FLAVONOIDS AS HORMONES

## A Perspective from an Analysis of Molecular Fossils

Michael E. Baker

Department of Medicine, 0623B
University of California, San Diego
9500 Gilman Drive
La Jolla, California 92093-0623

## 1. ABSTRACT

Although for centuries plants have been known to have hormone-like actions in humans, the mechanism(s) by which plant-derived compounds act in humans is still being elucidated, a goal that has assumed more importance due to interest in the protective actions of fruits and vegetables in diseases such as cancer. Here I use the "molecular fossil record" of amino acid sequences of proteins involved in regulating the actions steroids, retinoids, thyroid hormone and prostaglandins to propose some mechanisms by which flavonoids in fruits and vegetables can have hormone-like actions in humans. I focus on: i) hormone receptors that bind to DNA and regulate gene transcription and ii) the enzymes that regulate the concentrations of these hormones. Comparative analyses of amino acid sequences show that nuclear receptors for steroids, retinoids, thyroid hormone and prostaglandins in humans and insects are descended from a common ancestor. Similar analyses of dehydrogenases that regulate the concentrations of steroids, retinoids and prostaglandins reveal strong sequence similarity to enzymes in plants, insects, fungi, and bacteria. The similarity is sufficient to suggest that some compounds that bind receptors or enzymes in invertebrates, plants or unicellular organisms may also bind to mammalian homologs that are involved in endocrine physiology. Among the phytochemicals that are candidates for such activity are flavonoids because they are involved in plant-insect and plant-bacteria interactions and have some structural and chemical similarities to steroids, retinoids, thyroid hormone, prostaglandins and fatty acids. These similarities and the kinship of human, plant, insect and bacterial proteins involved in signal transduction provide a conceptual framework for investigating flavonoids for hormone-like actions in humans. Understanding these modes of action may be useful in developing protocols for preventing hormone-dependent diseases such as breast and prostate cancer.

*Flavonoids in the Living System*, edited by Manthey and Buslig
Plenum Press, New York, 1998.

## 2. INTRODUCTION

For thousands of years, plants have been used as medicines (Monder, 1991; Davis and Morris, 1991; Duke, 1992; Riddle and Estes, 1992; Baker, 1995). The Assyrians, Babylonians, and Egyptians used various parts of plants for treating diseases and ailments. The writings of Hippocrates, Theophrastus, Pliney the Elder and Galen and texts from China and India discuss the importance of plants for treating various ailments. For a long time the use of plants for treating diseases was the province of folk medicine. This has changed in the last decade due to interest in diet for prevention of diseases, such as cancer (Middleton and Kandaswami 1993; Baker, 1995; Ames et al., 1995; Clarkson et al., 1995).

Important evidence for hormone action of plants came in the 1950s from research in Australia showing that clover contained compounds that are estrogenic in sheep (Wong and Flux, 1962; Adams, 1990). Later studies showed that plant-derived compounds have hormone-like activity in rats (Markaverich et al., 1988; Whitten and Naftolin, 1992; Sharma et al., 1992), in tumor cells (Martin et al., 1978; Scambia et al., 1990; Mousavi and Adlercreutz, 1993) and even in fish (Pelissero et al., 1991). These reports stimulated epidemiological studies to analyze the diet of humans which revealed that the presence or absence of certain plants in the diet can influence the incidence of certain estrogen and androgen-dependent cancers (Adlercreutz, 1984; Setchell et al., 1984; Adlercreutz et al., 1987, 1992, 1993; Barnes et al., 1990; Peterson and Barnes, 1993). Concurrently, many laboratories reported that plant-derived compounds bind to receptors or enzymes that are important in regulating the actions of estrogens, androgens and other steroid hormones (Stewart et al., 1987; Edwards et al., 1988; Monder et al., 1989; Monder, 1991; Funder et al., 1988; 1990; Baker, 1991a; Adlercreutz et al., 1993; Peterson and Barnes, 1993) and prostaglandins (Degan, 1990; Baker, 1991a,b, 1994; Baker and Fanestil, 1991a,b) in humans, suggesting that these proteins could be targets for the endocrine effects of plants. This research showing that plant-derived compounds have hormone-like activity verifies many of the medicinal uses of plants reported by Greek and Roman writers, giving these early writings "scientific respectability".

In this paper, I use an evolutionary approach to discuss mechanisms for the actions of flavonoids found in fruits and vegetables on human endocrine function. This evolutionary perspective connects the action of flavonoids as signals between plants and bacteria involved in formation of nitrogen-fixing nodules (Nap and Bisseling, 1990; Fisher and Long, 1992), the synthesis of pigments in petunias, snapdragons and other flowers (Stafford, 1991; Koes et al., 1994), and ancestry of enzymes that regulate steroid hormone and prostaglandin action in humans (Baker, 1991a,b, 1992a,b). With this approach, I seek to provide another perspective on hormone-like actions of flavonoids in humans that may lead to understanding how to use fruits and vegetables for preventing and treating diseases in humans.

Throughout this paper, I focus on the steroid-like activities of flavonoids, and, in particular, their estrogenic actions because these have been most widely studied. In addition to the well known flavonoid binding to the estrogen receptor, I discuss how flavonoids may act via enzymes that regulate estrogen action. However, it is important to keep in mind that this model is a general one; it is valid for responses to thyroid hormone, retinoids, prostaglandins, fatty acids and other hormones.

## 3. ANALYSIS OF MOLECULAR FOSSILS

Molecular fossils are amino acid or nucleic acid sequences of homologous genes from different organisms, such as humans, birds, fish, insects, yeast, bacteria. Homologous genes are descended from a common ancestor. Sequence comparisons of homologous genes, such as the alpha and beta chains of hemoglobin in humans, birds, and frogs, are much like comparisons of bones from these animals. One looks for the features that are conserved with the thought that they are likely to be important in general functions; and one examines the differences with the thought that they may be important in functions unique to each organism. With site specific mutagenesis, it possible to change the amino acids in one protein to test their functional importance. Indeed, this is a major stimulus for interest in sequence analysis.

In this paper, I will consider only amino acid sequences of proteins, although the basic principles hold for comparisons of nucleic acid sequences.

## 3.1. Evolution of Steroid Hormone Receptors: A Common Ancestor with Retinoid, Thyroid Hormone, Prostaglandin, and Orphan Receptors (Kastner et al., 1995; Mangelsdorf and Evans, 1995; Manglelsdorf et al., 1995)

Steroids act by entering a target cell and binding to a receptor protein, which increases the affinity of the receptor for specific sites on DNA (Glass, 1994; Manglesdorf et al., 1995). Binding of the receptor to DNA regulates transcription of different gene products, evoking a physiological response characteristic of the hormone. Steroid hormone receptors are called nuclear receptors because they act in the nucleus. In the last decade, the sequences of receptors for various steroids were determined and found to have a common ancestor, which is not surprising. It is reasonable to think that a series of gene duplications of an ancestral steroid receptor followed by sequence divergence led to a family of proteins with specificities for different classes of steroids. What is surprising is the homology of steroid receptors to receptors for retinoids, thyroid hormone and prostaglandins (Glass, 1994; Forman et al., 1995; Kastner et al., 1995; Mangelsdorf and Evans, 1995; Mangelsdorf et al., 1995) because superficially the structures of their ligands are different (Figure 1). Sequence analysis also reveals the existence of a large group of nuclear receptors that are homologs of steroid receptors, but which have unknown ligands. These nuclear receptors are called orphan receptors (Mangelsdorf and Evans, 1995); recently, one was found to bind melatonin (Becker-Andre et al., 1994). Other orphan receptors may be activated by a flavonoid or one of its metabolites (Baker, 1991a). The nuclear receptor superfamily is an excellent example of gene duplication and divergence to yield proteins with specificities for different lipophilic molecules that regulate complex endocrine processes involving development, reproduction, metabolism, and homeostasis.

Figure 2 shows a phylogenetic tree (Feng and Doolittle, 1990) depicting the evolution of the nuclear receptor family based on an analysis of their hormone binding domains. The glucocorticoid, mineralocorticoid, progesterone and androgen receptors cluster closely with each other on one branch. The human mineralocorticoid and glucocorticoid receptors are 12 units and 23.5 units distant, respectively, from their common ancestor. Next closest is the estrogen receptor, which is 67.5 units distant from its common ancestor with these four receptors. More distantly related are the thyroid hormone, 1,25-dihydroxy-vitamin $D_3$, 15-deoxy-$\Delta^{12,14}$-prostaglandin $J_2$, ecdysone, retinoic acid and retinoid X receptors, which cluster on another branch, although these receptor sequences have diverged substantially from their common ancestor. For example, retinoid X receptor and ecdysone receptor are 66 units and 84.5 units distant, respectively, from their common ancestor.

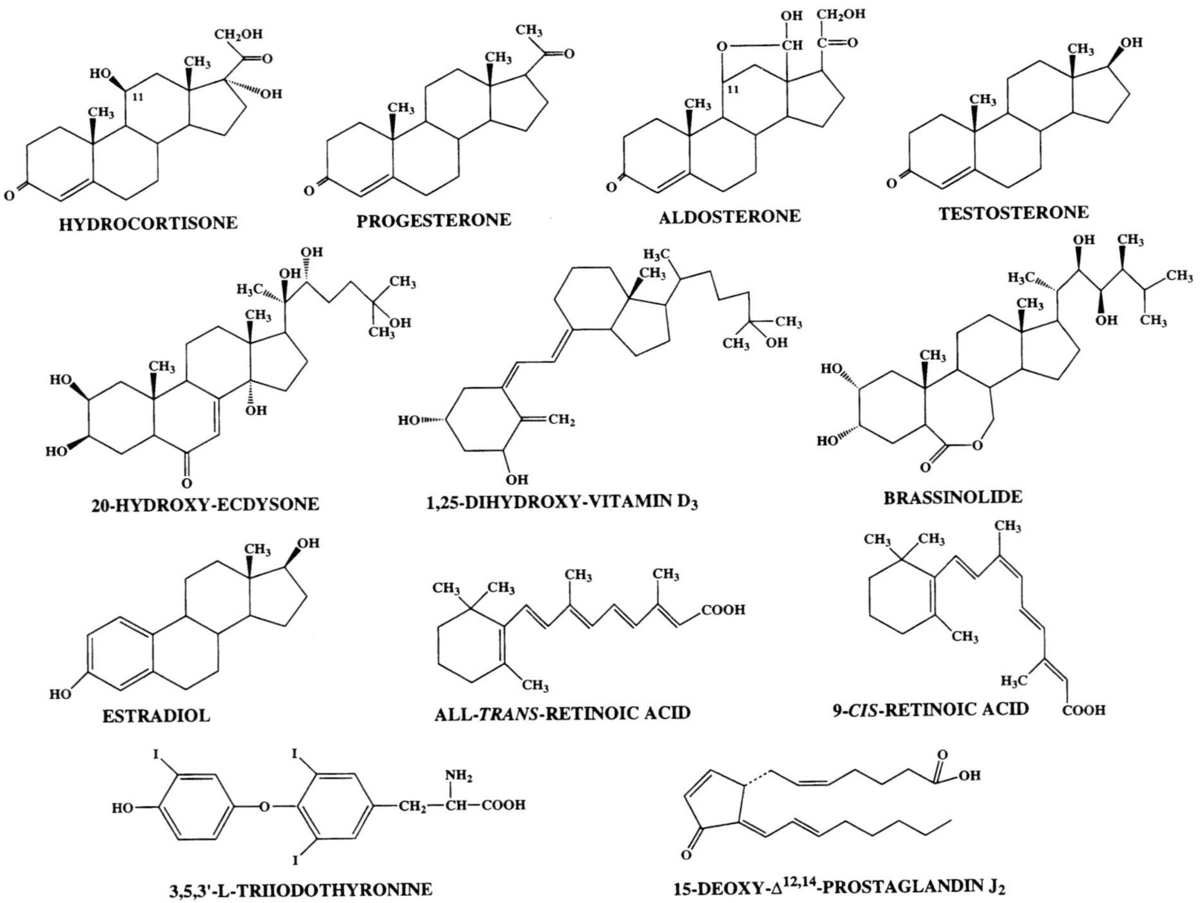

**Figure 1.** Structures of steroids, retinoic acid, 15-deoxy-$\Delta^{12,14}$-prostaglandin J$_2$, and thyroid hormone. Estradiol, testosterone, progesterone, hydrocortisone, aldosterone, and 1,25-dihydroxy-vitamin D$_3$ are active steroids in mammals. The steroid ecdysone regulates development of insect larvae. Ecdysteroids also are found in plants, where their functions are unknown. Brassinolide is a plant steroid. Receptors in plants for either ecdysteroids or brassinolide have not been identified. Although retinoids, thyroid hormone and 15-deoxy-$\Delta^{12,14}$-prostaglandin J$_2$ are not steroids, they regulate gene transcription by binding to receptors that have a common ancestor with receptors for mammalian and insect steroids.

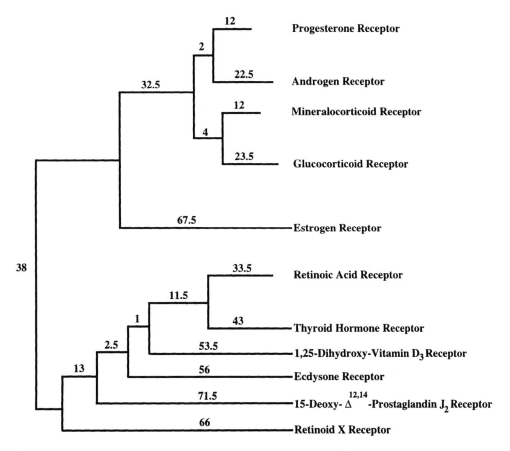

**Figure 2.** Evolution of receptors for steroids, thyroid hormone, 15-deoxy-$\Delta^{12,14}$-prostaglandin $J_2$ and retinoids. All receptor sequences except the *Drosophila melanogaster* ecdysone receptor sequence are from human genes. The phylogenetic tree was constructed with the Feng-Doolittle algorithm (Feng and Doolittle, 1990). In this method, the protein sequences first are progressively aligned using the Dayhoff PAM-250 scoring matrix to assess pairwise similarity of each sequence with the others, and the scores are assembled into a distance matrix. Then the branching order is calculated for the sequences. Branch lengths are calculated by linear regression analysis of the best fit of the pairwise distances and the branching order. The lengths of the branches are proportional to the relative distance between the sequences. For example, human progesterone and androgen receptors are 12 and 22.5 units distant, respectively, from their common ancestor.

## 3.2. A Common Ancestor for Enzymes That Regulate Steroid, Retinoid, and Prostaglandin Action in Humans and Signals between Plants, Bacteria, and Insects

Several years ago, we made the surprising discovery that the amino acid sequence of *R. meliloti* NodG is similar to human 17β-hydroxysteroid dehydrogenase (Baker, 1989; 1991a), the enzyme that metabolizes estrogens and androgens in humans (Figure 3). Analysis of the sequences of these two proteins indicated that the probability of these similarities occurring by chance was less than $10^{-24}$. Thus, instead of chance explaining the sequence similarity, more likely it is due to descent of the two proteins from a common ancestor. Both proteins are homologous to the animal 11β-hydroxysteroid dehydro-

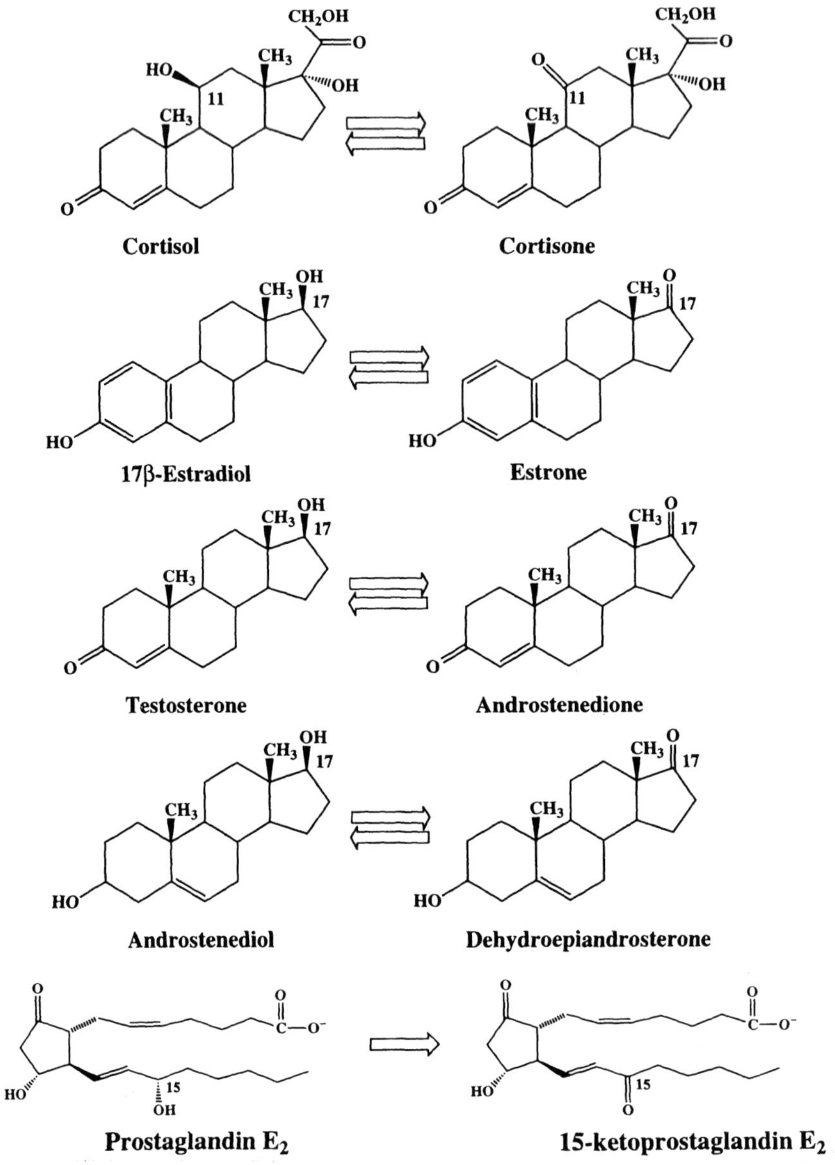

Figure 3. Metabolism of steroids and prostaglandins by dehydrogenases.

genase, 20β-hydroxysteroid dehydrogenase and 15-hydroxyprostaglandin dehydrogenase (Baker, 1990; 1991b; Persson et al., 1991), as well as to a plant reductase that catalyzes a key step in chlorophyll synthesis (Baker, 1994b) and a dehydrogenase that may regulate the action of a plant hormone (DeLong et al., 1993). Furthermore, these proteins have a common ancestor with β-ketoreductases that are part of the enzyme complex for synthesizing polyketide antibiotics in soil bacteria (Baker, 1990; 1991a,b; Hopwood and Sherman, 1990). Thus, this family goes back at least 2 billion years to the time when eukaryotes and prokaryotes diverged from a common ancestor.

11β-hydroxysteroid dehydrogenase catalyzes the interconversion of the active glucocorticoid hydrocortisone and the inactive steroid cortisone. 17β-hydroxysteroid dehydrogenase catalyzes the interconversion of the C17 alcohol and ketone groups on estradiol and testosterone. 15-hydroxyprostaglandin dehydrogenase inactivates prostaglandin $E_2$ by catalyzing the oxidation of the C15 alcohol to a ketone.

## 3.3. Flower Colors, Progesterone Synthesis, and Pox Viruses

Anthocyanins, pigments in flowers, are a class of flavonoids. Conversion of a flavanone to a leuco-anthocyanidin requires sequential modification by flavanone 3β-hydroxylase and dihydroflavonol 4-reductase (Stafford, 1991; Koes et al., 1994). Sequence analyses show that plant dihydroflavonol 4-reductases share a common ancestor with human 3β-hydroxysteroid dehydrogenase, an enzyme that converts pregnenolone to progesterone (Baker et al., 1990; Baker and Blasco, 1992). The reactions catalyzed by these enzymes are shown in Figure 4. Note the similarities in structures of the substrates for the plant and animal enzymes.

Other analyses show that vaccinia virus, a relative of the smallpox virus, contains a gene that is homologous to plant dihydroflavonol reductase and mammalian 3β-hydroxysteroid dehydrogenase (Baker and Blasco, 1992). Vaccinia virus ORF's amino acid sequence is about 35% identical to mammalian 3β-hydroxysteroid dehydrogenase (Baker and Blasco, 1992). Despite this sequence divergence, the vaccinia ORF has 3β-hydroxysteroid dehydrogenase activity (Moore and Smith, 1992); that is, the vaccinia protein catalyzes the conversion of pregnenolone to progesterone.

**Reaction Catalyzed by Mammalian 3β-hydroxysteroid Dehydrogenase**

**Reaction Catalyzed by Plant Dihydroflavonol Reductase**

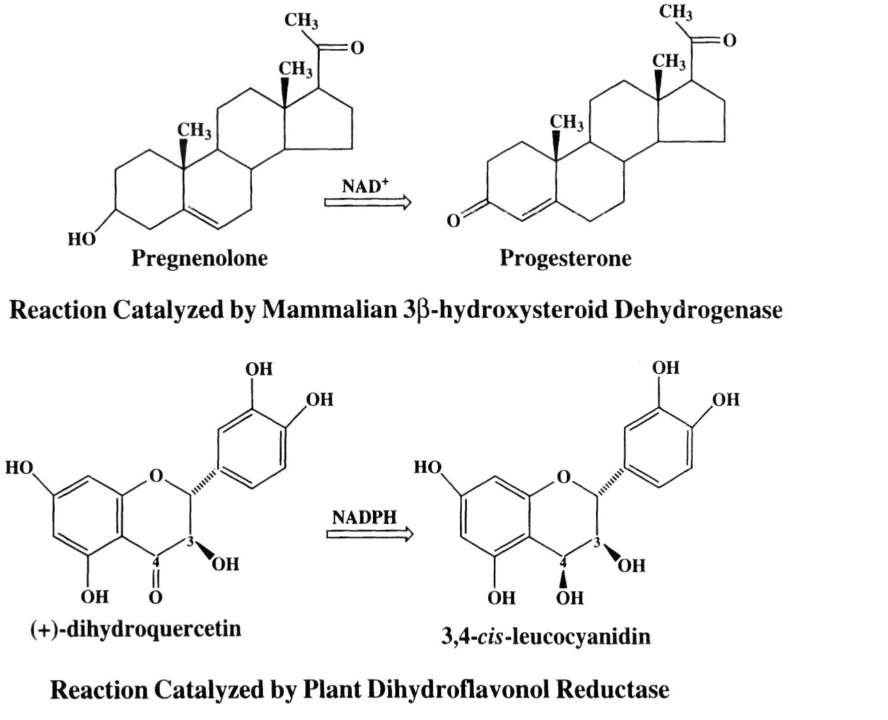

**Figure 4.** Reactions catalyzed by human 3β-hydroxysteroid dehydrogenase and plant dihydroflavonol reductase.

## 3.4. The Plant Steroid Brassinolide

Recently, enzymes important in the synthesis of the plant steroid brassinolide were cloned and sequenced and found to have strong sequence similarity to mammalian homologs. An *Arabidopsis* 5α-reductase has about 38% sequence identity to mammalian homologs (Li et al., 1996), sufficient to suggest that the animal enzyme could function in plants that are deficient in the plant 5α-reductase. Similarly, the *Arabidopsis* cytochrome P450 enzyme in the pathway for the synthesis of brassinolide (Szekeres et al., 1996) has 38% identity to animal homologs. The presence in plants of enzymes with this similarity to animal enzymes important in steroid hormone synthesis means that compounds that bind to the plant enzymes could bind to the animal enzymes and affect steroid hormone levels in animals.

## 4. COMMON STRUCTURES AND CHEMISTRY IN ANIMAL, PLANT, AND BACTERIA SIGNALS: WHAT IS REQUIRED FOR ESTROGENIC ACTIVITY?

The previous discussion has emphasized the common ancestry of mammalian proteins involved in hormone action with proteins in plants, insects, and bacteria. The similarities are sufficient in some cases to suggest that molecules that recognize a plant, insect or bacterial protein may interact with a mammalian relative. However, there is another factor that is important in "unexpected hormonal actions" of flavonoids during evolutionary divergence of mammals, insects and plants, Nature has taken certain chemical themes and reworked them for construction of intercellular signals. Signals usually contain hydrophobic aromatic and/or aliphatic groups with hydroxyl and methyl substituents. Combinations of these groups can yield signals with high specificity for target cells as found in the signaling between plants and bacteria for the formation of nitrogen fixing nodules in roots (Nap and Bisseling, 1990; Baker, 1992a,b; Fisher and Long, 1992; Koes et al., 1994).

A flavonoid may have hormonal activity if it contains a functionally important group of the human hormone; in which case, the flavonoid can compete with the hormone for binding to receptors or enzymes and have an endocrine action. The presence of these functional groups on pesticides that have been used for controlling insects results in hormone-like activity for pesticides. A good example of how functional groups in the proper 3-dimensional configuration on a compound can result in potent hormone activity is seen in non-steroidal compounds that act like estradiol. These compounds are discussed next.

## 4.1. Diethylstilbestrol and Tamoxifen

Synthetic non-steroidal compounds that lack the B and C rings of estradiol can have estrogenic activity. Comparison of estradiol, the biological estrogen, with diethylstilbestrol (DES), a synthetic estrogen, and tamoxifen, a synthetic compound with antiestrogen or estrogen action depending on the target tissue, reveals that all three compounds contain a phenolic A ring, which is crucial for estrogenic activity (Korach et al., 1989; Gantchev et al., 1994; Wiese et al., 1995) (Figure 5). Although DES lacks a B and C ring, DES is a more potent estrogen than estradiol, indicating that the presence on DES of two phenolic groups in the proper 3-dimensional structure suffices to activate the estrogen receptor.

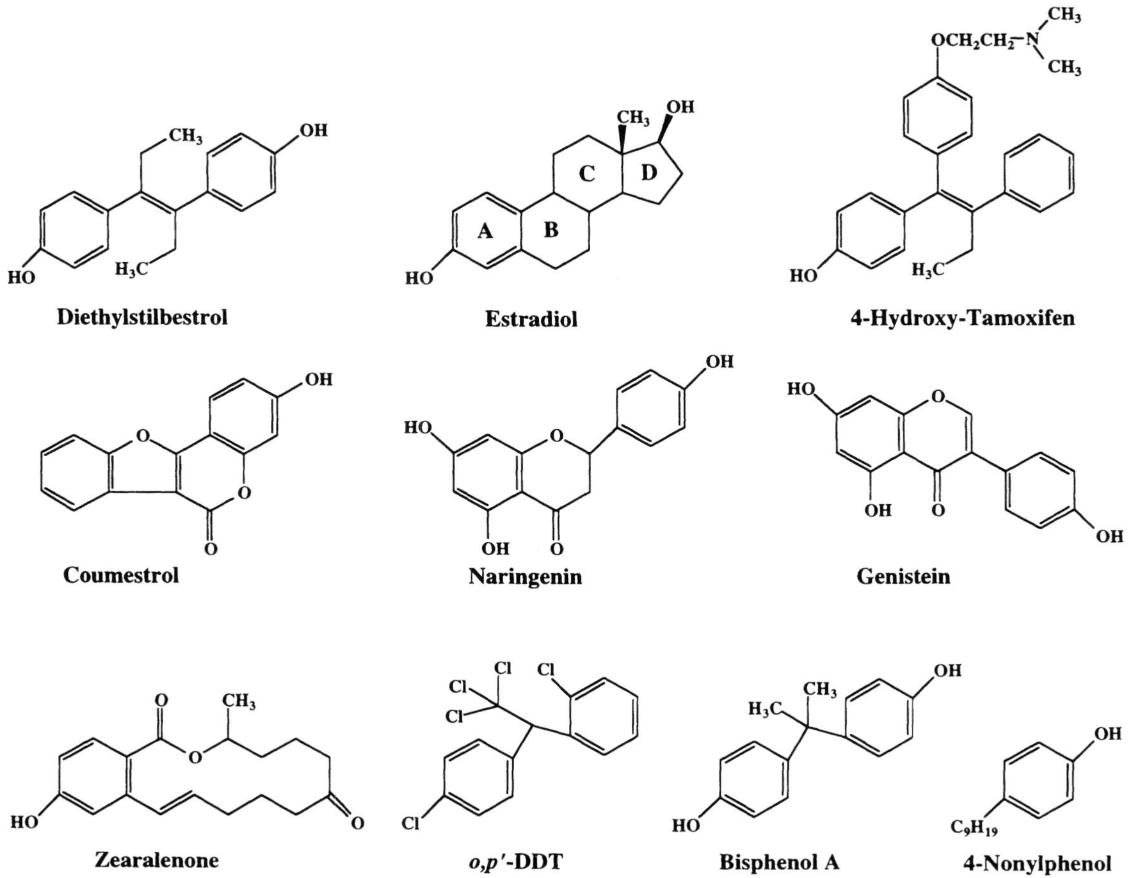

**Figure 5.** Comparison of estradiol with non-steroidal compounds that have estrogenic activity. Diethylstilbestrol (DES) is a synthetic estrogen that is more potent than estradiol, although DES lacks the B and C rings of estradiol. 4-Hydroxy-tamoxifen is a synthetic compound that can have estrogenic or anti-estrogenic activity, depending on the tissue. The flavonoids genistein and naringenin have some structural similarities to estradiol and can compete with estradiol for binding to its receptor. Also shown are the fungal compound zearalenone, $o,p'$-DDT, nonylphenol and bisphenol A, which have estrogenic activity.

## 4.2. Phenolics with Low Affinity for the Estrogen Receptor

The importance of phenolic group in estrogenicity is seen in reports that bisphenol A and nonylphenol (Figure 5) have estrogenic activity (Soto et al., 1991; Krishnan et al., 1993; White et al., 1994) although their affinity for the estrogen receptor is at least 10,000 fold lower than estradiol. Such compounds can be estrogenic if they are present in high concentrations or if they accumulate to high concentrations in fatty tissue.

## 4.3. DDT and Its Metabolites

Another illustrative example comes from the unexpected estrogenicity of *o,p'*-DDT, a metabolite of DDT (Figure 5), which binds to the estrogen receptor with affinity that is 1000 fold lower than estradiol (Kelce et al., 1995). Nevertheless, because *o,p'*-DDT accumulates in fatty tissue, it can achieve a concentration sufficient to activate the estrogen receptor. Another DDT metabolite, *p,p'*-DDE binds to the androgen receptor (Kelce et al., 1995). The unexpected hormonal actions of DDT metabolites and other insecticides have been an important stimulus for understanding structure-function activities of nonsteroidal compounds.

## 5. FLAVONOIDS AS HORMONES

Genistein, naringenin and luteolin (Figures 5 and 6) and other flavonoids contain a phenolic group and 6-membered rings with different degrees of desaturation in a structure that resembles, in part, either estradiol, other steroids, thyroid hormone, or retinoic acid. Considering the estrogenic activity of DES, bisphenol A and DDT metabolites, it is clear that flavonoids or a metabolite could bind to the estrogen receptor or another nuclear receptor.

Indeed, flavonoids have been shown to bind to the estrogen receptor. However, most flavonoids are active at $10^{-6}$M or higher (Miksicek, 1993; Ruh et al., 1995), which is $10^3$ to $10^4$ higher concentration than estradiol, raising the question of whether the flavonoid's *in vivo* concentration can be high enough to be biologically important. Flavonoid accumulation in fatty tissue could reach a concentration sufficient to activate a nuclear receptor. The flavonoid could be metabolized to a more active form. Another possibility is the binding of flavonoids to the recently cloned estrogen receptor-beta, which differs in some respects in estrogen specificity from the classical estrogen receptor-alpha (Kuiper et al., 1996). Flavonoid binding to estrogen receptor-beta has not been described.

Alfalfa secretes the flavonoid luteolin into the soil where it diffuses to *R. meliloti*, stimulating the synthesis of proteins that synthesize the Nod factor signal. The Nod factor diffuses to the root hair in alfalfa, stimulating the synthesis of proteins that permit entry of *R. meliloti* into the root hair and the subsequent formation of a nitrogen fixing nodule.

The point to keep in mind is that regulation of gene transcription in mammals by estrogens and other steroids is mediated by nuclear receptors that can also respond to nonsteroidal compounds that are lipophilic, even if the compounds have a much lower affinity than the endogenous hormone for its receptor. What is required is similarity at key structures of the hormone, such as having a phenolic group if the compound is to be an estrogen. Because phenolic groups, unsaturated 6-membered rings and aliphatic chains are widely used in construction of compounds that act as signals between plants, fungi, bacteria and insects (Lamb et al., 1989; Hopwood and Sherman, 1990; Nap and Bisseling, 1990; Baker 1991a,1992a,b; 1995, Fisher and Long, 1992; Koes et al., 1994) these signals are potentially active in human endocrine physiology.

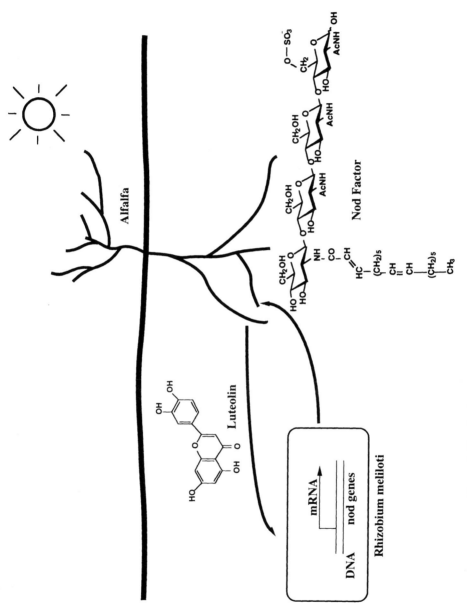

**Figure 6.** Signaling between Alfalfa and *Rhizobium meliloti*, a nitrogen-fixing bacterium.

## 5.1. Signals in Warfare and Peaceful Coexistence between Plants, Insects, and Bacteria

Long before placental mammals arose about 80 million years ago, plants, insects and bacteria were interacting with each other. Coevolution of plant, insect, and bacteria in an adversarial milieu has yielded a variety of strategies for survival. As with modern life, the focus is on consuming other organisms, defending against attacks, and forming alliances with other organisms to claim territory for growth and reproduction. To accomplish this, plants secrete signals, which include flavonoids, that interfere with the biochemistry of their adversary (Lamb et al., 1989; Gold et al., 1992; Byrne et al., 1995) or signal another organism to cooperate in a mutually beneficial alliance. As described next, there is a strong link between cooperative coevolution between plants and bacteria and enzymes that regulate human steroid and prostaglandin function.

## 5.2. Flavonoids as Signals between Plants and Nitrogen-Fixing Bacteria

The formation of nitrogen-fixing nodules in the roots of leguminous plants such as alfalfa, soybeans, peas and clover is a cooperative process between a plant and a soil bacterium. Communication between the plant and bacterium is mediated by flavonoid signals (Nap and Bisseling, 1990; Baker, 1992a,b; Fisher and Long, 1992; 1994). For example, alfalfa roots secret luteolin which binds to a specific protein in *Rhizobium meliloti*, inducing the synthesis of several nodulation proteins—one of which is NodG—one of several enzymes that the bacterium uses to synthesize a signal called a Nod factor (Figure 6). This signal permits the entry of the bacterium into the alfalfa root and the formation of a nodule that can fix atmospheric nitrogen into amino acids for the plant.

The NodG protein, which is induced by luteolin, is about 29% identical to human 15-hydroxyprostaglandin and 17β-hydroxysteroid dehydrogenases (Baker, 1989; 1990; 1991b). NodG is important in synthesis of the fatty acid part of the Nod factor; 17β-hydroxysteroid dehydrogenase is important in the synthesis of estradiol and testosterone; 15-hydroxyprostaglandin dehydrogenase is important in the inactivation of prostaglandins $E_2$ and $F_{2\alpha}$. Thus, these three proteins have a common activity of regulating the concentration of an intercellular signal. Other legumes such as soybeans, peas, beans and clover secrete other flavonoids to signal a specific rhizobium to form nitrogen-fixing nodules. This also is analogous to human endocrine function in which steroids with different structures are used to regulate gene transcription in specific target cells. A more detailed discussion of the analogies between the plant-rhizobia interaction and the human endocrine system is found in (Baker, 1991a; 1992a,b; 1995).

## 5.3. Signals That Control Insects

Since earliest times, plants and insects have been engaged in an arms race. The plant arsenal contains a variety of signals, including flavonoids, phytochemicals that mimic insect hormones and prostaglandin-like molecules, that interfere with normal insect physiology.

An example of a protective flavonoid is the flavone maysin, which is found in maize silks. Maysin provides resistance to corn earworm (Byrne et al., 1995).

Two examples of plant compounds that have hormonal activity in insects are ecdysone and juvenile hormone. Plants synthesize β-ecdysones; insects synthesize α-ecdysones. Plants and insects also synthesize juvenile hormone. Both ecdysone and juvenile hormone are important regulators of development of insect larvae. Thus, plant analogs that compete with either of these insect hormones can interfere with metamorphosis. Analogs of juvenile hormone such as methoprene are used as insecticides, for example, to control mosquitoes.

Recently, Harmon et al. (1995) reported that methoprene acid activates mammalian retinoid X receptor. Their work illustrates several points that are relevant to hormonal actions in humans of flavonoids, other phytochemicals and synthetic compounds. Harmon et al. (1995) found that methoprene does not activate retinoid X receptor. It is the metabolite, methoprene acid (Figure 7), which is formed from methoprene by hydrolysis by cellular esterases. Thus, the hormonally active compound is a metabolite. Although methoprene acid has about 300 fold lower affinity for retinoid X receptor than 9-*cis*-retinoic acid (Harmon et al., 1995), it mediates gene transcription via the receptor. Interestingly, methoprene has teratogenic effects during mouse embryogenesis (Unsworth et al., 1974). The effects include limb deformities, a characteristic of high doses of retinoids. Methoprene acid may accumulate in fatty tissues to a concentration sufficient for biological activity. While methoprene acid lacks a 6-membered group found in retinoids (Figures 1 and 7), metho-

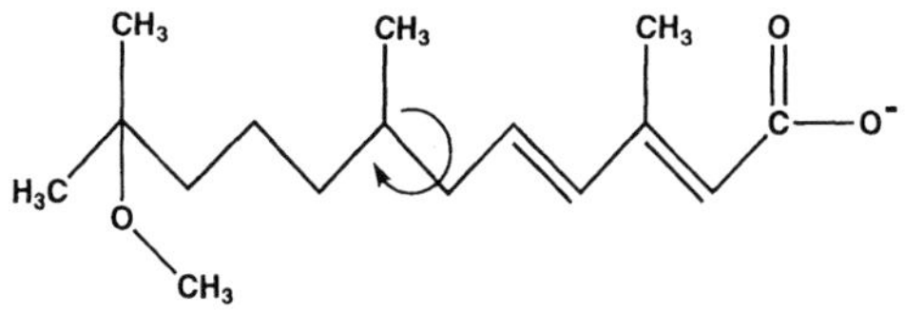

**Methoprene Acid**

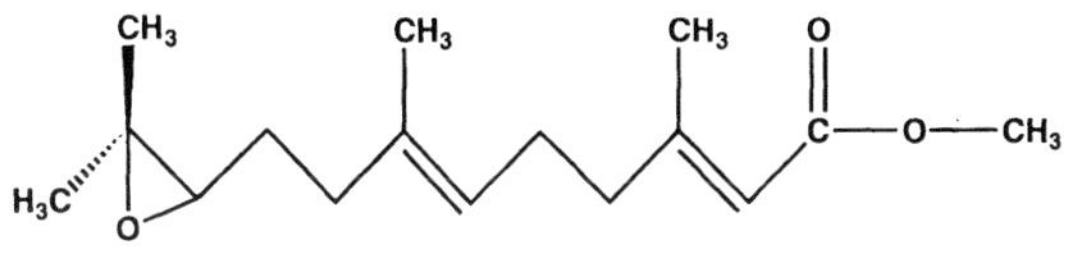

**Juvenile Hormone III**

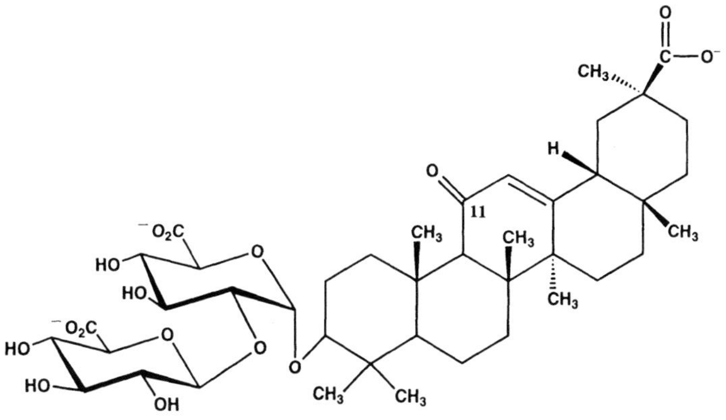

**Glycyrrhizic Acid**

**Figure 7.** Structure of methoprene acid, juvenile hormone, and glycyrrhizic acid.

prene acid can also assume a conformation that resembles sufficiently 9-*cis*-retinoic acid to activate a mammalian retinoid X receptor. This is important in view of the common ancestry of mammalian nuclear receptors with invertebrate receptors. It suggests that other phytochemicals or their metabolites that interact with an insect protein may have sufficient affinity for a human homolog to evoke an endocrine response.

Methoprene acid resembles juvenile hormone, and methoprene acid has juvenile hormone activity as well as 9-*cis*-retinoic acid activity. Juvenile hormone does not bind to retinoid X receptor, The double bond at C9 in juvenile hormone constrains this part from assuming a conformation like 9-*cis*-retinoic acid. Methoprene acid lacks this double bond, which may account for its binding to the retinoid X receptor. Glycyrrhizic acid is found in the extract of licorice root. This compound is metabolized by intestinal bacteria to the aglycone, which inhibits 11β-hydroxysteroid dehydrogenase and 15-hydroxyprostaglandin dehydrogenase.

## 6. FLAVONOID ACTIONS ON ENZYMES INVOLVED IN HORMONE ACTION

The previous discussion has focused on hormone receptors as sites for the endocrine effects of flavonoids in mammals. Flavonoid binding to enzymes may also be a mechanism for hormonal actions (Baker 1991a; 1992a,b; 1995). Structural similarities between flavonoids, steroids, retinoids, and thyroid hormone could lead to binding to enzymes involved in regulating the levels of these hormones. An important additional consideration is that enzymes usually have $K_m$s for steroid substrates at $10^{-6}$M, which is 1000 fold lower affinity than for steroid receptors. This means that flavonoids with $K_m$s of $10^{-6}$M to $10^{-4}$M for steroid metabolizing enzymes can be more effective in altering steroid metabolism. This also is relevant to retinol dehydrogenase, which is homologous to 11β- and 17β-hydroxysteroid dehydrogenases (Baker, 1996).

Plant-derived compounds that inhibit 17β-hydroxysteroid dehydrogenase will have important effects on female and male reproductive function and development (Makela et al., 1995). Compounds that inhibit the oxidation of cortisol to cortisone by 11β-hydroxysteroid dehydrogenase will affect blood pressure (Lee et al., 1996). Compounds that inhibit 15-hydroxyprostaglandin dehydrogenase will have prostaglandin-like activity. The effect of inhibiting these enzymes can mistakenly be interpreted as meaning that the compound acts by binding to a hormone receptor. The reality is that they are binding to hormone-inactivating enzymes and raising the local concentration of the hormone.

The best example of a steroid-like activity that is due to a phytochemical binding to an enzyme is the hormone-like actions of glycyrrhizic acid, a triterpenoid in a plant root that is the source of licorice. Although glycyrrhizic acid is not a flavonoid, it is a model for how a flavonoid can have hormone-like actions due to inhibition of an enzyme important in regulating steroid hormone concentrations.

### 6.1. Licorice Extract, King Tut, Addison's Disease, and Ulcers

Licorice extract provides an important example of an enzyme-mediated hormone activity (Stewart et al., 1987; Edwards et al., 1988; Funder et al., 1988, 1990; Monder et al., 1989; Monder, 1991a,b; Davis and Morris, 1991). Licorice is an extract of the roots of *Glycyrrhiza glabra*, a legume that is widely distributed throughout the Mediterranean and Orient. Over 2,000 years ago, licorice was used as an herbal for quenching thirst and promoting healing of ulcers (Monder, 1991a; Davis and Morris, 1991; Duke, 1992). Licorice root was considered to be so significant that it was placed in King Tutankhamen's tomb. As valuable as licorice was for the ancients, in the 20th Century licorice's principal use is as a flavoring agent; for the most part, its medicinal properties are ignored.

We owe our present interest and understanding of licorice's medicinal properties to the pioneering efforts of Reevers (1948) and Dr. S. Gottfried at Biorex (Jones, 1968), who were interested in herbal medicines. Reevers found that an herbal extract containing licorice helped people with Addison's disease, which is caused by a glucocorticoid deficiency. Dr. S. Gottfried had a strong interest in herbal medicines, which led him to read about licorice in a botanical book, *Historia Botanica Practica*, published in 1774. Dr. Gottfried and his colleagues found that the compound with steroid-like activity is glycyrrhizic acid (Figure 7), which is metabolized by intestinal bacteria to its aglycone glycyrrhetinic acid, the biologically active species. Comparison of glycyrrhizic acid with the steroid structures in Figure 1 reveals some structural resemblance to steroids. Glycyrrhetinic acid acts by inhibiting 11β- hydroxysteroid dehydrogenase, which increases hydrocortisone levels leading to both glucocorticoid and mineralocorticoid effects (Stewart et al., 1987; Edwards et al., 1988; Funder et al., 1988; 1990; Monder et al., 1989; Monder, 1991b; Baker, 1994a). It is the increase in glucocorticoid levels that explains the ability of licorice extract to help people with Addison's disease.

Folk medicine also describes the use of licorice to treat ulcers. This was a stimulus for Dr. Gottfried to collaborate with Dr. E.E. Turner and his colleagues at the University College to develop carbenoxolone, a synthetic analog of glycyrrhetinic acid. Carbenoxolone is more water soluble and more suitable for medical use. Carbenoxolone is thought to act by inhibiting 15-hydroxyprostaglandin dehydrogenase (Baker, 1994a), raising the local prostaglandin concentration and promoting healing of ulcers.

## 6.2. Hormonal Actions of Flavonoids Mediated by Other Enzymes

The evidence for hormone-like actions of plant-derived compounds due to inhibition of steroid and prostaglandin-metabolizing dehydrogenases suggests more intensive examination of other classes of enzymes that metabolize steroids, thyroid hormone and retinoids as sites of action for flavonoids. Flavonoids inhibit aromatases (Miller and O'Neill, 1990; Ibrahim and Abul-Hajj, 1990; Pelissero et al., 1996), which are important in synthesis of estrogens. Some flavonoids contain the resorcinol moiety, which makes them substrates for peroxidases (Lyttle et al., 1984; Divi and Doerge, 1994). These enzymes could activate flavonoids to form covalent adducts with receptors and enzymes in target tissues. Moreover, a flavonoid has been found to bind with high affinity to the ATP binding site on a cell cycle control kinase (Azevedo et al., 1996). A variety of mammalian enzymes have high affinity for either steroids or thyroid hormone (Kato et al., 1989; Vie et al., 1996), sufficient for these hormones to bind to these proteins under physiological conditions. Flavonoids may compete with steroids and thyroid hormone (Gaitan et al., 1989) for these proteins.

## 7. SUMMARY

Plants and insects have been under coevolutionary pressures for over 150 million years. Plant signals such as ecdysteroids, juvenile hormone and flavonoids and their metabolites have profound effects on insect physiology. Insect receptors for ecdysteroids have a common ancestor with human receptors for steroids, retinoids, thyroid hormone and prostaglandins. Moreover, enzymes important in human steroid, retinoid and prostaglandin function have a common ancestry with insect and plant enzymes. Flavonoids, steroids, thyroid hormone and retinoids also have some common chemical structures. Together this suggests that flavonoids or their metabolites that act in insects may also bind to related proteins in humans. Elucidation of flavonoid interactions with receptors and enzymes may yield important clues for the protective effects of diet in disease.

## 8. NOTE IN ADDED PROOF

Kuiper et al. (*Endocrinology* **1997**, 138, 863–870) report that flavonoids inhibit estrogen binding to estrogen receptor-β. Li et al. (*Proc. Natl. Acad. Sci. USA* **1997**, 94, 3554–3559) report that plant 5α-reductase can metabolize animal steroids and human 5α-reductase can act substitute for the plant homolog in the synthesis of brassinosteroids in mutant plants lacking the native reductase. Divi et al. (*Biochem. Pharmacol.* 1997, in press) report that isoflavones from soybean inhibit thyroid peroxidase, providing an explanation for the anti-thyroid effects of soybeans.

## ACKNOWLEDGMENT

I thank scientists at the National Center for Toxicological Research for stimulating discussion of mechanisms for physiological effects of flavonoids in humans.

## REFERENCES

Adams, N. R. Permanent infertility in ewes exposed to plant oestrogens. *Australian Vet. J.* **1990**, 67, 197–201.
Adlercreutz, H. Does fiber-rich food containing animal lignan precursors protect against both colon and breast cancer? An extension of the "fiber hypothesis". *Gastroenterology* **1984**, 86, 761–766.
Adlercreutz, H.; Hockerstedt, K.; Bannwart, C.; Bloigu, S.; Hamalainen, E.; Fotsis, T.; Ollus, A. Effects of dietary components, including lignans and phyto-oestrogens, on enterohepatic circulation and liver metabolism of oestrogens and on sex hormone binding globulin (SHBG). *J. Steroid Biochem.* **1987**, 27, 1135–1144.
Adlercreutz, H.; Mousavi, Y.; Clark, J.; Hockerstedt, K.; Hamalainen, E.; Wahala, K.; Makela, T.; Hase, T. Dietary phytoestrogens and cancer: *in vitro* and *in vivo* studies. *J. Steroid Biochem. Molec. Biol.* **1992**, 41, 331–337.
Adlercreutz, H.; Bannwart, C.; Wahala, K.; Makela, T.; Brunow, G.; Hase, T.; Arosemena, P. J.; Kellis, J. T.; Vickery, L. E. Inhibition of human aromatase by mammalian lignans and isoflavonoid phytoestrogens. *J. Steroid Biochem. Molec. Biol.* **1993**, 44, 147–153.
Ames, B. N.; Gold, L. S.; Willett, W. C. The causes and prevention of cancer. *Proc. Natl. Acad. Sci. USA* **1995**, 92, 5258–5265, 1995.
Azevedo, F. W. Jr., Mueller-Dieckmann, H-J.; Schulze-Gahmen, U.; Worland, P. J.; Sausville, E.; Kim, S-H. Structural basis for specificity and potency of a flavonoid inhibitor of human CDK2, a cell cycle kinase. *Proc. Natl. Acad. Sci. USA* **1996**, 93, 2735–2740.
Baker, M. E. Human placental 17β-hydroxysteroid dehydrogenase is homologous to NodG protein of *Rhizobium meliloti*. *Mol. Endocrinol.* **1989**, 3, 881–884.
Baker, M. E. A common ancestor for human placental 17β-hydroxysteroid dehydrogenase, *Streptomyces coelicolor* actIII protein, and *Drosophila melanogaster* alcohol dehydrogenase. *FASEB J.* **1990**, 4, 222–226.
Baker, M. E. Genealogy of regulation of human sex and adrenal function, prostaglandin action, snapdragon and petunia flower colors, antibiotics, and nitrogen fixation: functional diversity from two ancestral dehydrogenases. *Steroids* **1991a**, 56, 354–360.
Baker, M. E. Evolution of enzymatic regulation of prostaglandin action: novel connections to regulation of human sex and adrenal function, antibiotic synthesis and nitrogen fixation. *Prostaglandins* **1991b**, 42, 391–410.

Baker, M. E. Evolution of regulation of steroid-mediated intercellular communication in vertebrates: insights from flavonoids, signals that mediate plant-rhizobia symbiosis. *J. Steroid Biochem. Molec. Biol.* **1992a**, *41*, 301–308.

Baker, M. E. Similarities between legume-rhizobium communication and steroid-mediated intercellular communication in vertebrates. *Can. J. Microbiol.* **1992b**, *38*, 541–547.

Baker, M. E. Licorice and enzymes other than 11β-hydroxysteroid dehydrogenase. *Steroids* **1994a**, *59*, 136–141.

Baker, M. E. Protochlorophyllide reductase is homologous to human carbonyl reductase and pig 20β-hydroxysteroid dehydrogenase. *Biochem. J.* **1994b**, *300*, 605–607.

Baker, M. E. Endocrine activity of plant-derived compounds: an evolutionary perspective. *Proc. Soc. Exptl. Biol. Med.* **1995**, *208*, 131–138.

Baker, M. E. Unusual evolution of 11β- and 17β-hydroxysteroid and retinol dehydrogenases. *BioEssays* **1996**,*18*, 63–70.

Baker, M. E.; Luu-The, V.; Simard, J.; Labrie, F. A common ancestor for mammalian 3β-hydroxysteroid dehydrogenase and plant dihydroflavonol reductase. *Biochem. J.* **1990**, *269*, 558–558.

Baker, M. E.; Fanestil, D. D. Licorice, computer-based analyses of dehydrogenase sequences and regulation of steroid and prostaglandin action. *Mol. Cell. Endocrinol.* **1991a**, *78*, C99–102.

Baker, M. E.; Fanestil, D. D. Liquorice as a regulator of steroid and prostaglandin metabolism. *Lancet* **1991b**, *337*, 428–429.

Baker, M. E.; Blasco, R. Expansion of the mammalian 3β-hydroxysteroid dehydrogenase/plant dihydroflavonol reductase superfamily to include a bacterial cholesterol dehydrogenase, a bacterial UDP-galactose-4-epimerase, and open reading frames in vaccinia virus and fish lymphocystis disease virus. *FEBS Lett.* **1992**, *301*, 89–93.

Barnes, S.; Grubbs, C.; Setchell, K. D.; Carlson, J. Soybeans inhibit mammary tumors in models of breast cancer. *Progress in Clinical and Biological Research* **1990**, *347*, 239–253.

Becker-Andre, M.; Wiesenberg, I.; Schaeren-Wiemers, N.; Andre, E.; Missbach, M.; Saurat, J-H.; Carlberg, C. Pineal gland hormone melatonin binds and activates an orphan of the nuclear receptor superfamily. *J. Biol. Chem.* **1994**, *269*, 28531–28534.

Byrne, P. F.; McMullen, M. D.; Snook, M. E.; Musket, T. A.; Theuri, J. M.; Widstrom, N. W.; Wiseman, B. R.; Coe, E. H. Quantitative trait loci and metabolic pathways: genetic control of the concentration of maysin, a corn earworm resistance factor, in maize silks. *Proc. Natl. Acad. Sci. USA* **1995**, *93*, 8820–8825.

Clarkson, T. B.; Anthony, M. S.; Hughes, C. L. Jr. Estrogenic soybean isoflavones and chronic disease. *Trends Endocrinol. Metab.* **1995**, *6*, 11–16.

Davis, E. A.; Morris, D. J. Medicinal uses of licorice through the millennia: the good and plenty of it. *Mol. Cell. Endocrinol.* **1991**, *78*, 1–6.

Degen GH. Interaction of phytoestrogens and other environmental estrogens with prostaglandin synthase *in vitro*. *J. Steroid Biochem.* **1990**, *35*, 473–479.

DeLong, A.; Calceron-Urrea, A.; Dellaporta, S. L. Sex determination gene *TASSELSEED2* of maize encodes a short-chain-alcohol dehydrogenase required for stage-specific floral organ abortion. *Cell* **1993**, *74*, 757–768.

Divi, R. L.; Doerge, D. R. Mechanism-based inactivation of lactoperoxidase and thyroid peroxidase by resorcinol derivatives. *Biochemistry* **1994**, *33*, 9668–99674.

Duke, J. A. Handbook of Biologically Active Phytochemicals and Their Activities. Boca Raton: CRC Press, 1992.

Edwards, C. R. W.; Stewarts, P. M.; Burt, D.; Brett, L.; McIntyre, M. A.; Sutanto, W. S.; De Kloet E. R.; Monder, C. Localization of 11β-hydroxysteroid dehydrogenase-tissue specific protector for the mineralocorticoid receptor. *Lancet* **1988**, *ii*, 986–989.

Feng, D-F.; Doolittle, R. F. Progressive alignment and phylogenetic tree construction of protein sequences. *Meth. Enzymol.* **1990**, *183*, 375–387.

Fisher, R. F.; Long, S. R. Rhizobium-plant signal exchange. *Nature* **1992**, *357*, 655–660.

Forman, B. M.; Tontonoz, P.; Chen, J.; Brun, R. P.; Speigelman, B. M.; Evans, R. M. 15-deoxy-$\Delta^{12,14}$-prostaglandin $J_2$ is a ligand for the adipocyte determination factor PPAR. *Cell* **1995**, *83*, 803–812.

Funder, J. W.; Pearce, P. T.; Smith, R.; Smith, A. I. Mineralocorticoid action: Target tissue specificity is enzyme, not receptor, mediated. *Science* **1988**, *242*, 583–586.

Funder, J. W.; Pearce, P. T.; Myles, K.; Roy, L. P. Apparent mineralocorticoid excess, pseudohypoaldosteronism, and urinary electrolyte excretion: toward a redefinition of mineralocorticoid action. *FASEB J.* **1990**, *4*, 3234–3238.

Gaitan, E.; Lindsay, R. H.; Reichert, R. D.; Ingbar, S. H.; Cooksey, C.; Legan, J.; Meydrech, E. F.; Hill, J.; Kubota, K. Anti-thyroid and goitrogenic effects of millet: role of c-glycosylflavones. *J. Clin. Endocrinol. Metab.* **1989**, *68*, 707–714.

Gantchev, T. G.; Ali, H.; van Lier, J. E. Quantitative structure-activity relationships/comparative molecular field analysis (QSAR/CoMFA) for receptor-binding properties of halogenated estradiol derivatives. *J. Med. Chem.* **1994**, *37*, 4164–4176.

Glass, C. K. Differential recognition of target genes by nuclear receptor monomers, dimers, and heterodimers. *Endocrine Rev.* **1994**, *15*, 391–407.

Gold, L. S.; Slone, T. H.; Stern, B. R.; Manley, N. B.; Ames, B. N. Rodent carcinogenesis: setting priorities. *Science* **1992**, *258*, 261–265.

Harmon, M. A.; Boehm, M. F.; Heyman, R. A.; Mangelsdorf, D. J. Activation of mammalian retinoid X receptors by the insect growth regulator methoprene. *Proc. Natl. Acad. Sci. USA* **1995**, *92*, 6157–6160.

Hopwood, D. A.; Sherman, D. H. Molecular genetics of polyketides and its comparison to fatty acid biosynthesis. *Annu. Rev. Genet.* **1990**, 24, 37–66.

Ibrahim, A-R.; Abul-Hajj, Y. J. Aromatase inhibition by flavonoids. *J. Steroid Biochem. Molec. Biol.* **1990**, *37*, 257–260.

Jones, F.A. General introduction. In: A Symposium on Carbenoxolone Sodium. Robson, J. M.; Sullivan, F. M. (eds). Butterworths. London. pp 1–4, 1968.

Kastner, P.; Mark, M.; Chambon, P. Nonsteroid nuclear receptors: what are genetic studies telling us about their role in real life? *Cell* **1995**, *83*, 859–869.

Kato, H.; Fikuda, T.; Parkison, C.; McPhie, P.; Cheng, S. Y. Cytosolic thyroid hormone-binding protein is a monomer of pyruvate kinase. *Proc. Natl. Acad. Sci. USA* **1989**, *86*, 7861–7865.

Kelce, W. R.; Stone, C. R.; Laws, S. C.; Gray, L. E.; Keppalnen, J. A.; Wilson, E. M. Persistent DDT metabolite p,p'-DDE is a potent androgen receptor antagonist. *Nature* **1995**, *375*, 581–585.

Koes, R. E.; Quattrocchio, F.; Mol, J. N. M. The flavonoid biosynthetic pathway in plants: function and evolution. *BioEssays* **1994**, *16*, 123–132.

Korach, K. S.; Chae, K.; Levy, L. A.; Duax, W. L.; Sarver, P. J. Diethylstilbestrol metabolites and analogs: stereochemical probes for the estrogen receptor binding site. *J. Biol. Chem.* **1989**, *264*, 5642–5647.

Krishnan, A. V.; Stathis, P.; Permuth, S. F.; Tokes, L.; Feldman, D. Bisphenol-A: an estrogenic substance is released from polycarbonate flasks during autoclaving. *Endocrinology* **1993**, *132*, 2279–2286.

Kuiper, G. G. J. M.; Enmark, E.; Pelto-Huikko, M.; Nilsson, S.; Gustafsson, J-A. Cloning of a novel estrogen receptor expressed in rat prostate and ovary. *Proc. Natl. Acad. Sci. USA* **1996**, *93*, 5925–5930, 1996.

Lamb, C. J.; Lawton, M. A.; Dron, M.; Dixon, R. A. Signals and transduction mechanisms for activation of plant defenses against microbial attack. *Cell* **1989**, *56*, 215–224.

Lee, Y. S.; Lorenzo, B. J.; Koufis, T.; Reidenberg, M. M. Grapefruit juice and its flavonoids inhibit 11β-hydroxysteroid dehydrogenase. *Clin. Pharmacol. Ther.* **1996**, *59*, 62–71.

Li, J.; Nagpal, P.; Vitart, V.; McMorris, T. C.; Chory, J. A role for brassinosteroids in light- dependent development of *Arabidopsis*. *Science* **1996**, *272*, 398–401.

Lyttle, C. R.; Medlock, K. L.; Sheehan, D. M. Eosinophils as the source of uterine nuclear type II estrogen binding sites. *J. Biol. Chem.* **1984**, *259*, 2697–2700.

Makela, S.; Poutanen, M.; Lehtimaki, J.; Kostian, M-L.; Santii, R.; Vihko, R. Estrogen-specific 17β-hydroxysteroid oxidoreductase type I as a possible target for the action of phytoestrogens. *Proc. Soc. Experimental Biol. Med.* **1995**, *208*, 51–59.

Mangelsdorf, D. J.; Thummel, C.; Beato, M.; Herrlich, P.; Schutz, G.; Umesono, K.; Blumberg, B.; Kastner, P.; Mark, M.; Chambon, P.; Evans, R. M. The nuclear receptor superfamily: the second decade. *Cell* **1995**, *83*, 835–839.

Mangelsdorf, D. J.; Evans, R. M. The RXR heterodimers and orphan receptors. *Cell* **1995**, *83*, 841–850.

Markaverich, B. M.; Roberts, R. R.; Alejandro, M.A.; Johnson, G.A.; Middleitch, B.S.; Clark, J. H. Bioflavonoid interaction with rat uterine type II binding sites and cell growth inhibition. *J. Steroid Biochem.* **1988**, *30*, 71–78.

Martin, P. M.; Horwitz, K. B.; Ryan, D. S.; McGuire, W. L. Phytoestrogen interaction with estrogen receptors in human breast cancer cells. *Endocrinology* **1978**, *103*, 1860–1867.

Middleton, E. Jr.; Kandaswami, C. The impact of plant flavonoids on mammalian biology: implications for immunity, inflammation and cancer. pp 619–652. The Flavonoids: Advances in research since 1986. Ed. Harborne, J. B., Chapman & Hall London, 1993.

Miksicek, R. J. Commonly occurring plant flavonoids have estrogenic activity. *Molec. Pharmacology* **1993**, *44*, 37–43, 1993.

Miller, W. R.; O'Neill, J.S. The significance of steroid metabolism in human cancer. *J. Steroid Biochem. Molec. Biol.* **1990**, *37*, 317–325.

Monder, C. Corticosteroids, kidneys, sweet roots and dirty drugs. *Mol. Cell. Endocrinol.* **1991a**, *78*, C95-C98.

Monder, C. Corticosteroids, receptors, and the organ-specific functions of 11β-hydroxysteroid dehydrogenase. *FASEB J.* **1991b**, *5*, 3047–3054.

Monder, C.; Stewart, P. M.; Lakshmi, V.; Valentino, R.; Burt, D.; Edwards, C.R. Licorice inhibits corticosteroid 11β-dehydrogenase of rat kidney and liver: *in vivo* and *in vitro* studies. *Endocrinology* **1989**, *125*, 1046–1053.

Moore, J. B.; Smith, G. L. Steroid hormone synthesis by a vaccinia enzyme-a new type of virus virulence factor. *EMBO J.* **1992**, *11*, 1973–1980.

Mousavi, Y.; Adlercreutz, H. Genistein is an effective stimulator of sex hormone-binding globulin production in hepatocarcinoma human liver cancer cells and suppresses proliferation of these cells in culture. *Steroids* **1993**, *58*, 301–304.

Nap, J-P.; Bisseling, T. Developmental biology of a plant-prokaryote symbiosis: the legume root nodule. *Science* **1990**, *250*, 948–954.

Pelissero, C.; Bennetau, B.; Babin, P.; Le Menn, F.; Dunogues, J. The estrogenic activity of certain phytoestrogens in the Siberian sturgeon *Acipenser baeri*. *J. Steroid Biochem. Mol. Biol.* **1991**, *38*, 293–299.

Pelissero, C.; Lenczowski, M. J. P.; Chinzi, D.; Davail-Cuisset, B.; Sumpter, J. P.; Fostier, A. Effects of flavonoids on aromatase activity, an *in vitro* study. *J. Steroid Biochem. Molec. Biol.* **1996**, *57*, 215–223.

Persson, B.; Krook, M.; Jornvall, H. Characteristics of short-chain alcohol dehydrogenases and related enzymes. *Eur. J. Biochem.* **1991**, *200*, 537–543.

Peterson, G.; Barnes, S. Genistein and biochanin A inhibit the growth of human prostate cancer cells but not epidermal growth factor receptor tyrosine autophosphorylation. *Prostate* **1991**, *22*, 335–345.

Reevers, F.E. De behandeling von ulcus ventriculi en ulcus duodenei met succus liquiritiae. *Ned. Tijdschr. Geneesk* **1948**, *92*, 2968–2973.

Riddle, J. M.; Estes, J. W. Oral contraceptives in ancient and medieval times. *Amer. Scientist* **1992**, *80*, 226–233.

Ruh, M. F.; Zacharewski, T.; Connor, K.; Howell, J.; Chen, I.; Safe, S. Naringenin: a weakly estrogenic bioflavonoid that exhibits antiestrogenic activity. *Biochem. Pharmacology* **1995**, *50*, 1485–1493.

Scambia, G.; Ranelletti, F. O.; Panici, P. B.; Piantelli, M.; Rumi, C.; Battaglia, F.; Larocca, L. M.; Capelli, A.; Mancuso, S. Type-II estrogen binding sites in a lymphoblastoid cell line and growth-inhibitory effect of estrogen, anti-estrogen and bioflavonoids. *Int. J. Cancer* **1990**, *46*, 1112–1116.

Setchell, K. D. R.; Borriello, S. P.; Hulme, P.; Kirk, D. N.; Axelson, M. Non-steroidal estrogens of dietary origin: possible role in hormone-dependent disease. *Am. J. Clin. Nutr.* **1984**, *40*, 569–578.

Sharma, O. P.; Adlercreutz, H.; Strandberg, J. D.; Zirkin, B. R.; Coffey, D. S.; Ewing, L. L. Soy of dietary source plays a preventive role against the pathogenesis of prostatisis in rats. *J. Steroid Biochem. Molec. Biol.* **1992**, *43*, 557–564.

Soto, A. M.; Justicia, H.; Wray, J. W.; Sonnenschein, C. p-Nonylphenol: an estrogenic xenobiotic released from "modified" polystyrene. *Environ. Health Perpect.* **1991**, *92*, 167–173.

Stafford, H. A Flavonoid evolution - an enzyme approach. *Plant Physiol.* **1991**, *96*, 680–685.

Stewart, P. M.; Valentino, R.; Wallace, A. M.; Burt, D.; Shackleton, C. H. L.; Edwards, C. R. W. Mineralocorticoid activity of liquorice: 11-beta-hydroxysteroid dehydrogenase deficiency comes of age. *Lancet* **1987**, *ii*, 821–823.

Szekeres, M.; Nemeth, K.; Koncz-Kalman, Z.; Mathur, J.; Kauschmann, A.; Altmann, T.; Redei, G. P.; Nagy, F.; Schell, J.; Koncz, C. Brassinosteroids rescue the deficiency of CYP90, a cytochrome P450, controlling cell elongation and de-etiolation in *Arabidopsis*. *Cell* **1996**, *85*, 171–182.

Unsworth, B.; Hennen, S.; Krishnakumaran, A.; Ting, P.; Hoffman, N. Teratogenic evaluation of terpenoid derivatives. *Life Sci.* **1974**, *15*, 1649–1655.

Vie, M-P.; Blanchet, P.; Samson, M.; Francon, J.; Blondeau, J-P. High affinity thyroid hormone-binding protein in human kidney: kinetic characterization and identification by photoaffinity labeling. *Endocrinology* **1996**, *137*, 4563–4570.

White, R.; Jobling, S.; Hoare, S. A.; Sumpter, J. P.; Parker, M. G. Environmentally persistent alkylphenolic compounds are estrogenic. *Endocrinology* **1994**, *135*, 175–182.

Whitten, P. L.; Naftolin, F. Effects of a phytoestrogen diet on estrogen-dependent reproductive processes in immature female rats. *Steroids* **1992**, *57*, 56–61.

Wiese, T. E.; Dukes, D.; Brooks, S. C. A molecular modeling analysis of diethylstilbestrol conformations and their similarity to estradiol-17β. *Steroids* **1995**, *60*, 802–808.

Wong, E.; Flux, D. S. The oestrogenic activity of red clover isoflavones and some of their degradation products. *J. Endocrinol.* **1962**, *24*, 341–348.

# INDEX

Abscisic acid, 79
Acacetin, 186, 191, 210, 222
   acacetin-7-$O$-β-D-galactopyranoside, 191, 210
*Acaulospora*, 25
Acetylsalicylic acid, 78
Adhesion molecules, 4, 115, 179, 183, 184, 188, 189
*Afraegle paniculata*, 95, 96
*Agrobacterium*-mediated transformation, 72, 82
Aldosterone, 252
Alfalfa, 6, 27, 31–33, 45, 46, 48, 49, 52–56, 58, 60–62, 64, 65, 70, 81, 258, 260
Allelochemicals, 56, 150
Allergy, 181, 182, 189
   allergic response, 4
*Allium cepa*, 13, 32
*Amorpha fructicosa*, 193
Amorphigenin, 193
   12αβ-hydroxyamorphigenin, 193
Amorphispironone, 193, 204, 223, 225
*Anacardiaceae*, 193
Analogue(s), 3, 82, 191–193
Androgen(s), 6, 250, 251, 253, 258, 266
Anther, 35, 37, 42
Anthocyanin(s), 43, 63, 70, 77, 80, 82–84, 157, 163, 255
Anti-allergic, 192, 222
Anti-HIV, 3, 191, 192, 207, 209–213, 216, 220–223, 225
Anti-inflammatory, 68, 106, 183, 187–189, 192, 209, 210, 222, *see also* Antiinflammatory
Anti-invasive agents, 201
Antiadhesive activity, 171, 172
Antiatherogenic activity, 177
Antibacterial, 151, 209
Anticancer, 3, 85, 100, 106, 115, 149, 175, 176, 180, 207, 219, 223–225, 227, 238, 245
   anticancer activity, 100, 106, 115, 175
   anticarcinogenic activity, 180
Antiestrogen, 28, 256
Antifungal, 32, 128, 129, 209, 224
Antihepatotoxic, 192
Antiinflammatory, 3, 151, 175, 176, 207
   antiinflammatory activity, 176
Antileukemic, 191, 192, 200, 223
Antimetastatic activity, 180
Antimicrobial agents, 13, 62
Antimitotic agent(s), 194, 216, 223
Antioxidant(s), 3, 5, 68, 85, 103, 124, 149–164, 175–177, 181, 190, 192, 223, 224, 237, 247
   antioxidant activity, 5, 151, 153, 159–161, 163, 164, 177, 181, 223
   antioxidative, 128, 129
Antiproliferative, 6, 115, 180, 199, 222, 228, 230, 238, 247
Antispirochetal, 209
Antitumor, 3, 124, 180, 191, 192, 198, 200, 201, 208, 216, 221, 223–225
Antiviral, 115, 151, 153, 175, 176, 181, 182, 192, 207–209, 211, 213, 221–225
Apigenin, 4, 17, 25, 27, 28, 76, 94, 97, 98, 168, 179, 183, 185–189, 191, 193, 203, 205, 209, 225
   apigenin-7-$O$-β-D-glucopyranoside, 191, 209, 225
   tri-$O$-methyl-apigenin, 168
Aplysiatoxin, 203
Apoptosis, 115, 201, 222
   flavonoid-induced apoptosis, 201
Apples, 154, 157–159
Arachidonic acid, 78, 176, 189
Arbuscular mycorrhiza, 9, 32
   arbuscular mycorrhizal fungi, 2, 9, 30–32
   AM, 9, 10, 12, 13, 17, 25–28, 30, 32, 162, 163, 181, 189, 224, 235, 246–248, 267
   AM symbiosis, 10, 13, 17, 28, 30
   AMF, 2, 9–19, 26, 27, 30
*Arnica*, 201, 225
Arthritis, 192
Ascorbic acid, 85, 160, 161, *see also Vitamin C*
*Aspergillus*, 117, 118, 128, 129
*Asteraceae*, 194
Asthma, 176–179, 204
*Atalantia*, 72, 96
Atherogenic, 155, 156, 166, 167, 172

Atherosclerosis, 153–155, 163, 166, 168, 172, 181
*Aurantioideae*, 85, 86–90, 93, 94, 97, 100, 101
Auxin transport regulators, 53, 68, 82
Axillarin, 206, 209

Baicalein, 137, 141, 200, 211–213, 224, 232, 248
Baicalin, 185, 200, 201, 224, 234, 245, 248
*Balsamocitrinae*, 94, 95
*Balsamocitrus*, 94
Basophil histamine release, 6, 116, 178, 181, 203, 223, 224
Basophils, 115, 175, 177–179
Beans, 157
Benzyladenine, 78
Beverages, 151, 156–158, 160–162
Biochanin A, 24, 26–28, 213, 243, 248, 267
Bioflavonoid(s), 6, 31, 103, 106, 115, 163, 189, 222, 266, 267
Biological activities, 4, 56, 71, 85, 91, 105, 124, 131, 192, 216
Biosynthesis, 2–4, 32, 36, 37, 43, 44, 46, 53, 54, 56, 58, 60, 64–66, 68–74, 80–83, 85, 97–101, 266
Black swallowtail butterfly, 141, 143, 145, 150
Black tea, 154, 157–163
Blastogenesis, 200, 201
Blood, 4, 6, 7, 103, 105–107, 116, 153, 154, 158, 162, 166–169, 171–173, 177, 178, 180, 189, 200, 201, 244, 245, 262
    blood cell aggregation, 105, 166–169, 172, 173
    blood cell clumping, 106, 169
    blood flow, 103, 166, 171
    blood viscosity, 4, 105, 171
    red blood cell aggregation, 105
Bombesin, 210, 224
*Bradyrhizobium*, 17, 28, 31
*Brassicaceae*, 10
Brassinolide, 252, 256
Broccoli, 159, 163
Butein, 140

Cabbage looper, 141, 143, 144, 146, 147, 149
E-cadherin, 107, 115
Caffeic acid, 58, 65, 157, 222, 240
    caffeic acid 3-*O*-methyltransferase, 58, 65
Calamondin, 74, 80, *see also C. mitis*
Calcium channel blockers, 105
Callus, 72, 75, 76, 78, 80, 81, 84
*Calycopteris floribunda*, 197
Calycopterone, 197
Cancer, 3, 4, 6, 105–107, 112, 113, 115, 116, 149, 150, 153, 162, 163, 180–182, 192, 194, 197, 199–201, 207, 221, 222, 223–225, 227, 228, 230–232, 235–238, 242, 243, 246–250, 264–267
    breast cancer, 107, 115, 227, 228, 230–232, 235, 236, 248, 264–266
    MCF-7 human breast cancer cells, 107, 233–236
    non-small cell lung cancer, 194, 207, 224

Cancer (*cont.*)
    small cell lung cancer, 194, 207, 224
    central nervous system cancer, 194
    ovarian cancer, 194
    pancreatic cancer, 200
    prostate cancer, 197, 200, 224, 248, 249, 267
    renal cancer, 194
Capillary, 1, 85, 103, 113, 151, 165, 166, 172, 173, 246
    capillary fragility, 103, 113, 165, 166, 173
    capillary permeability, 103, 151
    capillary resistance, 166, 172
Carbenoxolone, 263, 266
Carcinogenic, 68, 106, 131, 150, 192, 205, 207, 223
Carcinogens, 180, 222, 238
    anthracene, 228, 236
        2-aminoanthracene, 206
        benz[a]anthracene, 206
        7,12-dimethylbenz(a)anthracene, 228, 236
    cholanthrene, 238, 239
        3-methyl cholanthrene, 238, 239
    chrysene, 206
    fluorene
        acetylaminofluorene, 206
    pyrene, 205, 206
        benzo[a]pyrene, 205, 206
    benzo[c]phenanthrene, 206
Carcinoma
    colon adenocarcinoma, 199, 207
    ileocecal carcinoma, 216
    oral carcinomas, 105
    oral epidermoid carcinoma, 197
    squamous cell carcinoma, 6, 115, 199, 222, 223
Cardiovascular problems, 197
Carotene, 105, 116, 153, 177, 238
    α-carotene, 238
    β-carotene, 177
Carotenoid(s), 155, 237, 238, 247, 248
Catechin, 105–107, 133, 140, 152, 156, 160, 163, 176, 201, 202, 205, 209, 213, 221, 237, 240
    (+)-catechin, 201, 202, 221
    D-catechin, 205
    (–)-*epi*catechin, 161, 211
    (–)-*epi*catechin-3-*O*-gallate, 211
    *epi*-catechin, 213
    *epi*-gallocatechingallate, 157, 161, 241
    *epi*catechin, 105, 106, 158, 211
    *epi*gallocatechin gallate, 157, 158, 160, 163
Catechol, 5, 133, 143, 156
Cell(s), 2–6, 9, 10, 13, 17, 31, 35, 36, 38, 39, 43–46, 53, 54, 56, 60–67, 69–84, 86, 92, 100, 101, 105–108, 112, 113, 115, 116, 166–169, 171–173, 175–177, 179–181, 183, 184, 187–190, 193, 194, 197, 199–202, 204, 205, 207, 213, 216, 221–225, 228, 230, 232, 236, 245–248, 251, 263–267
    activated cells, 4–6, 175, 177, 187
    blood cell aggregation, 105, 166–169, 172, 173
    blood cell clumping, 106, 169
    cell activation, 4, 175, 177, 179, 184, 199

Cell(s) (*cont.*)
  cell adhesion, 4–6, 100, 107, 115, 116, 172, 179, 181, 183, 184, 188, 189, 201
  cell culture, 3, 67, 69, 70, 72, 74, 77, 79–81, 197, 221, 246
  cell damage, 3
  cell proliferation, 4, 180, 202, 236
  cell types, 4, 175, 179, 183, 184, 232
  cell-cell adhesion, 107
  cell-cell interaction, 2
  endothelial cells, 179, 187
  epithelial cells, 107
  leukemic cells, 84, 101, 108, 116, 200, 224
  mast cells, 4, 175, 177, 179
  MCF-7 human breast cancer cells, 107, 235, 236
  MO4 malignant mouse tumor cells, 106
  non-small cell lung cancer, 194, 207, 224
  red blood cell aggregation, 105
  small cell lung cancer, 194, 207, 224
  squamous cell carcinoma, 6, 115, 199, 222, 223
Cellular phagocytic activity, 197
*Centaurea urvillei*, 194, 225
*Cephalocereus senilis*, 57, 62, 65, 82
Chalcone, 2, 6, 7, 36, 43–45, 53, 54, 57–61, 63–65, 70, 74, 81, 82, 101, 133, 196, 203, 222
  chalcone 2′-*O*-methyltransferase, 57, 58, 61
  chalcone 4′-*O*-methyltransferase, 45
  chalcone flavanone isomerase, 74
  chalcone isomerase, 45, 50, 54, 57, 58, 65, 81
  chalcone reductase, 45, 53, 54, 57, 58, 60, 64
  chalcone synthase, 2, 6, 7, 36, 43–45, 51, 53, 54, 57, 59, 63–65, 74, 81, 82
  4,4′-dihydroxy-2′-methoxychalcone, 26, 46
  naringenin chalcone, 45, 70
Chemical functionality, 67
Chemometric, 117, 118
  cluster analysis, 117, 119
  composition patterns, 118, 120
  discriminant analysis, 117, 119
  GA-LDA, 117, 119, 121
  genetic algorithm, 117, 119, 121, 129
  linear discriminant analysis (LDA), 117, 119–121
  pattern recognition, 117–120, 127
  soft independent modelling of class analogy (SIMCA), 117, 119–121
  SPSS, 119
  UNSCRAMBLER, 119
Chemotaxonomy, 92
*Chenopodiaceae*, 10
CHI transcript, 50, 52
Chinese herbs, 210
Chlorogenic acid, 59, 157, 161
Cholecystitis, 210
Cholesterol, 153–155, 163, 168, 247, 265
Cholic acid, 168
Chondroitin sulfate A (CSA), 168, 173
2-phenylthiochromen-4-ones, 191, 214, 216, 218–221, 225
  5,6,7,8-substituted-2-phenylthio-chromen-4-ones, 216

$\gamma$-chromones, 212, 213
*Chrysanthemum morifolium*, 191, 209, 222
Chrysin, 24, 185, 191, 205, 209, 240
Chrysoeriol, 31, 185, 209
Chrysoplenol, 209
Chrysoplenoside, 209
CHS genes, 46, 48
CHS transcript, 46, 48–52
Cinnamic acid(s), 57, 58, 60, 62, 65, 70, 77, 152
  cinnamate 4-hydroxylase, 57, 65, 74
  cinnamic acid esters, 77
*Citrinae*, 86, 91–94, 96
Citromitin, 91
  5-*O*-desmethylcitromitin, 91
Citron, 71, 93, see also *C. medica*
*Citrus* relatives, see Aurantioideae
  *C. aurantifolia*, 76, 80, see also lime
  *C. aurantium*, 80, 97, 100, see also sour orange
  *C. grandis*, 92, 93, 101, see also pummelo
  *C. limon*, 80, see also lemon
  *C. limonia*, 100
  *C. medica*, 82, 83, 93, see also citron
  *C. mitis*, see also Calamondin
  *C. paradisi*, 80, 82, 83, 100, 101, see also grapefruit
  *C. reticulata*, 91–93, 100, 196, see also mandarin or tangerine
  *C. sinensis*, 14, 81–83, 101, see also sweet orange
Citrus, 1, 3, 4, 6, 31, 43, 67, 68, 71–86, 90–95, 97, 99–101, 103, 105–107, 113, 115, 116, 159, 160, 162, 165, 166, 171, 172, 173, 196, 201, 222, 224, 227, 230, 231, 235, 236, 240, 246
  zinc-deficient citrus, 97, 99
*Clauseneae*, 91, 94
*Clymenia*, 95
Coenzyme Q10 155
Colchicine, 194, 216
Cold pressed peel oil solids, 103, 109, 110, 112
*Combretaceae*, 197
Combretastatin, 216
*Compositae*, 206, 209, 210
Conditional male fertility, 7, 36, 43, 44
Coronary thrombosis, 106, 166
Cough, 192, 204
*P*-coumaric acid, 13, 57, 65
  *p*-coumaryl-CoA, 70
  4-coumarate:CoA ligase, 57, 59
  4-coumaroyl CoA, 45, 57, 60
Coumestrol, 13, 17, 28, 243, 247
Cromolyn, 177
*Cryptocarya kurzii*, 196, 222
*Cudrania tricuspidata*, 197, 198, 223
Curcumin, 105, 116
Cyanide, 134–136, 146–150
Cyanidin, 152, 157, 161, 185
Cyanogenic glycosides, 148
Cyanohydrin, 135
Cyclic voltammetry, 140
Cycloheximide, 201

Cytochrome P450 60, 65, 74, 81, 105, 115, 248, 256, 267
Cytokine, 4, 100, 179, 181, 183, 184, 187–189, 245
Cytotoxic, 42, 68, 131, 150, 179, 182, 191–194, 196–198, 201, 202, 207, 212, 216, 221–225, 228, 230, 236

Daidzein, 13, 17, 28, 57, 61, 62, 117, 123–125, 128, 238, 240, 242–248
Dancy tangerine, 100, 103, 108, 109, 112, 113, 115
*Daucus carota*, 12, 32, 82, 83
*p,p'*-DDE, 258, 266
DDT, 258, 266
*o,p'*-DDT, 258
Delphinidin, 134, 140, 141, 205
*O*-desmethylangolensin, 243
Diabetes, 197
2,4-dichlorophenoxyacetic acid, 78
Dicoumarol, 152
Diethylstilbestrol (DES), 256–258, 266, 267
Diets, 79, 124, 154, 166–168, 180, 227, 228, 243
Dihydrofisetin, 209
2,3-dihydroxysuccinic acid, 127, see also Tartaric acid
Diosmetin, 76, 97, 98, 205, 209
Diosmin, 73, 76, 97, 99, 105
Disaggregation, 169
DMBA, 115, 228, 234, 236

Ecdysone, 251, 253, 260
 α-ecdysones, 252, 260
 β-ecdysones, 260
*Egletes viscosa*, 210, 223
Electron transport, 132, 134, 137
Enzymes, 2–6, 37, 43, 45, 55–59, 61–66, 69, 70, 72, 74, 81, 84, 105, 124, 131, 134, 141, 148, 150, 175–177, 180, 211, 213, 228, 249, 250, 253, 255, 256, 260, 262–265, 267
 acetyl CoA carboxylase, 56, 57
 ATPase(s), 149, 177
 caffeic acid 3-*O*-methyltransferase, 58, 65
 catalase, 135, 137, 141, 144, 148
 chalcone 2'-*O*-methyltransferase, 58, 61
 chalcone 4'-*O*-methyltransferase, 45
 chalcone flavanone isomerase, 74
 chalcone isomerase, 45, 54, 58, 65, 81
 chalcone reductase, 45, 53, 54, 58, 60, 64
 chalcone synthase, 2, 6, 7, 36, 43–45, 53, 54, 57, 59, 63–65, 74, 81, 82
 cinnamate 4-hydroxylase, 65, 74
 4-coumarate:CoA ligase, 59
 cyclic nucleotide phosphodiesterase, 177
 cyclooxygenase, 4, 151, 177, 183, 184, 188
 cytochrome oxidase, 134
 flavanone 3β-hydroxylase, 255
 flavanone 7-*O*-glucosyltransferase, 74
 flavonol 3-*O*-galactosyl transferase, 38
 dihydroflavonol 4-reductase(s), 255
 dihydroflavonol reductase, 255, 265
 GalTase, 41, 43

Enzymes (*cont.*)
 glutathione reductase, 131, 141, 144, 145, 148, 149, 151
 integrase, 181, 192, 213, 214, 222, 224
 isoflavone *O*-methyltransferase, 61, 64
 isoflavone reductase, 45, 46, 54, 58
 isoflavone synthase, 62
 β-ketoreductases, 254
 lipoxygenase(s), 4, 151, 177, 183, 188, 203, 222, 224
 methyltransferase, 45, 54, 58, 61, 62, 64–66, 74, 80
 mitochondrial succinoxidase, 131, 133, 140, 149
 NADH-oxidase, 4, 131, 133, 134, 149
 phenylalanine ammonia lyase (PAL), 45, 52, 53, 58–60, 62, 63, 74
 L-phenylalanine ammonia-lyase, 58, 64, 83, 101
 phosphatase A2 4
 phosphatidyl inositol kinase, 180
 phospholipase, 151, 175–177, 188, 189
 phospholipase A2 151, 175–177, 189
 phospholipase C, 177
 phospholipases, 4, 183
 protein kinase, 4, 6, 175, 177, 178, 181, 184, 188, 189, 202, 222–224, 232, 234–236
 protein kinase C, 4, 6, 175, 177, 178, 181, 202, 222–224, 232, 234, 235
 protein tyrosine kinase, 4, 177, 188
 proteinase, 181, 192, 212, 213, 222
 reverse transcriptase, 176, 192, 210–212, 214, 224, 225
 S-adenosyl methionine synthase, 56
 3β-hydroxysteroid dehydrogenase, 255
 11β-hydroxysteroid dehydrogenase, 255, 262, 265, 266
 17β-hydroxysteroid dehydrogenase, 253, 255, 260, 262, 264
 20β-hydroxysteroid dehydrogenase, 254, 265
 superoxide dismutase, 135, 144, 150
 topoisomerase, 4, 180, 188, 189, 214, 216, 228, 235
 tyrosine protein kinase, 232, 236
Epidemiological studies, 153, 227, 250
Equol, 243
*Eremocitrus*, 95
Eriocitrin, 73, 76, 103
Eriodictyol, 75, 80
*Erwinia*, 52
Erythrocyte, 3, 4, 114, 168, 169, 172, 173
 erythrocyte adhesion, 4
 erythrocyte aggregation, 3, 4, 168, 172, 173
 erythrocyte sedimentation rate (ESR), 138, 140, 168, 169, 171
Estradiol, 27, 28, 231, 255, 256, 258, 260, 266, 267
 17β-estradiol, 27, 28
Estrogen, 27, 28, 33, 107, 230–232, 235, 236, 248, 250, 251, 256, 258, 264, 266, 267
 estrogen receptor, 107, 230–232, 235, 236, 250, 251, 256, 258, 264, 266
 estrogenic, 31, 128, 250, 256, 258, 265–267

# Index

*Fabaceae*, 206
Felodipine, 105, 115, 228, 235
Ferulic acid, 152
N-feruloyltyramine, 13
Fisetin, 26, 140, 178, 180, 202, 203, 205, 213
Flavane, 245
Flavanone(s), 1, 4, 17, 25, 46, 54, 60, 62, 70, 71, 73, 74, 79–83, 86, 90–92, 94–97, 99–101, 103, 106, 109, 111, 112, 152, 159, 176, 180, 184, 188, 196, 203, 204, 211, 212, 222, 255
  flavanone 3β-hydroxylase, 255
  flavanone 7-*O*-glucosyltransferase, 74
  flavanone aglycone, 71, 73
  flavanone glycosides, 4, 71, 73, 81, 86, 90, 91, 94–97, 103, 106
  flavanone neohesperidosides, 71, 97
  flavanone rutinosides, 71
  biflavanones, 203
  5,6,7,8,3′,4′-hexamethoxyflavanone, 111, 114
  5,7,8,3′,4′-pentamethoxyflavanone, 111, 114
  prenyl (3-methyl-2-butenyl), 204
  prenylated flavanones, 95
Flavans, 196, 211, 224
  biflavans, 196
Flavone(s), 3, 4, 7, 17, 25, 26, 32, 70, 73, 75, 81, 86, 90–92, 94, 97–101, 105–109, 111–113, 115, 133, 136, 150, 152, 154, 156, 162, 168, 169, 171–173, 181, 185, 186, 189, 196–199, 201, 203–205, 207, 209–214, 216, 221–225, 235, 236, 240, 245, 247, 260
  flavone 8-acetic acid, 207
  flavone acetic acid ester, 207
  flavone-xanthone C-glucoside, 210
  amentoflavone, 177, 185, 211
  5,6-benzoflavone, 205
  7,8-benzoflavone, 205
  4′,7-dihydroxyflavone, 17, 26, 27
  7,4′-dihydroxyflavone, 46
  7,8-dihydroxy flavone, 136
  methoxylated flavones, 75, 168, 169, 171–173, 197
  3-methoxyflavones, 209, 222, 225
  3′,4′,3,4,6,7,8-heptamethoxyflavone, 169
  3′,4′,5,6,7,8-hexamethoxyflavone, 169
  3′,4′,5,6,7-pentamethoxyflavone, 169
  4′,5-dihydroxy-3,3′,7,8-tetramethoxyflavone, 210
  5,2′-dihydroxy-6,7,8,6′-tetramethoxyflavone, 194, 198
  5,4′-dihydroxy-3,6,7,8,3′-pentamethoxyflavone, 194
  5′,7′-dihydroxy-6,8,3′,4′-tetramethoxyflavone, 111
  3,7-dimethoxyflavone, 201, 202, 224
  7,8-dihydroxy-3′,4′-dimethoxylflavone, 199
  7-chloro-3,5,6,8,3′,4′-hexamethoxyflavone, 112
  dihydroxypentamethoxyflavones, 191
  heptamethoxyflavone, 106–108, 168, 169
  7″-*O*-methylrobustaflavone, 197
  5,7,2′-trihydroxy, 204
  polymethoxylated flavones, 3, 4, 86, 94, 99, 100, 105–109, 112, 113, 115

Flavonoid(s), 1–7, 9, 10, 13, 17, 25–28, 30–32, 37, 43–46, 49–56, 58–62, 64–86, 90–94, 97, 99, 100, 101, 103, 105–107, 111, 114–116, 124, 131, 133–138, 140, 141, 148–156, 158, 159, 162, 163, 165, 166–169, 171–173, 175–184, 186–192, 194, 196, 198–205, 207, 209–213, 216, 221–225, 227, 228, 230–232, 235–243, 245–251, 255, 256, 258, 260–267
  flavonoid absorption, 140
  flavonoid binding protein, 6
  flavonoid biosynthesis, 2, 3, 32, 44, 54, 56, 60, 65, 69, 70, 72–74, 81–83, 97, 98, 101
  flavonoid esters, 90
  flavonoid-induced apoptosis, 201
  methoxylated flavonoids, 168
  prenyl (3-methyl-2-butenyl), 204
  prenylated flavonoids, 90, 94
Flavonol(s), 3, 6, 7, 17, 25, 30, 35–44, 70, 76, 84, 90, 95, 100, 101, 116, 133, 152, 154–156, 163, 164, 173, 176, 181, 182, 184, 194, 204, 221, 222, 224, 225, 236, 247
  flavonol 3-*O*-galactosyl transferase, 36, 38
  flavonol aglycones, 35–37, 42
  flavonol glycoside, 84
  flavonol receptor, 42
  flavonol uptake, 42
  flavonol-deficient pollen, 35, 36
  dihydroflavonol 4-reductase(s), 255
  dihydroflavonol reductase, 255, 265
  dihydroflavonols, 70, 197, 223
Formononetin, 17, 26, 28, 57, 62, 243
*Fortunella*, 86, 90, 95, 100
Fremontin, 206, 223
Fremontone, 206, 223
French paradox, 153, 162, 163, 177
Fucosylated surface glycopeptides, 202
Fungi, 2, 9, 13, 28, 30–32, 57, 68, 69, 79, 117, 118, 124, 249, 258
Furanocoumarin, 105, 149
Fustin, 140

Gallic acid, 152
Galangin, 24, 140
Gastritis, 197
Gastroprotection, 210
Gene activation, 28, 58, 83
Gene expression, 3–6, 35, 49–52, 68, 100, 181, 183, 186–189
Genistein, 26, 62, 105, 117, 123–125, 180, 185, 200, 205, 213, 224, 228, 230–232, 234–236, 238, 240–248, 257, 258, 267
  8-hydroxygenistein, 117, 125
  5-*O*-methylgenistein-7-*O*-β-D-glucopyranoside, 209
*Gentianaceae*, 210
Gericudranins, 197
Gibberellic acid, 78
*Gigaspora*, 12, 14, 15, 25, 26, 30–32
Ginsenoside-Rb1 200
Glioblastoma, 197

*Glomus*, 10, 11, 14, 15, 25, 26, 28, 31–33
Glucose, 38, 43, 44, 71, 73, 82, 90, 118, 153, 157
  *C*-glucosides, 90
  *O*-glycosides, 86, 94, 99, 152
*Glycine max*, 13, 17, 28, 31, 66, 247
*Glycosmis*, 90
Glycyrrhetinic acid, 204, 263
*Glycyrrhiza glabra*, 262
Glycyrrhizic acid, 261–263
Golgi apparatus, 43
*Gramineae*, 13
Grape juice, 161
Grapefruit, 67, 71–75, 78–83, 97, 99–101, 103, 105, 115, 116, 159, 160, 171–173, 227, 228, 230, 234, 235, 247, 266, see also *C. paradisi*
  grapefruit juice, 80, 105, 115, 160, 161, 228, 230, 234, 247, 266
Green tea, 157, 158, 163
*Gutierrezia microcephala*, 194, 222

Heart disease, 5, 7, 151, 153, 154, 156, 162–164, 172
Heated aroma, 118
Hepatitis, 197, 209
Hepatogenous jaundice, 210
Hepatoprotection, 210
Hesperetin, 24, 26, 27, 71, 73, 75, 80, 90, 97, 100, 103, 105, 106, 157, 160, 161, 176, 202, 203, 209, 230, 232, 234, 235, 237, 240
  hesperetin 7-*O*-glucoside, 73, 80, 97, 100
  hesperetin triglycoside, 92
Hesperidin, 1, 71, 73–76, 78, 85, 90, 91, 97, 99, 103, 105–107, 115, 116, 157, 159, 160–162, 165–169, 172, 181, 227, 237, 240
Hexa-*O*-methyl gossypetin, 106
High density lipoprotein, 155
High-performance liquid chromatography (HPLC), 36, 39, 59, 68, 77, 95, 108, 109, 113, 114, 117–121, 124, 127, 154, 156, 158, 159, 237, 242, 243, 245–248
Hinokiflavone, 185, 191, 192, 223
Hispidulin, 194
Histamine, 6, 106, 115, 116, 177, 178, 181, 203, 223, 224
  histamine release, 6, 106, 115, 116, 177, 178, 181, 203, 223, 224
*Hoffmanosseggia intricata*, 206, 225
Homoisoflavones, 206
Homoisoflavonoids, 206, 225
Hormone(s), 2, 5, 30, 33, 44, 52, 79, 180, 224, 232, 249–251, 254, 256, 258, 260–267
Hydrocortisone, 252, 255, 263
Hydroxyfarrerol, 245, 248
Hydroxyl radical, 138
Hymenoxin, 194
*Hypericum*, 77, 82
Hyperplastic lesions, 105
Hyphae, 9, 12, 17, 28
  hyphal growth, 2, 10, 12, 16, 17, 20–23, 25–28, 30–32
  hyphal growth stimulation, 12, 28

ICAM-1 179, 183, 184, 186, 188, 189
Immunocytes, 201
Immunosuppression, 204
Indoleacetic acid, 78
Indolebutyric acid, 78
*In vivo* absorption, 5, 151, 158
Inflammation, 3–5, 150, 163, 179, 181, 183, 187, 236, 248, 266
Insect model, 143
Inflammatory genes, 4, 183
Interleukin-1 183
Interspecies hybrids, 72
Intricatin, 206, 225
Intricatinol, 206, 225
Isochamaejasmin, 203
Isocryptomerin, 197
Isoflavone(s), 17, 25, 31, 45, 46, 54, 58, 61, 62, 64–66, 70, 105, 117, 118, 123–125, 127–129, 177, 180, 184, 191, 194, 206, 222, 223, 228, 237, 240, 242–248, 264, 265, 267
  isoflavone aglycones, 128
  isoflavone derivatives, 117, 125, 128
  isoflavone *O*-methyltransferase, 57, 61, 64
  isoflavone reductase, 45, 46, 54, 57, 58
  isoflavone synthase, 57, 62
  isorhamnetin, 94, 209
  isorhoifolin, 76, 97, 99
  (5′-α,α-dimethylallyl)-5,7,2′,4′-tetrahydroxyisoflavone, 206
  6,7,4′-trihydroxyisoflavone, 128, 129

Jaceosidin, 201
Juglone, 152
Juvenile hormone, 261

Kaempferid, 205
Kaempferol, 24–26, 36–41, 43, 44, 76–79, 83, 94, 140, 157, 159, 191, 192, 203–205, 209, 223
  kaempferol 3-*O*-galactoside, 39–41, 43
  kaempferol 3-*O*-glucoside, 37, 40
  kaempferol glycosides, 38, 40, 77, 78, 83
  kaempferol-3-*O*-β-D-glucopyranoside, 191, 192
  kaempferol-3-*O*-galactoside, 41
  dihydrokaempferol, 25
*Kummerowia striata*, 191, 209, 225
Kurzichalcolactones, 196
Kurziflavolactones, 196
Kuwanon, 210, 224

*Lauraceae*, 196
LDL oxidation, 156, 159, 160
Leaf, 59, 69, 71, 72, 77, 86, 90, 95–97, 99, 101, 124, 149
Leaves, 43, 59, 60, 71, 73–76, 78, 80–82, 86, 97, 99, 100, 103, 109, 112, 113, 153, 194, 197, 204
*Leguminoseae*, 9, 13
Lemon, 67, 73–75, 82, 84, 101, 103, 105, 165
Leucicyanidin, 209

# Index

Leukemia, 107, 194, 197, 200, 201, 211, 216
  human acute promyelocytic leukemia, 197
  lymphocytic leukemia, 200, 201
  mouse myeloid leukemia, 197
Lignin, 58–60, 64, 65
*Liliaceae*, 13
Lipoprotein, 7, 154, 155, 160–164
Liquiritigenin, 26, 57, 61
*Lotus corniculatus*, 63
Low density lipoprotein, 155
Low molecular weight dextran (LMD), 168
Lung function, 154, 162
*Lupinus polyphyllus*, 28
Lutein, 238
Luteolin, 17, 24, 26, 27, 31, 94, 140, 152, 178, 185, 202, 203, 205, 209, 258–260
  luteolin-7-*O*-glucoside, 26

Macrophage(s), 4, 155, 175, 177, 179, 189
Magniferin, 152
Maillard reaction, 118
Malonyl-CoA, 45, 70
Mandarin, 71, 75, 77, 82, 91, 92, 103, 105, 107, *see also C. reticulata*
Mauritianin, 204
Maysin, 260, 265
MDA-MB-435 estrogen receptor, 107, 233, 234
*Medicago sativa*, 13, 17, 45, 46, 53, 54, 64, 65
*Medicago truncatula*, 31, 63
Medicarpin, 17, 46, 53, 56, 57, 61, 62, 65
  medicarpin malonyl glucoside, 17
Medulloblastoma, 194
Melanoma, 105, 115, 194, 199, 207, 216, 223
Metabolic
  metabolic compartments, 58
  metabolic engineering, 2, 55
  metabolic flux, 41, 55, 58, 63
Metastasis, 4, 6, 115, 201, 221
Methoprene, 261, 262, 266
  methoprene acid, 261, 262
Mexican lime, 77, 78
*Microcitrus*, 95
Microtubule assembly, 178
Mitochondria, 7, 101, 132, 134, 137
  mitochondrial FeS centers, 132, 134
  mitochondrial respiration, 92, 101, 131–133, 148–150
  mitochondrial succinoxidase, 131–133, 137, 140, 149
Mitogen, 177, 184, 188, 189, 200, 201
  mitogen-activated protein kinase, 184
  mitogenic stimuli, 199
Monocytes, 4, 177, 184
*Moraceae*, 197, 204
Morin, 24, 26, 105, 140, 203, 205, 209
*Morus alba*, 204, 210, 223
Morusin, 210
  morusin-4'-glucoside, 210

Mouse, 6, 106, 131, 150, 180, 187, 197, 201, 203, 204, 221, 223, 238, 248, 261
  mouse fibroblast, 238
  mouse myeloid leukemia, 197
Mulberry trees, 204
*Muntingia calabura*, 195, 196, 223
*Murraya*, 90, 96, 115
*Murraya koenigii*, 96
Mutagenic, 68, 131, 192, 205–207, 223
Myeloblastosis, 211
Myelocytes, 201
Myricetin, 26, 27, 134–137, 140, 141, 205, 206, 211, 212
Myricitrin, 157, 204, 205

Naphthaleneacetic acid, 78
Naringenin, 25–27, 45, 70, 71, 73, 75, 81, 83, 90, 97, 101, 103, 105, 106, 115, 152, 159, 185, 188, 204, 205, 228, 230, 232, 234, 235, 247, 258, 267
  naringenin 7-*O*-glucoside, 73, 97
  naringenin chalcone, 45, 70
  naringenin triglycoside, 92
Naringin, 71, 73–75, 79–83, 90, 91, 97, 99–101, 103, 105–107, 115, 159, 162, 166–169, 171, 176, 209, 227, 228, 230
Narirutin, 71, 75, 76, 90, 91, 103
Neochamaejasmin A, 203
Neodiosmin, 73, 80, 97, 100
Neohesperidin, 71, 73–75, 79, 80, 100
Neoplastic transformation, 242
Neutrophil(s), 4, 131, 150, 175–77, 182, 184, 188, 189
  neutrophil activation, 176
*Nicotiana tabacum*, 10
Nifedipine, 228, 235
Nobiletin, 4, 99, 106–108, 165, 168, 169, 172, 199, 203, 205, 227, 230–232, 240, 242
*Nod* gene-inducing flavonoids, 46, 57
*Nod* genes, 2, 31, 46, 52, 57, 65, 259
Nod-factor, 52, 259
Norwogonin, 137
Nuclear magnetic resonance (NMR), 37, 38, 108, 109, 111–114, 117, 120, 122, 124–127, 128

Onions, 154, 157–159
Orange, 67, 71, 73–75, 80–82, 85, 86, 97–101, 103, 105–107, 113, 116, 159–161, 165, 173, 228, 230, 234, 235
  orange juice, 105, 113, 159–161, 165, 173, 228, 230, 234, 235
  orange peel oils, 106
  sour orange, 71, 73–75, 97, *see also C. aurantium*
  sweet orange, 74, 75, 81, *see also C. sinensis*
Orobol, 180
Oxidation, 3, 70, 135–138, 140, 141, 143, 148–150, 155–157, 159–162, 164, 177, 210, 255, 262
Oxidative stress, 5, 131, 141, 148, 150
Oxyayanin, 209

*Pamburus*, 90, 94
*Papeda*, 75
*Papilionaceae*, 209
Parainfluenza, 105, 176, 209
*Paramignya monophylla*, 94
Peanuts, 128
Pectolinarigenin, 209
Peel oil, 100, 103, 109, 110, 112, 115
Pelargonidin, 209
Pentose phosphate pathway, 56, 64
Petunia, 6, 7, 35–37, 43, 44, 264
Phenyl benzo-γ-pyrone, 7, 172, 173
Phenylalanine, 45, 53, 56–58, 62, 64, 65, 68, 70, 73, 74, 81–84, 101, 178
   phenylalanine ammonia lyase (PAL), 45, 52, 53, 57–60, 62, 63, 74
   L-phenylalanine ammonia-lyase, 58, 64, 83, 101
Phenylpropanoid biosynthetic pathway, 68, 72
Phloretin, 159, 203
Photo-oxidation, 210
Phylogenetic tree, 253
Phytoalexin(s), 2, 30, 31, 33, 46, 49, 54, 56–58, 60–62, 64, 66, 81–83
Phytoestrogen(s), 124, 235, 243, 246–248, 264–267
Pinocembrin, 57
*Pisum sativum*, 12, 30
Plant, 2, 3, 5–7, 9, 10, 12, 13, 17, 26, 28, 30–33, 35, 36, 43–46, 49, 52–57, 61, 63–72, 74, 77–86, 91, 97, 99–101, 116, 141, 150, 152, 162–164, 172, 173, 175, 177, 181, 182, 192, 194, 196, 205, 206, 209, 216, 221–225, 227, 235–237, 247–250, 254–256, 260, 262–267
   plant growth regulators, 67
   plant-fungus signaling, 17
Plantagoside, 180, 182
Platelet(s), 4, 6, 106, 151, 163, 171–173, 175, 177, 180, 189, 228
   platelet adhesion, 4
   platelet aggregation, 151, 173, 180, 228
Podophyllotoxin, 216
*Polanisia dodecandia*, 194
*Polanisia dodecandra*, 191
Polio, 105, 176
Pollen, 2, 3, 6, 7, 35–45, 56
   pollen germination, 7, 35–40, 42, 44, 56
   pollen tube, 2, 3, 36, 43, 44
Polyphenol absorption, 158
*Poncirus*, 83, 86, 95
Potatoes, 157
Pro-oxidants, 141, 148, 150
Proanthocyanidins, 70, 177
Prodifferentiation, 180
Progesterone, 251, 251, 255
Prostaglandin(s), 6, 176, 179, 249–255, 260, 262–265
   15-deoxy-$\Delta^{12,14}$-prostaglandin J$_2$, 251, 265
   15-hydroxyprostaglandin dehydrogenase, 254, 255, 260, 262, 263
   15-hydroxyprostaglandin dehydrogenase, 254, 255, 260, 262, 263

Protein(s), 4–6, 35, 39, 58, 61, 64, 74, 75, 100, 107, 143, 148, 150, 151, 154, 156, 163, 175, 177–179, 181, 183, 184, 188, 189, 200–202, 222–224, 232, 234–236, 246, 248, 251, 255, 256, 260, 262, 264–267
   protein kinase, 4, 6, 175, 177, 178, 181, 184, 188, 189, 202, 222–224, 232, 234–236
   protein kinase C, 4, 6, 175, 177, 178, 181, 202, 222–224, 232, 234, 235
   protein tyrosine kinase, 4, 177, 188
   proteinase, 181, 192, 212, 213, 222
Protoplast(s), 65, 72, 81–84
   protoplast fusion, 72, 83
Prunin, 73, 75, 80, 100
*Pseudomonas*, 52, 53, 64
Pseudorabies, 209
*Psorothamnus fremontii*, 206, 223
Pummelo, 71, 72, 74, 75, 92, 227
Pyrogallol, 5, 133

Quercetagetin, 134, 135–137, 141, 211, 212
Quercetin, 24–28, 37–39, 77, 94, 105, 134, 137, 138, 140, 141, 143, 144–150, 152, 155, 157, 159, 160–162, 168, 169, 176, 177, 178–182, 189, 194, 199, 202, 203, 205–207, 209, 211–213, 222, 228, 232, 234, 236, 247
   quercetin 3-*O*-β-glucopyranosyl-7-*O*-α-L-rhamnopyranoside, 194
   quercetin 3-*O*-galactoside, 39
   3,7-dimethylquercetin, 194
   7-methoxyquercetin, 205
   dihydroquercetin, 25, 176, 178, 205, 209
   pentamethoxyquercetin, 205
Quinine, 168
Quinoline
   4-nitroquinoline 1-oxide, 105
Quinolone(s), 3, 191, 216, 223
   2-phenyl-4-quinolones, 3, 191, 214–217, 223
Quinones, 134

Rabies, 207
Reactive oxygen species, 5, 131, 119–142, 149
Receptor, 2, 5, 6, 28, 42, 43, 107, 154, 155, 178, 179, 184, 210, 224, 230–232, 235, 236, 248, 250, 251, 256, 258, 261, 262, 264–267
   receptor protein, 154, 251, 253
   receptor-ligand binding, 2
   receptor-mediated endocytosis, 155
Red wine, 153, 158–163, 177, 181, 222
Resorcinol, 263, 265
Respiratory, 134, 135, 137, 140, 148, 176, 209
   respiratory burst, 134–137, 148
Response, 2, 4, 6, 26, 31–33, 41, 42, 46, 48–52, 54, 56, 58, 64, 65, 68, 71, 80, 83, 141, 150, 156, 157, 176, 179, 180, 182–184, 187–189, 243, 251, 262
   allergic response, 4
   hypersensitive response, 49, 52
   immune response, 6, 179, 182

Response (cont.)
 inflammatory response, 183
 Ini response, 46, 51
Resveratrol, 152, 157, 161
Retinoic acid, 251, 252, 258, 261, 262
 9-*cis*-retinoic acid, 261, 262
Retinoid(s), 6, 239, 249–251, 253, 261–264, 266
 retinoid X receptor(s), 251, 261, 262, 266
Rhamnetin, 136, 137, 141
Rhamnose, 39, 43, 71, 73, 80, 100, 152
Rhizobia, 2, 46, 49, 51, 52, 260, 265
*Rhizobium*, 2, 6, 9, 10, 13, 17, 31, 32, 45–48, 50, 53, 54, 65, 259, 260, 264, 265
*Rhizobium meliloti*, 31, 32, 45, 48, 53, 54, 65, 259, 260, 264
Rhizobium-legume symbiosis, 2, 45, 46
Rhizosphere, 2, 12, 32, 56
Rhoifolin, 97, 99
*Rhus succedanea*, 191, 223
Robinetin, 137, 141, 205, 206
Robinin, 205
Root exudates, 10, 12, 13, 17, 26–28, 30–32, 46, 53, 54
Rotenoids, 204, 223, 225
*Rutaceae*, 82, 86, 100, 101, 113, 196
Rutin, 24, 25, 59, 60, 76, 77, 90, 105, 106, 157, 161, 166–169, 178, 202, 205, 209, 241
Rutinose, 71, 152, 213

Saikosaponin-a, 200, 201
Saikosaponin-b2 200
Saikosaponin-d, 200
*Salmonella typhimurium*, 205, 206
Salicylic acid, 78, 152
*Scoparia dulcis*, 194, 222
*Scrophulariaceae*, 194
Scutellarein, 106, 211
 tetra-*O*-methylscutellarein, 168
*Scutellaria baicalensis*, 194, 204, 223
*Scutellospora*, 25
Secondary metabolites, 55, 56, 63, 67–70, 77, 83, 227
*Selaginella*, 197, 224
*Selaginella willdenowii*, 197, 198, 224
*Selaginellaceae*, 197
E-selectin, 179, 183, 184, 186, 189
Semiquinone, 138
Sequestration, 42
Serum antioxidant activity, 159
*Severinia*, 72, 81, 96
Signal transduction, 56, 58, 177, 178, 180, 181, 236, 249
Sinensetin, 99, 106–108, 168, 169, 172
Skin diseases, 197
Skullcapflavone, 194
Somatic hybrid, 81, 82
Southern Armyworm, 141, 143, 144, 150
Soy, 3, 117–121, 124, 180, 236–238, 240, 242, 243, 245–248, 267
 soy foods, 128, 237, 238, 243, 246

Soy (cont.)
 soy sauce, 117–122, 128, 129
  dark-colored soy sauce, 117–120
  light-colored soy sauce, 117–121
 soybeans, 33, 105, 118, 128, 129, 228, 243, 260, 264, 265
*Spinacea oleracea*, 28
Spore germination, 2, 10–12, 16, 20–23, 25, 26, 28, 31, 32
*Stellera chamaejasme*, 203
Steroid(s), 5, 6, 27, 30, 31, 33, 231, 236, 249–256, 258, 260, 262–267
 steroid hormone receptors, 5, 251, 253, 257
 3β-hydroxysteroid dehydrogenase, 255
 11β-hydroxysteroid dehydrogenase, 255, 262, 265, 266
 17β-hydroxysteroid dehydrogenase, 253, 255, 260, 262, 264
 20β-hydroxysteroid dehydrogenase, 254, 265
Stigma, 35, 36
Stinging nettle, 10
Strawberries, 153
Structure-activity relationships, 176, 178, 181, 189, 213, 222, 235, 266
Swertifrancheside, 210
*Swertia franchetiana*, 210, 225
*Swinglea glutinosa*, 95–97
Syphilis, 192

T & B Lymphocytes, 175
Tamoxifen, 115, 231, 232–236, 256
Tangelo, 75
Tangeretin, 4, 6, 99, 106–108, 115, 165, 167, 168, 185, 199, 201, 202, 205, 221, 222, 227, 230–233, 240, 242
Tangerine(s), 100, 103, 106, 108, 109, 112, 113, 115, 227, 230, *see also C. reticulata*
 tangerine peel oils, 106
Tannins, 70, 152, 163
Tartaric acid, 117, 127–129
 caffeoyl ester of tartaric acid, 128
 ethers of tartaric acid, 117
Taxifolin, 140, 176, 178, 179, 199, 202, 203; *see also* Dihydroquercetin
 6-p-hydroxybenzyltaxifolin, 197
 8-p-hydroxybenzyltaxifolin, 197
 6,8-di-p-hydroxybenzyltaxifolin, 197
Teleocidin, 203
Tephrosin, 193, 204
Ternatin, 210
Testosterone, 252, 255, 260
Thiouracil, 168
Thrombogenesis, 3–5, 169, 172
Thrombogenic, 166, 167, 172
Thrombosis, 4, 106, 166, 167, 169
*Thymelaeaceae*, 192, 203
Thyroid hormone, 249–253, 258, 262–264, 266, 267
Tobacco, 10, 55, 59, 60, 64, 65, 83, 124
Tocopherol, 155, 160, 161, 232, 246

Tocotrienol(s), 5, 107, 115, 231–236
　γ-tocotrienol, 232
　δ-tocotrienol, 232
Transcription, 58, 64, 68, 80, 183, 184, 188, 189, 249, 251, 258, 260, 261
　transcription factor complexes, 184
Transformation assay, 239
Transgenic tobacco plants, 55, 59, 60, 64
Tricin, 191, 192, 223
*Trifolium subterraneum*, 51, 54, 66
Trimodal action, 116, 171, 172
*Triphasia*, 94
*Triphasiinae*, 94
Tubulin, 194, 216, 223
　tubulin polymerization, 194, 216, 223
Tumor necrosis factor, 183, 187
Tumorigenesis, 105, 115, 228, 230, 234, 236
Two-stage carcinogenesis, 203, 204

UDP, 40
Urinary excretion rates, 244
Urinary tract infections, 197
*Urtica dioica*, 10, 32
Urticaria, 176

*Vancouveria hexandra*, 77, 84
Vasodilatory actions, 151
VCAM-1 179, 183, 184
Vegetables, 103, 151–154, 156–159, 162, 163, 175, 176, 180, 227, 228, 235, 236, 238, 246–250
Very low density lipoprotein, 155
Vestitone, 57
Vicinal diphenol, 237, 240, 245

Viral infection, 3, 176
Viruses, 105, 115, 176, 181, 207, 209, 223, 255
　adenovirus, 209
　Epstein-Barr virus, 204, 223
　herpes simplex, 105, 176, 209, 221
　herpes virus type, 1 176
　human immunodeficiency virus, 176, 181, 212
　poliovirus, 209
　potato virus, 63, 209
　respiratory syncytial virus, 176, 209
　rhinovirus, 209
　Sindbis virus, 209
Vitamins, 7, 101, 116, 153, 154, 157, 159, 163, 166, 173, 236, 237
　vitamin C, 148, 153, 154, 157, 166–168, 172, 181
　vitamin C sparing activity, 181
　vitamin D, 168
　　1,25-dihydroxy-vitamin $D_3$, 251, 252
　vitamin E, 153–157, 159, 160, 177, 231, 236, 239, 240
　vitamin P, 7, 85, 100, 101, 116, 151, 165, 166, 172, 173

Whooping cough, 192
*Wikstroemia indica*, 191, 223

Xanthone, 210, 225
　1,5,8-trihydroxy-3-methoxy-7-(5',7',3'',4''-tetrahydroxy-6'-C-β-D-glucopyranosyl-4'-oxy-8'-flavyl)-xanthone, 210
Xenografts, 207

Zeatin, 78

**UCSF LIBRARY MATERIALS MUST BE RETURNED TO:**
**THE UCSF LIBRARY**
530 Parnassus Ave.
University of California, San Francisco

This book is due on the last date stamped below.
Patrons with overdue items are subject to penalties.
Please refer to the Borrower's Policy for details.
Items may be renewed within five days prior to the due date.
For telephone renewals -- call (415) 476-2335
Self renewals -- at any UCSF Library Catalog terminal in the Library, or
renew by accessing the UCSF Library Catalog via the Library's web site:
http://www.library.ucsf.edu
All items are subject to recall after 7 days.

## 28 DAY LOAN

28 DAY

MAY 1 1 1999

RETURNED

MAY 2 6 1999

28 DAY

NOV 2 3 1999